Electrical and Electronic Engineering: Concepts and Applications

Edited by Jeremy Giamatti

CLANRYE
INTERNATIONAL
www.clanryeinternational.com

Clanrye International,
750 Third Avenue, 9ᵗʰ Floor,
New York, NY 10017, USA

ISBN: 978-1-63240-624-8

Cataloging-in-Publication Data

Electrical and electronic engineering : concepts and applications / edited by Jeremy Giamatti.
  p. cm.
Includes bibliographical references and index.
ISBN 978-1-63240-624-8
1. Electrical engineering. 2. Electronics. 3. Electronic instruments. 4. Engineering instruments.
I. Giamatti, Jeremy.
TK145 .E44 2017
621.3--dc23

For information on all Clanrye International publications visit our website at www.clanryeinternational.com

𝓒LANRYE
𝓘NTERNATIONAL

Printed in the United States of America.

Contents

Permissions

List of Contributors

Index

Preface

Electrical engineering is a field that studies the principles and applications of electricity and the technology that has been developed around it. This book elucidates new techniques and their applications in a multidisciplinary approach. It consists of contributions made by international experts. It seeks to provide comprehensive information dealing with the various sub-disciplines of electrical engineering and the technological advancements in these areas of study. Detailed information is provided in a simple and analytical manner. For all readers who are interested in electrical and electronic engineering, the case studies included in this book will serve as an excellent guide to develop a comprehensive understanding.

After months of intensive research and writing, this book is the end result of all who devoted their time and efforts in the initiation and progress of this book. It will surely be a source of reference in enhancing the required knowledge of the new developments in the area. During the course of developing this book, certain measures such as accuracy, authenticity and research focused analytical studies were given preference in order to produce a comprehensive book in the area of study.

This book would not have been possible without the efforts of the authors and the publisher. I extend my sincere thanks to them. Secondly, I express my gratitude to my family and well-wishers. And most importantly, I thank my students for constantly expressing their willingness and curiosity in enhancing their knowledge in the field, which encourages me to take up further research projects for the advancement of the area.

Editor

A least square support vector machine-based approach for contingency classification and ranking in a large power system

Bhanu Pratap Soni[1]*, Akash Saxena[2] and Vikas Gupta[1]

*Corresponding author: Bhanu Pratap Soni, Department of Electrical Engineering, Malaviya National Institute of Technology, Jaipur, Rajasthan 302017, India
E-mail: er.bpsoni2011@gmail.com

Reviewing editor: Kun Chen, Wuhan University of Technology, China

Abstract: This paper proposes an effective supervised learning approach for static security assessment of a large power system. Supervised learning approach employs least square support vector machine (LS-SVM) to rank the contingencies and predict the system severity level. The severity of the contingency is measured by two scalar performance indices (PIs): line MVA performance index (PI_{MVA}) and Voltage-reactive power performance index (PI_{VQ}). SVM works in two steps. Step I is the estimation of both standard indices (PI_{MVA} and PI_{VQ}) that is carried out under different operating scenarios and Step II contingency ranking is carried out based on the values of PIs. The effectiveness of the proposed methodology is demonstrated on IEEE 39-bus (New England system). The approach can be beneficial tool which is less time consuming and accurate security assessment and contingency analysis at energy management center.

Subjects: Electrical & Electronic Engineering; Power Engineering; Systems & Controls

Keywords: critical line outage; sensitive lines; power system stability; artificial neural network; contingency analysis; performance index (PI); static security assessment; support vector machines (SVMs)

1. Introduction
Modern power system is a complex interconnected network having multiple utilities of different nature at generation, transmission, and distribution ends. This diverse nature of the devices makes

ABOUT THE AUTHORS
Bhanu Pratap Soni is pursuing Ph.D. from Malaviya National Institute of Technology Jaipur, India. His research interests include development and application of optimization algorithm, power system dynamic stability.

Akash Saxena is an associate professor of Electrical Engineering at Swami Keshvanand Institute of Technology, Management and Gramothan, Jaipur, India. His research interests include power system operations and control, automatic generation control, and application of natural inspired meta-heuristic algorithms.

Vikas Gupta is an associate professor and head in Department of Electrical Engineering, Malaviya National Institute of Technology Jaipur, India. His research interests include power system dynamics, electrical machines and drives, and non-conventional energy sources.

PUBLIC INTEREST STATEMENT
With the increase in population and ongoing demand of the electricity, power utilities are working near to the operating limits. The relevant information of the power system's state is beneficial for the operation and control of the power system at energy management center. This paper presents a supervised learning approach for ranking the contingencies in three states namely not critical, critical and most critical. The supervised learning model is developed by the offline studies and two standard performance indices are used. After reading this paper, the readers will be able to know about the static security assessment and online contingency ranking of a large power system.

the operation of power system more complex as compared with earlier days. This complexity is also increasing due to exponential increase in population and escalating load demands. Under these conditions a question mark appears on the reliable operation of the power system (Souza, Do Coutto Filho, & Schilling, 2002). The utilities at different ends are operating at their operating and security limits. To make system extremely reliable, the offline studies under different operating scenarios for occurrence of probable contingencies is the field of interest. Contingency analysis is carried out by screening and sensitivity-based ranking methods (Devaraj, Yegnanarayana, & Ramar, 2002; Niazi, Arora, & Surana, 2004; Patidar & Sharma, 2007; Refaee, Mohandes, & Maghrabi, 1999; Shanti Swarup & Sudhakar, 2006; Singh & Srivastava, 2007; Singh, Srivastava, & Sharma, 2000; Srivastava, Singh, & Sharma, 2000; Verma & Niazi, 2012). In ranking methods, calculation of standard indices is carried out. These calculations are based online MVA, bus voltage, and reactive power generation of the system. However, in screening methods different load flow methods are employed namely DC load flow, Ac load flow, and local solution methods. Sensitivity-based methods are efficient but inaccurate, on the other hand screening methods are less efficient.

In past few decades, this area invited the interest of researchers to develop a foolproof contingency evaluation scheme. Souza et al. (2002) proposed a fast contingency selection based on pattern search analysis in this approach authors employed Multi Layer Perceptron (MLP). However, this method is tested for small IEEE 24 bus system. Srivastava et al. (2000) proposed a fast voltage contingency screening through a hybrid neural network. This neural network is obtained by the combination of filter module and ranking modular network. Radial Basis Function Neural Networks (RBFNN) was used in the approaches (Devaraj et al., 2002; Singh & Srivastava, 2007; Srivastava et al., 2000). These networks exploited as a supervised agent to estimate the line loadings and bus voltages of different power systems. The RBFNN was employed in this problem, due to simplicity and training efficiency. Some more approaches employed RBFNN for the calculation of the indexes (Refaee et al., 1999). In Singh and Srivastava (2007) mutual information method is used for selecting the feature of the neural net. This method was employed to define the relationship between independent variables and dependent variables. Method of correlation coefficients has been discussed and employed in Verma and Niazi (2012). In the work reported in Verma and Niazi (2012), 11 best features were chosen as input features. Euclidean-based clustering technique was applied by Jain, Srivastava, & Singh, (2003) to select the appropriate number of hidden layers for the RBFNN for voltage contingency screening. For selection of features, class seperability index and correlation coefficients based approach were employed in many researches (Devaraj et al., 2002; Patidar & Sharma, 2007; Singh et al., 2000; Verma & Niazi, 2012).

Supervised learning approaches, namely feed forward neural network (FFNN) (Shanti Swarup & Sudhakar, 2006; Verma & Niazi, 2012), RBFNN (Devaraj et al., 2002; Singh & Srivastava, 2007; Srivastava et al., 2000), cascaded neural network (CNN) (Niazi et al., 2004; Singh et al., 2000) have been presented to estimate and classify the critical contingencies for many models of power networks. The most important part of these learning approaches is input feature selection and choice of the parameters which determine the micro and macro structure of neural nets. In literature bus injections, state variables associated with generating and loading conditions were employed to generate a large database. To aggregate research in a more promising way, two major thrust areas are identified and those are as follows: firstly, development of an intelligent feature selection algorithm which can map dependent and independent variables and secondly to employ fast and accurate supervised learning model to contemporary power system for accurate contingency ranking. Recent years LS-SVM is used as a classifier in many approaches (Ekici, 2012; Erişti, Yıldırım, Erişti, & Demir, 2013; Jain et al., 2003). Erişti et al., (2013) presented a study based on wavelet transform to classify power quality events into fault events. These events were self regulating faults, line energizing events, and non-fault interruption events. Nine different features were extracted for this study. Similar work is reported by Sami Ekici (Erişti et al., 2012 to classify the power system disturbances. Power load forecasting along with Ant Colony Optimization is presented by Niu, Wang, & Wu, (2010); In Niu et al. (2012) optimal feature selection is performed by Ant Colony Optimization. Different Neural topologies presented in the work (Goyal & Goyal, 2011; Ghosh & Lubkeman, 1995; Niu et al., 2010; Toha & Osman Tokhi, 2008;

Williams & Zipser, 1989). The size reduction of the data and optimal feature selection are the key issues addressed in these approaches.

In view of the above literature review, following are the objectives of this research paper.

(i) To develop a supervised learning based model which can predict the performance indices based on MVA power flow and line voltage reactive power flow for a large interconnected standard IEEE 39 bus test system under a dynamic operating scenario.

(ii) To develop a classifier which can screen the contingencies of the power system into three states namely not critical, critical, and most critical.

(iii) To present the comparative analysis of the reported approaches with the proposed approach based on accuracy in prediction of the PIs.

The paper is organized as follows, Section 2 contains the details and mathematical formulation of the performance indices, in Section 3 philosophy of support vector machine is discussed, in Sections 4 and 5 proposed methodology and simulation results are discussed. Section 6 conclusion enlists the main finding of this work.

2. Contingency analysis

Contingency evaluation is an essential practice to know the emergency situations in power networks. Without knowing the severity and the impact of a particular contingency, preventive action cannot be initiated by the system operator at energy management center (Devaraj et al., 2002). Contingency analysis is an important tool for security assessment. On the other hand, prediction of the critical contingencies at earlier stage (which can present a potential threat to the system stability (voltage or rotor angle)) helps system operator to operate the power system in a secure state and initiate the corrective measures. In this paper, line outages at every bus in New England system are considered as a potential threat to the system stability. Performance Indices (PI) methods are widely used for contingency ranking (Jain et al., 2003; Singh & Srivastava, 2007; Srivastava et al., 2000; Verma & Niazi, 2012). Following subsection presents definition of performance indices used for contingency ranking.

2.1. Line MVA performance index (PI$_{MVA}$)

On the basis of literature review, it can be judged that the contingency ranking performed by the performance indices. System loading conditions in a modern emerging power system are dynamic in nature and impose a great impact on the performance of the power system. An index based on Line MVA flow is determined to estimate the extent of overload. Equation (1) shows the mathematical representation.

$$PI_{MVA} = \sum_{1}^{Nl} \left(\frac{W_{Li}}{M} \right) \left[\frac{S_i^{post}}{S_i^{max}} \right]^M \tag{1}$$

where S_i^{post} is the post contingency MVA flow of line i, S_i^{max} is the MVA rating of the line i, N_L is the number of lines in the system in this study (N_L = 46), W_{Li} is the weighting factor(=1). M (=2n) is the order of the exponent of penalty function (Verma & Niazi, 2012). To avoid misranking high value of exponential order (n = 4) is chosen in this paper. In order to classify the power system security states, on the basis of PIs calculation the status of power system is subdivided into three categories and indicated in Figure 1. Class A non-critical contingencies, Class B critical contingencies, and Class C most critical contingencies. Class B contingencies are related to the violation of the loading limits or voltage limit violations. However, the Class C contingencies indicate that they are not safe under any operating condition.

0<PI<0.2	0.2<PI<0.8	0.8<PI<1
Class A (Non-Critical)	Class B (Critical)	Class C (Most-Critical)

Figure 1. Classification criterion.

2.2. Line voltage reactive performance index (PI$_{VQ}$)
The system stress is measured in terms of bus voltage limit violations and transmission line over loads. An index based on Line VQ flow is determined to estimate the extent of overload.

$$PI_{VQ} = \sum_1^{N_B} \left(\frac{W_{Vi}}{M}\right)\left[\frac{V_i - V_i^{sp}}{\Delta V_i^{Lim}}\right]^M + \sum_1^{N_G}\left(\frac{W_{Gi}}{M}\right)\left[\frac{Q_i}{Q_i^{max}}\right]^M \tag{2}$$

where $\Delta V_i^{Lim} = V_i - V_i^{max}$ for $V_i > V_i^{max}$, $V_i^{min} - V_i$ for $V_i < V_i^{min}$, V_i is the post contingency Voltage at the ith bus, V_i^{sp} the specified (base case) voltage magnitude at the ith bus, V_i^{max} the maximum limit of voltage at the ith bus, V_i^{min} the minimum limit of voltage at the ith bus, N_B the number of buses in the system, W_{Vi} the real non-negative weighting factor (=1), and M(=2n) is the order of the exponent for penalty function. The first summation is a function of only the limit violated buses chosen to quantify system deficiency due to out-of limit bus voltages. The second summation, penalizes any violations of the reactive power constraints of all the generating units, where Q_i is the reactive power produced at bus i, Q_i^{max} the maximum limit for reactive power production of a generating unit, N_G the number of generating units, W_{Gi} is the real non-negative weighting factor (=1). The determination of the proper value of "n" is system specific. The optimum integer value "n" for this paper is taken as 4. In following section the basic details of Least Square Support Vector Machines (LS-SVMs) are interwoven to understand the role of this supervised learning model as a regression agent and classifier (Figure 2).

3. Support vector machine
Recently the mappings and classification problems are handled well by the artificial neural networks (ANNs). Two basic properties of neural nets make themselves different from other conventional approaches. These properties are:

(a) Learning from the training samples

(b) To adapt according to new environment

In LS-SVMs the input data are mapped with high-dimensional feature space with the help of kernel functions. Using kernel functions, the problem can be mapped in linear form. The least square loss function is used in LS-SVM to construct the optimization problem based on equality constraints.

The least squares loss function requires only the solution of linear equation set instead of long and computationally hard quadratic programming as in case of traditional SVMs. LS-SVM equation for function estimation can be written as shown in Equation (3).

$$y(x) = \sum_{k=1}^N \alpha_k K(x, x_k) + b \tag{3}$$

where α_k is the weighting factor, x are the training samples and x_k are the support vectors, b represents the bias, and N is the training samples. The architecture of LS-SVM is shown in Figure 3.

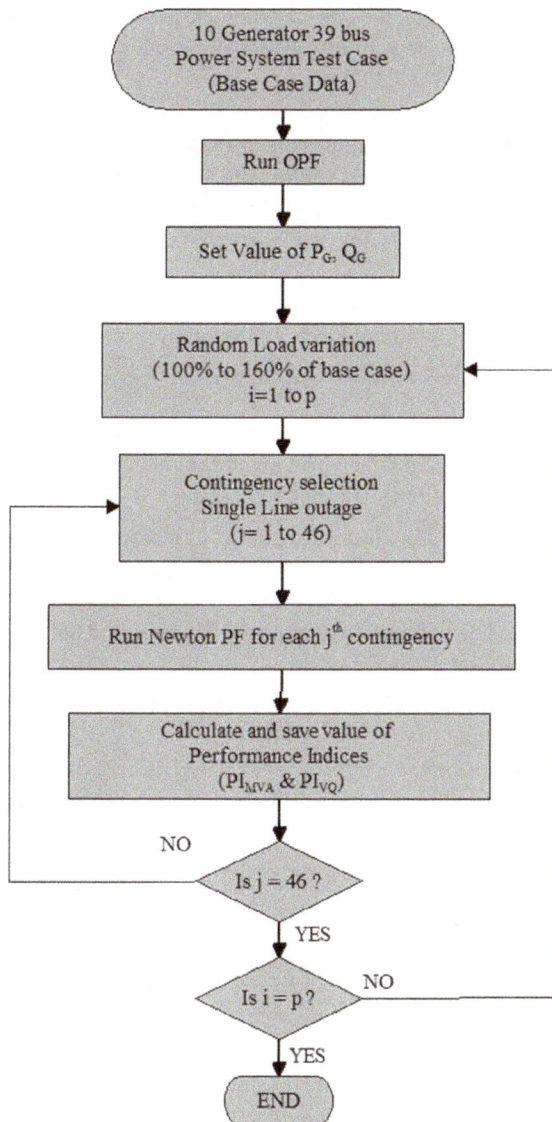

Figure 2. Data generation for contingency analysis.

The RBF kernel function for the proposed SVM tool can be written by Equation (4).

$$K(x, x_k) = \exp\left(-\frac{\|x - x_k\|^2}{2\sigma^2}\right) \qquad (4)$$

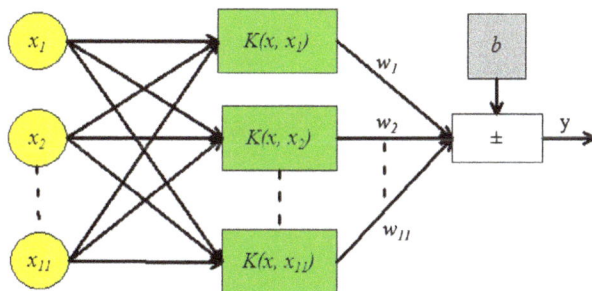

Figure 3. The structure of LS-SVM network.

In the present simulation work LS-SVM is interfaced with MATLAB software (*Matpower 4.0 User's Manual*, 2015; *Power system test cases*, 2010) and data-set of 13,800 different operating conditions along with line outages of each line is considered.

The values of PIs obtained from standard Newton Raphson methods are used for training purpose. The least square estimator uses the optimize values of σ (kernel width). The larger the value, the more will be the width of the kernel. This value indicates that system is global and near to a linear system. Unlike neural network SVM trains in less time and possess no hidden layers.

4. Proposed methodology

Data generation is an important task in supervised learning approach. In this study, a rich data of 13,800 samples are employed to train, test, and validate the networks. Following are the steps involved in the process.

(i) A large number of load patterns are generated by randomly perturbing the real and reactive loads on all the buses and real and reactive generation at the generator buses.

(ii) The features are selected as per (Verma & Niazi, 2012). Total 11 features as indicated in work are chosen for training purpose. These features are $P_{g10}, Q_{g1}, Q_{g2}, Q_{g3}, Q_{g4}, Q_{g5}, Q_{g7}, Q_{g8}, Q_{g9}, Q_{g10}$, and Q_{d14}. A contingency set for all credible contingencies are employed. N−1 contingencies are the most common event in power system. Single line outages are considered for each load pattern and the value of index is stored for each iteration of the simulation.

(iii) The obtained values of the index are normalized between 0.1–0.9 to train the SVM. Further the binary classification is done to train the classifier.

The system operating state contingency type and the regression performance of the network is stored for each operating scenarios (Figure 4).

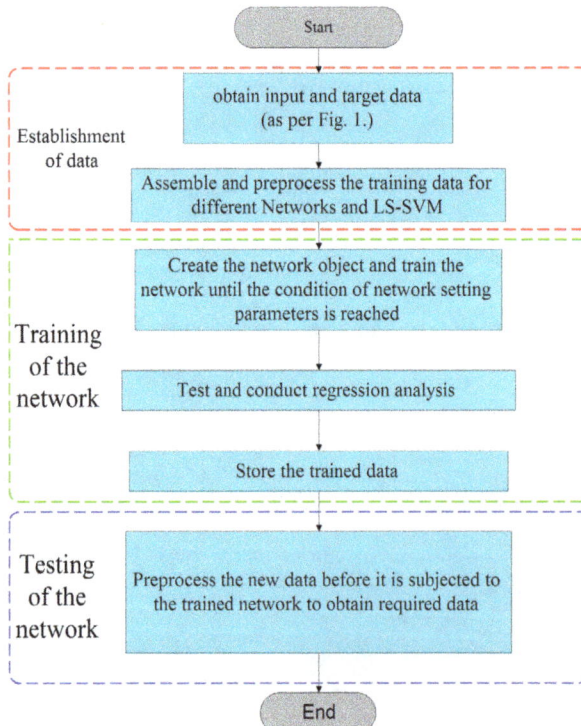

Figure 4. Implementation procedure of contingency classification.

5. Simulation results

The Simulink implementation of proposed approach has been implemented in MATLAB and tested over IEEE 39 bus test system (New England) (Lin, Horne, Tiňo, & Lee Giles, 1996) shown in Figure 5. The modeling of the system and simulation studies are performed over Intel ® core ™, i7, 2.9 GHz 4.00 GB RAM processor unit. Bus no. 39 has taken as slack bus. For line contingency 13,800 patterns are generated, which includes 46 line outages and different loading patterns (300). Out of these 200 patterns are those that the Newton Raphson (NR) method failed to converge.

From Table 1 it can be judged that the LSSVM possess lower values of mean square error (*MSE*) and high values of regression coefficient (*R*). It is empirical to judge that often the performance of the ranking methods is questioned due to wrong detection or misranking of a critical contingency.

The comparative results for the performance of the neural networks for determination of PIs are shown in Figures 6 and 7 based on value of mean square error (*MSE*) and percentage R^2, respectively.

The values of calculated indices for different mentioned contingencies are shown in Table 2. Number of samples is exhibited to show the efficacy of the different methods. From Table 2, it can be judged that the line outage 6–7 during loading condition 1345 is the critical one as the values of the indices are higher for every method.

For sample no. 2984, the values of PI_{MVA} by NR method is 0.1102 and the values predicted by Elman Backdrop, NARX, and Cascaded FBNN are around 0.15. Higher values can be clustered near the classifier boundaries and a crisp classifier will not be able to classify the state of the power system by these values.

On the other hand, the values calculated by the LS-SVM method possess lower values. It is important to mention here that often the performance of the ranking methods is questioned due to wrong detection or misranking of a critical contingency.

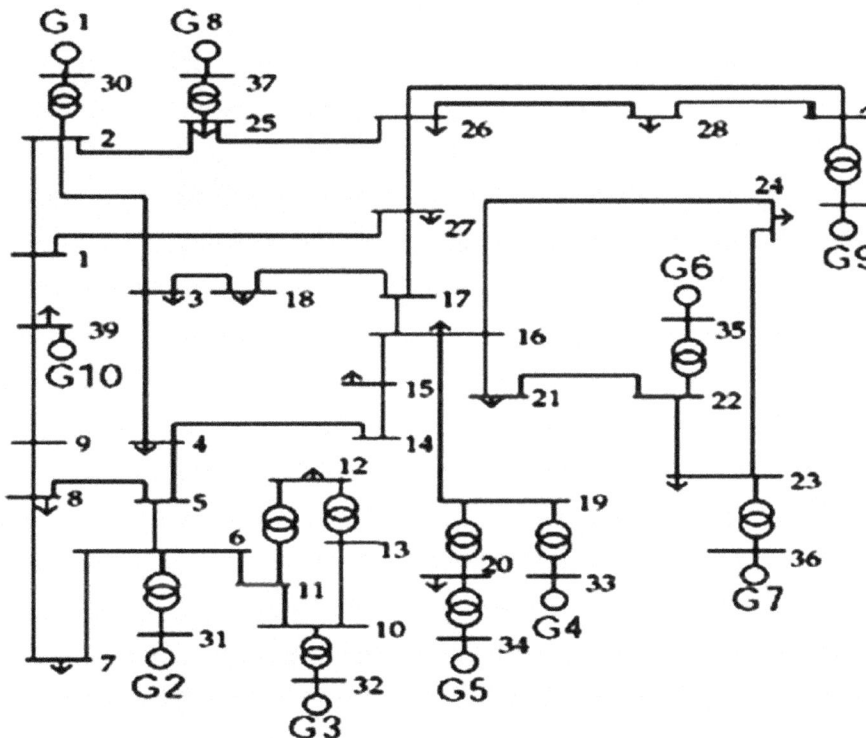

Figure 5. Single line diagram of New England system.

Table 1. Comparative performance of different neural networks (PI_{VQ} & PI_{MVA})				
Type of regression agent	PI_{MVA}		PI_{VQ}	
	MSE	R	MSE	R
Elman backdrop (Niu et al., 2010)	0.0006769	0.97453	0.0007062	0.96253
Cascaded FNN (Toha & Osman Tokhi, 2008)	0.0008139	0.9847	0.0008091	0.9837
FFNN (Verma & Niazi, 2012)	0.0008845	0.9856	0.0008755	0.9822
FFDTDNN (Goyal & Goyal, 2011)	0.0006829	0.98693	0.000683	0.98457
Layer recurrent (Ghosh & Lubkeman, 1995)	0.0007585	0.9837	0.0007752	0.9844
NARX (Williams & Zipser, 1989)	0.000659	0.98624	0.000646	0.98846
LS-SVM (Proposed)	0.000559	0.9912	0.000575	0.9915

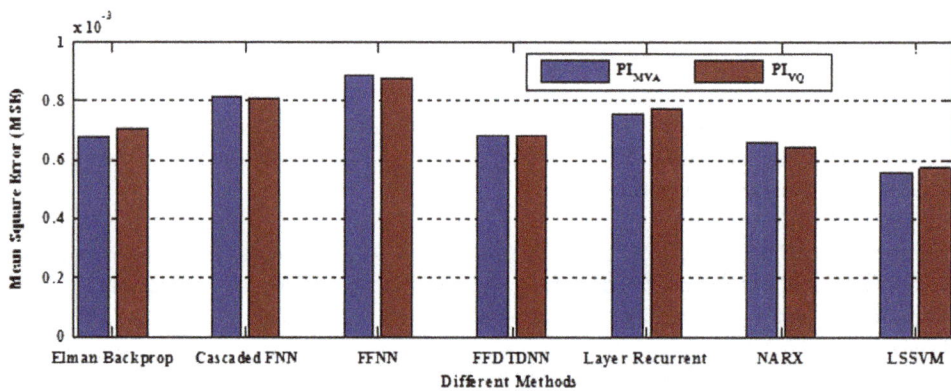

Figure 6. Comparative performance of different neural networks (*MSE*).

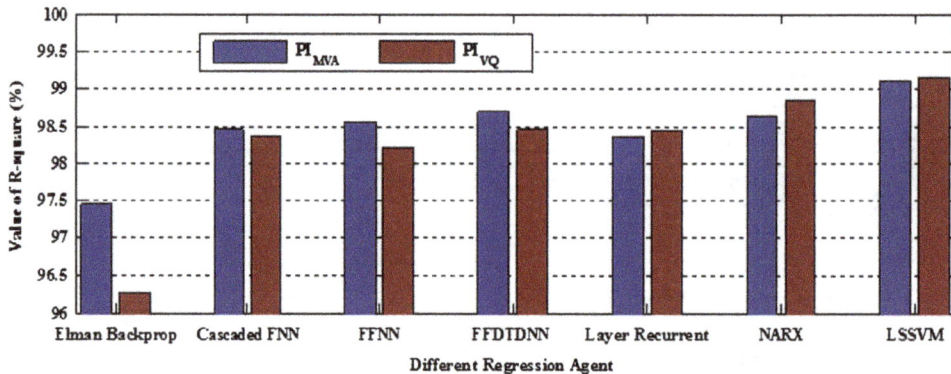

Figure 7. Comparison of the performance of different regression agents based on R^2.

The LS-SVM outperformed over the recent available topologies of neural networks (NNs) in prediction of performance indices. Classifications of contingencies are compared with the NR method and it is observed that LS-SVM can classify the contingencies well. For the ease of simplicity and understanding the excel plots are also included with the analysis. It can be observed from Figure 6 that values of MSE for FFNN are the highest. This shows the incapability of FFNN to predict the contingencies. MSE is the residual mean square, in statistical interpolation the value closer to zero indicates that the fit is more useful for prediction. From these values, it can be concluded that LS-SVM proven as a best regression agent for the prediction of both indices. LS-SVM method is suitable for prediction of contingencies. Values of R^2 are found minimum for Elman backdrop as shown in Figure 7. In statistical studies these values are the indication of how successful the fit is in explaining the variation of the data. The values which approach near to 1 as in case of LS-SVM shows that the machine learning model is able to predict the data very well. The value of adjusted R^2 is highest in the case of LS-SVM for calculation of both indices.

Following points are emerged from Table 2 (PI$_{MVA}$):

(i) It is observed that for line outage 6–7 the values of PI$_{MVA}$ from NR method is 0.8683. For this operating scenario, this contingency is not only a potential threat to the system stability but also most critical in nature. The value predicted by backdrop method is 0.7438; Cascaded FBNN is 0.7844 and 0.77859 FFDTDNN. LS-SVM predicted the value which is very near to the NR method and that is 0.8588.

(ii) It is also observed that for line 20–34 the contingency is neither severe nor critical hence the values predicted by all network topologies fall in the same range. However, the value predicted by LSSVM is quite close to the original values.

(iii) The classes identified by the NR method are same with the class identified by LS-SVM. It is also important to mention here that in this simulation study total no. of 13,800 cases were simulated. Out of these cases 1,232 no. of cases identified as critical contingencies and 10,243 were identified as most severe critical contingencies. The classification handled by LS-SVM is verified through NR and presented in a lucid manner for some contingencies.

(iv) Confusion matrix is an error matrix. The rows represent the instances in predicted class and column represents instances in actual class. Confusion matrix for the index detection is shown in Figure 8. Confusion matrix is a classical way to determine the accuracy of the classifier. From Figure 8, it can be observed that for class C, 100% cases were identified. For class B, 92% cases were identified, and 8% remaining cases were identified as a contingency A and C. Classification Efficiency for Class A and C is 100%. This shows an efficacy of the proposed approach to identify critical contingencies more accurately.

(v) It is concluded that both classification and prediction parts are performed by SVM with selected features. The efficacy of the proposed method is compared with contemporary types of neural networks. It is pragmatic to say that LS-SVM proved as a better supervised approach for real power system problem.

Table 2. Sample result of PI$_{MVA}$ and PI$_{VQ}$ calculation and contingency analysis

Outage No.		1345	7811	9014	587	2984
Line No.		6–7	8–9	9–39	25–26	20–34
PI$_{MVA}$	NR	0.8683	0.5438	0.4971	0.4094	0.1102
	Elman backprop (Toha & Osman Tokhi, 2008)	0.7483	0.7250	0.4635	0.3744	0.1565
	Cascaded FBNN (Goyal & Goyal, 2011)	0.7884	0.5030	0.3654	0.4649	0.1557
	FFNN (Verma & Niazi, 2012)	0.8236	0.4457	0.3972	0.3107	0.1270
	FFDTD (Ghosh & Lubkeman, 1995)	0.7785	0.5447	0.4482	0.3174	0.1567
	Layer recurrent (Williams & Zipser, 1989)	0.8330	0.7349	0.3381	0.2192	0.1150
	NARX (Lin et al., 1996)	0.8177	0.3869	0.2877	0.3399	0.1465
	LS-SVM	0.8588	0.5432	0.4865	0.4014	0.1098
PI$_{VQ}$	NR	0.8421	0.5846	0.3876	0.4232	0.1201
	Elman Backprop (Toha & Osman Tokhi, 2008)	0.8143	0.6510	0.4231	0.3647	0.1345
	Cascaded FBNN (Goyal & Goyal, 2011)	0.8001	0.4322	0.3870	0.4515	0.1141
	FFNN (Verma & Niazi, 2012)	0.8436	0.5561	0.4015	0.3484	0.1220
	FFDTD (Ghosh & Lubkeman, 1995)	0.8015	0.5334	0.4312	0.3486	0.1546
	Layer recurrent (Williams & Zipser, 1989)	0.8451	0.5457	0.3342	0.2247	0.1340
	NARX (Lin et al., 1996)	0.8245	0.3475	0.3015	0.3846	0.1426
	LS-SVM	0.8425	0.5901	0.3870	0.4231	0.1210
Class	LS-SVM	C	B	B	B	A
	NR	C	B	B	B	A

Confusion Matrix for Performance Index PI$_{MVA}$

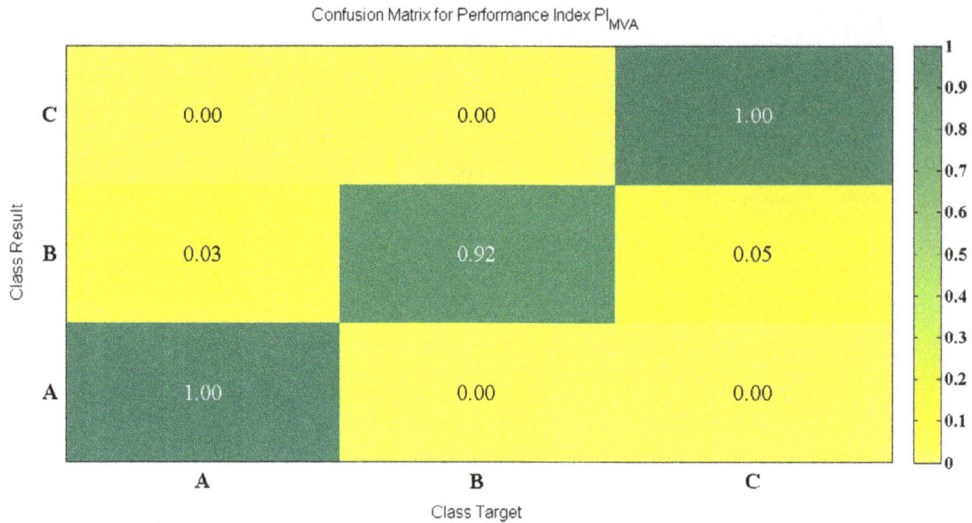

Figure 8. Confusion matrix for PI$_{MVA}$ of LS-SVM network.

It is also observed that most contingency ranked correctly by LSSVM shows the efficacy of the proposed approach over the recent conventional supervised learning models. In few cases it is observed that neural networks detect the class wrongly. To draw the fair comparison between all the neural nets, the hidden layers are kept 2 and number of neurons is kept 4. The neural nets are trained several times and the final results are taken from the best performed networks. It is observed that SVM is able to classify the contingency efficiently. It is also observed that the estimation of the PIs under different outages is also carried out in a very effective manner by SVM as compared with other approaches.

Following points are emerged from Table 2 (PI$_{VQ}$):

(i) It is observed that for line outage 6–7 the values of PI$_{VQ}$ from NR method is 0.8421. For this operating scenario, this contingency not only a potential threat to the system stability but also most critical in nature. The value predicted by backdrop method is 0.8413; Cascaded FBNN is 0.8001 and 0.8015 FFDTDNN. LS-SVM predicted the value which is very near to the NR method and that is 0.8425.

(ii) It is also observed that for line 20–34 the contingency is neither severe nor critical hence the values predicted by all network topologies fall in the same range. However, the value predicted by LSSVM is quite close to the original values.

(iii) The classes identified by the NR method are same with the class identified by LS-SVM. It is important to mention here that in this simulation study total 13,800 cases were simulated. Out of these cases 1,232 cases identified as critical contingencies and 10,243 were identified as most severe critical contingencies. The classification is handled by LS-SVM is verified through NR and presented in a lucid manner for some contingencies.

(iv) It is concluded that both classification and prediction parts are performed by SVM with selected features. The efficacy of the proposed method is compared with contemporary types of neural networks. It is pragmatic to say that LS-SVM proved as a better supervised approach for real power system problem. Confusion matrix for the index detection is shown in Figure 9. Confusion matrix is a classical way to determine the accuracy of the classifier. From Figure 9 it can be observed that for Class C, 100% cases were identified. For class B, 93% cases were identified i.e. from 1,232 cases 1,146 cases are identified by the classifier. 7% remaining cases were identified as a contingency A.

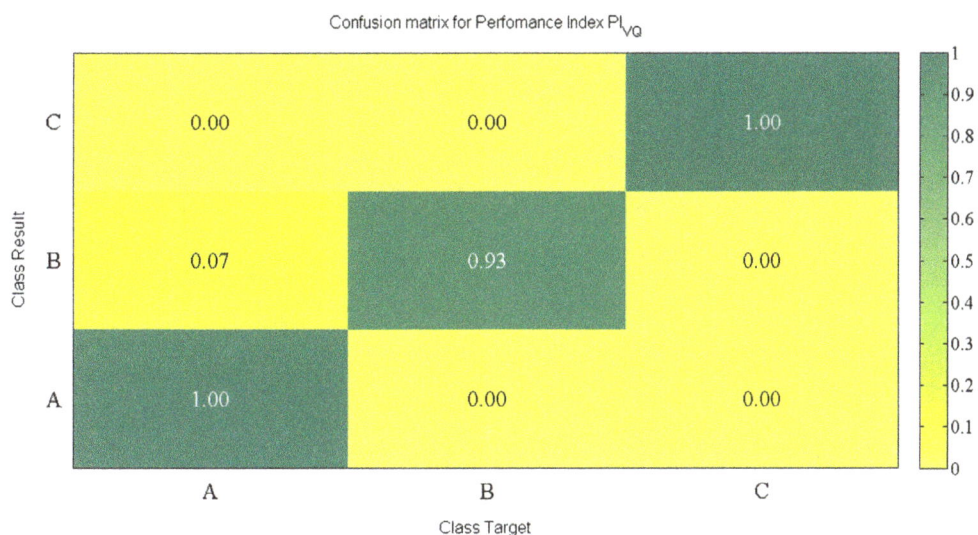

Figure 9. Confusion matrix for PI$_{VQ}$ of LS-SVM network.

It is observed that SVM is able to classify the contingency efficiently. It is also observed that the estimation of the PIs under different outages is also carried out in a very effective manner by SVM as compared with other approaches.

6. Conclusions

This paper proposes a supervised learning model based on least square loss function with RBF Kernel function to estimate the contingency ranking in a standard IEEE 39 bus system. Following are the main highlights of this work.

(a) Comparative analysis of existing learning-based approaches for contingency ranking through standard performance indices is carried out on a large interconnected power system while considering dynamic operating conditions. It is observed that neural nets of different topologies exhibit their quality to act as a regression agent. However, the best regression results are based on *MSE* and *R* are exhibited by LS-SVM. The numerical results obtained for the indices calculation advocated the efficacy of the proposed approach.

(b) In second part, the classification of the contingencies are carried out by LS-SVM. A binary classifier is obtained with three binary classes based on the values of Performance Indices. The performance of the SVM as a classifier is exhibited through the comparison of the results with NR method. It is concluded that SVM shows a satisfactory response to classify the contingencies.

(c) The proposed approach is suitable for online application. The operator at energy management center can easily get the details of the contingency and severity of the same with the help of these offline tested results. The study on larger system with multiple contingencies lays in the future scope.

Funding
The authors received no direct funding for this research.

Author details
Bhanu Pratap Soni[1]
E-mail: er.bpsoni2011@gmail.com
ORCID ID: http://orcid.org/0000-0001-8138-5392
Akash Saxena[2]
E-mail: akashvitjpr@gmail.com
Vikas Gupta[1]
E-mail: vgupta.ee@mnit.ac.in

[1] Department of Electrical Engineering, Malaviya National Institute of Technology, Jaipur, Rajasthan 302017, India.
[2] Department of Electrical Engineering, Swami Keshvanand Institute of Technology, Jaipur, Rajasthan 302017, India.

References

Devaraj, D., Yegnanarayana, B., & Ramar, K. (2002). Radial basis function networks for fast contingency ranking. *International Journal of Electrical Power & Energy Systems, 24*, 387–393. http://dx.doi.org/10.1016/S0142-0615(01)00041-2

Ekici, S. (2012). Support Vector Machines for classification and locating faults on transmission lines. *Applied Soft Computing, 12*, 1650–1658, ISSN 1568-4946. doi:10.1016/j.asoc.2012.02.011

Erişti, H., Yıldırım, Ö., Erişti, B., & Demir, Y. (2013). Optimal feature selection for classification of the power quality events using wavelet transform and least squares support vector machines. *International Journal of Electrical Power & Energy Systems, 49*, 95–103, ISSN 0142-0615. doi:10.1016/j.ijepes.2012.12.018

Ghosh, A. K., & Lubkeman, D. L. (1995). The classification of power system disturbance waveforms using a neural network approach. *IEEE Transactions on Power Delivery, 10*, 109–115. http://dx.doi.org/10.1109/61.368408

Goyal, S., & Goyal, G. K. (2011). Cascade and feed forward back propagation artificial neural networks models for prediction of sensory quality of instant coffee flavoured sterilized drink. *Canadian Journal on Artificial Intelligence, Machine Learning and Pattern Recognition, 2*, 78–82.

Jain, T., Srivastava, L., & Singh, S. N. (2003). Fast voltage contingency screening using radial basis function neural network. *IEEE Transactions on Power Systems, 18*, 1359–1366. http://dx.doi.org/10.1109/TPWRS.2003.818607

Lin, T., Horne, B. G., Tiňo, P., & Lee Giles, C. (1996). Learning long-term dependencies in NARX recurrent neural networks. *IEEE Transactions on Neural Networks, 7*, 1329–1338.

Matpower 4.0 User's Manual. (2015). Retrieved from http://www.pserc.cornell.edu/matpower/manual

Niazi, K. R., Arora, C. M., & Surana, S. L. (2004). Power system security evaluation using ANN: Feature selection using divergence. *Electric Power Systems Research, 69*, 161–167. http://dx.doi.org/10.1016/j.epsr.2003.08.007

Niu, D., Wang, Y., & Wu, D. D. (2010). Power load forecasting using support vector machine and ant colony optimization. *Expert Systems with Applications, 37*, 2531–2539.

Patidar N., & Sharma J. (2007). A hybrid decision tree model for fast voltage screening and ranking. *International Journal of Emerging Electric Power Systems, 8*, Article No. 7. doi:10.2202/1553-779X.1610

Power system test cases. (2010). Retrieved from http://www.ee.washington.edu/pstca/

Refaee, J. A., Mohandes, M., & Maghrabi, H. (1999). Radial basis function networks for contingency analysis of bulk power systems. *IEEE Transactions on Power Systems, 14*, 772–778. http://dx.doi.org/10.1109/59.761911

Shanti Swarup, K. S., & Sudhakar, G. (2006). Neural network approach to contingency screening and ranking in power systems. *Neurocomputing, 70*, 105–118. http://dx.doi.org/10.1016/j.neucom.2006.05.006

Singh, R., & Srivastava, L. (2007). Line flow contingency selection and ranking using cascade neural network. *Neurocomputing, 70*, 2645–2650. http://dx.doi.org/10.1016/j.neucom.2005.11.024

Singh, S. N., Srivastava, L., & Sharma, J. (2000). Fast voltage contingency screening and ranking using cascade neural network. *Electric Power Systems Research, 53*, 197–205. http://dx.doi.org/10.1016/S0378-7796(99)00059-0

Souza, J. C. S., Do Coutto Filho, M. B., & Schilling, M. T. (2002). Fast contingency selection through a pattern analysis approach. *Electric Power Systems Research, 62*, 13–19, ISSN 0378-7796. doi:10.1016/S0378-7796(02)00016-0

Srivastava, L., Singh, S. N., & Sharma, J. (2000). A hybrid neural network model for fast voltage contingency screening and ranking. *International Journal of Electrical Power & Energy Systems, 22*, 35–42. http://dx.doi.org/10.1016/S0142-0615(99)00024-1

Toha, S. F., & Osman Tokhi, M. (2008). MLP and Elman recurrent neural network modelling for the TRMS. In *7th IEEE International Conference on Cybernetic Intelligent Systems* (pp. 1–6), London.

Verma, K., & Niazi, K. R. (2012). Supervised learning approach to online contingency screening and ranking in power systems. *International Journal of Electrical Power & Energy Systems, 38*, 97–104.

Williams, R. J., & Zipser, D. (1989). A learning algorithm for continually running fully recurrent neural networks. *Neural Computation, 1*, 270–280. http://dx.doi.org/10.1162/neco.1989.1.2.270

2

A novel method of support vector machine to compute the resonant frequency of annular ring compact microstrip antennas

Ahmet Kayabasi[1]* and Ali Akdagli[2]

*Corresponding Author: Ahmet Kayabasi, Department of Electronic & Automation, Silifke-Tasucu Vocational School, Selcuk University, 33900, Silifke, Mersin, Turkey
E-mail: ahmetkayabasi@selcuk.edu.tr
Reviewing editor: Duc Pham, University of Birmingham, UK

Abstract: An application of support vector machine (SVM) to compute the resonant frequency at dominant mode TM_{11} of annular ring compact microstrip antennas (ARCMAs) is presented in this paper. ARCMAs have some useful features; resonant modes can be adjusted by controlling the ratio of the outer radius to the inner radius. The resonant frequencies of 100 ARCMAs with varied dimensions and electrical parameters in accordance with UHF band covering GSM, LTE, WLAN, and WiMAX applications were simulated with IE3D™ which is a robust numerical electromagnetic computational tool. Then, the SVM model was built with simulation data and 88 simulated ARCMAs were operated for training and the remaining 12 ARCMAs were used for testing this model. The proposed model has been confirmed by comparing with the suggestions reported elsewhere via measurement data published earlier in the literature, and it has further validated on an ARCMA operating at 3 GHz fabricated in this study. The obtained results show that this technique can be successfully used to compute the resonant frequency of ARCMAs without involving any sophisticated methods. The novelty of the approach described here is to offer ease of designing the process using this method.

ABOUT THE AUTHOR

Ahmet Kayabasi was born in 1980. In 2001, he received B.S. degree in Electrical and Electronics Engineering from Selcuk University, Turkey. In 2005, he received M.S. degree in Electrical and Electronics Engineering from Selcuk University, Turkey. He has been working as a lecturer at Electronics and Automation Department of Silifke-Tasucu Vocational School of Selcuk University since 2001. He has been studying toward the PhD degree at Electrical and Electronics Engineering from Mersin University since 2009. His current research interests include antennas, microstrip antennas, computational electromagnetic, artificial intelligent, and applications of optimization algorithms to electromagnetic problem such as radiation, resonance, and bandwidth.

PUBLIC INTEREST STATEMENT

The wireless communication devices such as laptop, netbook, and smart phone are moving toward the miniaturization very rapidly. Therefore, the antennas within these mobile devices should also be reduced in size with high performance, as well. Thus, compact microstrip antennas can be one of the best choices, since they allow to easily modify their geometry in order to achieve the desired characteristics. Annular ring compact microstrip antennas are miniaturized by loading a circular slot in the center of the circular patch. The resonant frequency determination of compact microstrip antennas is important, because these antennas inherently suffer from narrow bandwidth. The support vector machine is a recent and effective artificial intelligent technique as artificial neural network and adaptive neuro-fuzzy inference system. In this study, a method based on support vector machine model has been successfully used for determining the resonant frequency of annular ring compact microstrip antennas.

Subjects: Artificial Intelligence; Communications & Information Processing; Electromagnetics & Microwaves; Machine Learning - Design; Neural Networks

Keywords: annular ring compact microstrip antenna; resonant frequency; support vector machine (SVM)

1. Introduction

Microstrip antennas (MAs) are preferred in a wide range of applications such as aircraft antennas, missile guidance antennas, mobile radios, and array antennas due to their advantageous features as being low profile, having low fabrication costs, and ease of integration with microwave circuits. However, MAs also have some disadvantageous characteristics like having narrow bandwidth and low power handling capacity. Therefore, there has been a vast amount of researches on proposing alternative MA configurations that could eliminate the drawbacks of conventional MAs. As alternative antenna configurations, introduction of parasitic patches and modifying the shape of the microstrip patch are proposed (Wong, 2002).

Present portable communication and handheld devices inherently need miniaturized MAs. Using the substrate materials with high dielectric constant, the smaller antennas can be achieved, but this gives rise to decrease the bandwidth and efficiency performances (Kumar & Ray, 2003; Wong, 2002). Thus, it is difficult to carry out the requirements of mobile communication devices by using the traditional MAs. The compact geometry has been proved as an alternate methodology to design miniature MAs. The compact microstrip antennas (CMAs) are obtained by applying some modifications such as slot-loading and shorting-pin/wall on traditional MA structures (Wong, 2002). Several slot-loaded CMA configurations such as C (Deshmukh & Kumar, 2007), E (Akdagli, Toktas, Kayabasi, & Develi, 2013; Deshmukh, Phatak, Nagarbovdi, & Ahuja, 2013), H (Deshmukh & Kumar, 2007; Kayabasi, Bicer, Akdagli, & Toktas, 2011), L (Chen, 2000), rectangular ring (Deshmukh & Kumar, 2007), and annular ring (Chew, 1982) shapes have been presented in the literature as an alternative and effective method to reduce the physically size of antenna.

Annular ring compact microstrip antennas (ARCMAs) are miniaturized antenna constructed by loading a circular slot in the center of the circular patch. The size of the ARCMA is substantially smaller than circular microstrip antenna (CMA) at the same operating frequency (Chew, 1982). It can be appreciated that the average path length traveled by the current in the annular-ring patch is much longer than the corresponding circular patch for the lowest order mode (Chew, 1982). Also, by choosing the inner and outer radius of the ring properly, both bandwidth broadening (Chew, 1982) and controlling the separation of resonant modes can be managed (Dahele, Lee, & Wong, 1987). Due to these useful properties, it is one of the most studied MAs. In the literature, the ARCMA was theoretically investigated by its resonator model in (Bahl, Stuchly, & Stuchly, 1980; Wolff & Knoppik, 1971; Pintzos & Pregla, 1978; Wu & Rosenbaum, 1973). The mathematical tools such as vector Hankel transform, Galerkin's method, and Green functions were greatly utilized in the analysis of the ARCMAs (Ali, Weng, & Kong, 1982; Fan & Lee, 1991; Gurel & Yazgan, 2010; Liu & Hu, 1996a, 1996b; Motevasselian, 2011). Methods based on cavity model and transmission line model were presented to investigate some parameters such as the resonant frequency, input impedance, and bandwidth (Bahl & Stuchly, 1992; Bhattacharyya & Garg, 1985; El-khamy, El-Awadi, & El-Sharrawy, 1986; Gomez-Tagle & Christodoulou, 1997; Kumar & Dhubkarya, 2011; Richards, Jai-Dong, & Long, 1984; Sathi, Ghobadi, & Nourinia, 2008). The experimental studies concerning the ARCMA were also performed to confirm the theoretical calculations in Dahele et al. (1987), Bahl et al. (1980), Fan and Lee (1991), Liu and Hu (1996a), Kumar and Dhubkarya (2011), Dahele and Lee (1982), Lee, Dahele, and Ho (1983), Row (2004), and Shinde, Shinde, Kumar, Uplane, and Mishra (2010). It can be seen from the literature that these methods include rigorous calculation of Hankel and Fourier transforms and Bessel functions.

Analytical methods seem to be easier but they result in accurate solutions only for regular shapes of the patch, whereas the numerical electromagnetic computation methods are suitable for all shapes of the MA. However, the numerical methods require much more time in solving Maxwell's equations including integral and/or differential computations. So, it becomes time consuming since it repeats the same mathematical procedure even if a minor change in geometry is carried out. On the other hand, antenna designers prefer the easier approaches without requiring much rigorous computations and consuming time.

Due to the rapid development of computer technology in recent years, several robust and alternative methods based on nature-inspired optimization algorithms and artificial intelligent techniques (AITs) have emerged for solving the different kind of engineering problems. The most well-known AITs are the artificial neural network (ANN) (Kumar & Shukla, 2012), the adaptive neuro-fuzzy inference system (Dadgarnia & Heidari, 2010), and the support vector machine (SVM) (Bertsekas, 1995; Christodoulou, Martinez-Ramon, & Balanis, 2006; Cristianini & Shawe-Taylor, 2000; Tokan, 2008; Tokan & Gunes, 2008; Vapnik, 1998). AITs, which are effective and high speedy approaches, are used frequently for the solution of microwave and electromagnetic problems in recent years.

In machine learning, SVM is a new generation supervised learning model which is used for classification and regression analysis. In another terms, SVM is a classification and regression prediction tool that uses machine learning theory to maximize predictive accuracy while automatically avoiding over-fit to the data. The SVM is an advanced nonlinear learning machine (so-called Vapnik–Chervonenkis theory) (Vapnik, 1998). SVM is a machine learning method used for classification and regression implementations and also run in supervised or semi-supervised way. In the nonlinear problems such as ours, SVM depends on the principle which is separation of two classes with a hyperplane that is occurred by transforming data into the higher dimensions. The functions that have various features are used during the transformation into the high dimension and these functions are called as Kernel functions. Some parameters in the mathematical expression of these functions need to be defined by the user for using Kernel functions. SVM has been formed on powerful theoretical foundations (Tokan & Gunes, 2008). A distinct advantage of SVM is that it bypasses the repeated use of complex formulations or process for a new case given to it after proper training.

In this study, an application of SVM model is presented to compute accurately resonant frequencies of ARCMAs. The resonant frequency values of 100 ARCMAs corresponding most of UHF band covering GSM, LTE, WLAN, and WiMAX applications were determined by the electromagnetic simulator IE3D™ using method of moment (Harrington, 1993). The simulation parameters of 88 ARCMAs representing the overall problem space were used for training and the remaining 12 were then employed to test the accuracy. The results of the SVM model obtained in this study were confirmed by comparing with the measurement results published earlier in the literature (Bahl et al., 1980; Dahele & Lee, 1982; Dahele et al., 1987; Fan & Lee, 1991; Kumar & Dhubkarya, 2011; Lee et al., 1983; Liu & Hu, 1996a; Row, 2004; Shinde et al., 2010). Furthermore, the accuracy and validity of the proposed method of the SVM model was also verified on an ARCMA prototyped in this work.

2. SVM Modeling for computation the resonant frequency of ARCMAs

2.1. Support vector machine
In machine learning, SVM is a new generation supervised learning model (Christodoulou et al., 2006). N-dimension optimum hyperplane that separated two groups into data is occurred by the SVM. The SVM model is closely associated with ANN and it has an artificial network consisting of the two layers and feed-forward. An empirical risk minimization derived by minimizing the squares of the error on the data-set is not used by the SVM. It separates two groups into data by using the structural risk minimization in statistical learning theory. The SVM can be used in the classification and regression problems. The basic idea in the SVM regression method is determined by a linear discriminant function reflecting the characteristics of data-set. This function used for classification, regression, or other tasks is called hyperplane or margin. The SVM effectively constructs a maximum hyperplane (maximum margin) having equidistant from both of the data in a high or infinite dimensional space. The SVM can solve a problem having a linear or nonlinear structure. It easily finds a solution for

linearly separable problems. However, this method used to solve the linearly separable problem is not sufficient for nonlinear problem. One way to solve this problem is to map the data on to a higher dimensional space and then to use a linear classifier in the higher dimensional space. This mapping to a higher dimensional space in SVM is performed by using kernel function (Vapnik, 1998).

In this paper, the SVM was used to predict the resonant frequency (f_r) of ARCMAs. The problem related to the estimation of the f_r can be stated as follows. In the training process, the training data have been taken as a set of m training pairs {(x_0, f_{r0}), (x_1, f_{r1}), ... (x_m, f_{rm})} by considering x as the input variable vector and f_r as the output. x is physical dimensions (a_o, a_j, and h) and dielectric constant values (ε_r) of the simulated ARCMAs and f_r is also resonant frequencies of ARCMAs. Given a set of observed discrete data {(x_i, f_{ri}), $x_i \in R, f_{ri} \in R, i = 1, 2, ..., m$}, the SVM learning method in its basic form creates an $f_r^*(x)$ function. Starting from these samples of the input/output values of f_r, the goal is to find a function $f_r^*(x)$, which approximates, as well as is a possible unknown function $f_r(x)$. Using the support vector regression, f_r^* is defined as:

$$f_r^*(x) = \langle w, \varphi(x) \rangle + b \tag{1}$$

where $\langle ... \rangle$ denotes the inner product, and φ is a nonlinear mapping vector that performs a transformation of the input vector to a high dimensional space. The patterns that are not linearly separable are extended to a new higher dimension space where it is possible to separate them with a linear hyperplane. w and b are the weighting vector and bias, respectively, which are obtained by minimizing the primal convex objective function (Regression Risk) defined as:

$$R_{reg} = \frac{1}{2}\|w\|^2 + C \sum_{i=1}^{m} m^\varepsilon(x, f_r, f_r^*) \tag{2}$$

$$\xi_i = m^\varepsilon(x, f_{ri}, f_{ri}^*) \tag{3}$$

$$\min_{w, b, \xi} = \frac{1}{2} w^T w + C \sum_{i=1}^{m} \xi_i$$
$$\text{subject to } f_{ri}(w.x_i - b) \geq 1 - \xi_i$$
$$\xi_i \geq 0 \tag{4}$$

where ξ_i are called slack variables and they are used for measuring the occurred error at point (x_i, f_{ri}). C is a deterministic parameter that shows the change rate of the size of the margin and amount of error in training process and $m^\varepsilon(x, f_r)$ is a general loss function. Since the given objective function Equation 2 has no local minima and it guarantees the global minimum, which is one superiority of SVM on the other pattern recognition methods, particularly neural networks.

In our work, the so-called ε-insensitive loss function developed by Vapnik (1998) is used:

$$\xi_i = \begin{cases} 0, & \text{if } |f_{ri} - f_{ri}^*(x_i)| \leq \varepsilon \\ |f_{ri} - f_{ri}^*(x_i)| - \varepsilon, & \text{else} \end{cases} \tag{5}$$

This defines an ε tube so that if the predicted value is within the tube, the loss is zero, while if the predicted point is outside the tube, the loss is the magnitude of the difference between the predicted value and the radius of the tube.

This primal problem is transformed into dual problem, and then it can be solved by using Lagrange multipliers method (Bertsekas, 1995). The Lagrange function, J can be given as;

$$J(w, b, \alpha) = \frac{1}{2} w^T.w - \sum_{i=1}^{m} \alpha_i \{f_{ri}(w.x_i - b) - 1 + \xi_i\} \tag{6}$$

This primal problem is not preferred for solving. Therefore, this primal problem is transformed into dual problem. At this point, there are usually two main reasons mentioned for solving this problem in the dual:

(1) The duality theory provides a convenient way to deal with constraints.

(2) The dual optimization problem can be written in terms of dot product, thereby making it possible to use kernel functions.

The dual problem has the same optimal value as the primal problem. To describe the dual problem, Equation 6 is expanded as:

Maximize

$$Q_2(\alpha) = \sum_i \alpha_i - \frac{1}{2} \sum_i \sum_j \alpha_i \alpha_j f_{ri} f_{rj} x_i x_j \tag{7}$$

subject to

$$\sum_i \alpha_i f_{ri} = 0 \tag{8}$$

$$C \geq \alpha \geq 0 \tag{9}$$

The acquired dual problem can be solved by quadratic programming techniques. Here, $\alpha_1, \alpha_2, ... \alpha_i$ are Lagrange multipliers that correspond each sample. The Karush–Kuhn Tucker theorem is important for this optimization problem (Cristianini & Shawe-Taylor, 2000). The theorem is essential for solving the w, b, and α parameters. While the values of α are greater than zero, the expression in Equation 11 is applied for each training point. When the values of α are equal to zero, the points are excluded for not affecting the hyperplane. The values of the α^* are assumed to be the best solution for the dual problem. The expressions for the solution of w and b are given, respectively, as follows:.

$$w^* = \sum_i^m \alpha_i^* f_{ri} \, x_i = \sum_{\text{support vectors}} \alpha_i^* f_{ri} \, x_i \tag{10}$$

$$b^* = 1 - w^* x_i \tag{11}$$

x vector can be expressed as in Equation 12:

$$f_r^*(x) = \text{sign} \left(\sum_{i=1}^m \alpha_i^* f_{ri} \, x_i . x + b^* \right) \tag{12}$$

The decision function obtained by using the kernel function is given in Equation 13:

$$f_r^*(x) = \text{sign} \left(\sum_{i=1}^m \alpha_i^* f_{ri} \, K(x_i, x) + b^* \right) \tag{13}$$

Some common kernel functions are given below:

(1) Linear kernel function $\rightarrow K(x, y) = x^T y + c$

It is given by the inner product $<x, y>$ plus an optional constant c.

(2) Polynomial kernel function $\rightarrow K(x, y) = (x^T y + c)^d$

Adjustable parameters are the slope α, the constant term c, and the polynomial degree d.

(3) Gaussian kernel function $\rightarrow K(x, y) = \exp\left(-\frac{\|x-y\|^2}{2\sigma^2}\right)$

The adjustable parameter σ plays a major role in the performance of the kernel, and should be carefully tuned to the problem at hand.

(4) Exponential kernel function $\rightarrow K(x, y) = \exp\left(-\frac{\|x-y\|}{2\sigma^2}\right)$

(5) Laplacian kernel function $\rightarrow K(x, y) = \exp\left(-\frac{\|x-y\|}{2\sigma}\right)$

Different kernel functions have been tried for our problem in SVM model. The best results have been obtained with Gaussian kernel function. So, Gaussian kernel function was used for SVM model in this work. After several trials, value of σ was found as 28, which yielded satisfactorily results, and was employed in this work.

2.2. Geometry of ARCMAs and simulation phase

As shown in Figure 1, ARCMA has an annular ring patch formed by loading a circular slot with radius a_i on a circular patch of radius a_o on the substrate having overall relative dielectric constant ε_r on the ground plane.

In order to determine the resonant frequency values, simulations were performed by means of the IE3D™ software for 100 ARCMAs having different dimensions and various substrate dielectric constants which are tabulated in Table 1. The antennas operate over the frequency range 0.55–3.71 GHz corresponding to UHF band. In the simulations, the antennas were supposed to a probe feed with 50 Ω. For meshing process, cell/wavelength rate values were assumed as 40 in limit of 4 GHz. The built-in optimization module of the IE3D™ was utilized to determine the feed point that gives the best return loss value with the objective function S_{11} (dB) < -10 for the resonant frequencies at TM_{11} mode.

2.3. Training and test phases of the SVM model

The physical dimensions (a_o, a_i, and h) and dielectric constant values (ε_r) of the simulated ARCMAs were given as inputs and their respective resonant frequency values of IE3D™ were given as output to the SVM model. For this model, 88 of ARCMAs are employed for training phase, while 12 of ARCMAs are used for test phase. The block diagram of SVM model is shown in Figure 2. The parameters of the SVM model used in this work are tabulated in Table 2. To evaluate the performances of the constructed model for both training and test, the following average percentage error (APE) equation is assigned:

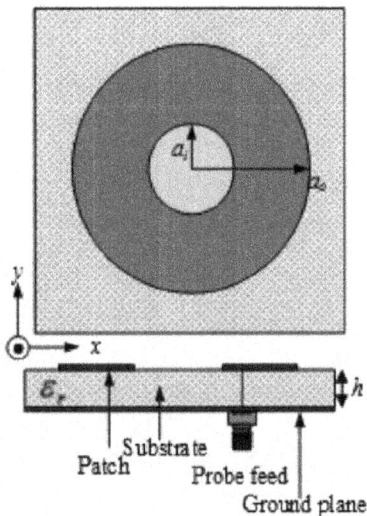

Figure 1. The geometry of ARCMA.

Table 1. Physical and electrical parameters of simulated ARCMAs

Simulation numbers	Antenna dimensions (mm)			ε_r
	a_o	a_i	h	
	15	2, 4, 6, 8, 10	0.640	4.50
	20	3, 6, 9, 12, 15	0.640	4.50
	25	4, 8, 12, 16, 20	0.640	4.50
	30	5, 10, 15, 20, 25	0.640	4.50
	35	6, 12, 18, 24, 30	0.640	4.5
	15	2, 4, 6, 8, 10	1.570	2.33
	20	3, 6, 9, 12, 15	1.570	2.33
	25	4, 8, 12, 16, 20	1.570	2.33
	30	5, 10, 15, 20, 25	1.570	2.33
	35	6, 12, 18, 24, 30	1.570	2.33
4×25	15	2, 4, 6, 8, 10	2.500	9.80
	20	3, 6, 9, 12, 15	2.500	9.80
	25	4, 8, 12, 16, 20	2.500	9.80
	30	5, 10, 15, 20, 25	2.500	9.80
	35	6, 12, 18, 24, 30	2.500	9.80
	15	2, 4, 6, 8, 10	3.175	2.20
	20	3, 6, 9, 12, 15	3.175	2.20
	25	4, 8, 12, 16, 20	3.175	2.20
	30	5, 10, 15, 20, 25	3.175	2.20
	35	6, 12, 18, 24, 30	3.175	2.20

$$APE = \sum \left| \frac{f_{IE3D} - f_{SVM}}{f_{IE3D}} \right| \times 100 \qquad (14)$$

As shown in Figure 3, resonant frequency values, which are computed by SVM, and results obtained from simulations are in a good harmony and APE was calculated as 0.952% for the training data.

To test the performances of the SVM model constructed here, 12 simulated ARCMAs whose electrical and psychical parameters listed in Table 3 are employed. The predicted resonant

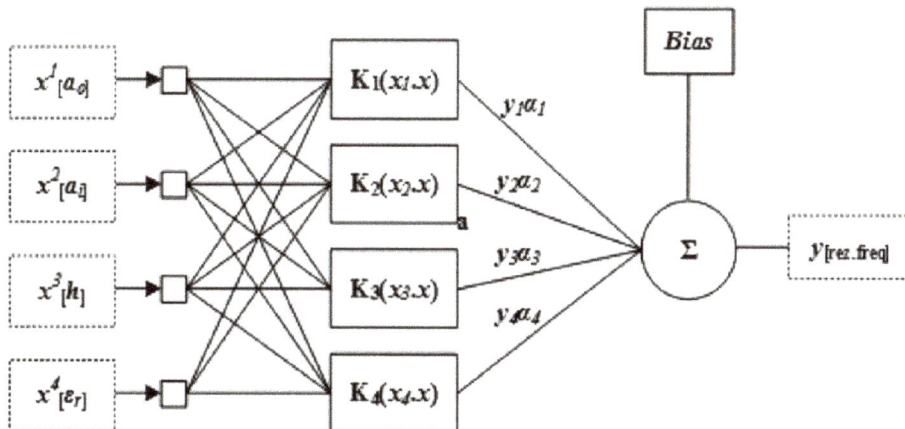

Figure 2. Block diagram of the SVM Model.

Table 2. The SVM parameters	
Parameters	Set type/value
Kernel function	Gaussian
Kernel function coefficient (σ)	30
Penalty weight (C)	100000
Slack variables (ξ)	0.001
Number of input	4
Number of output	1

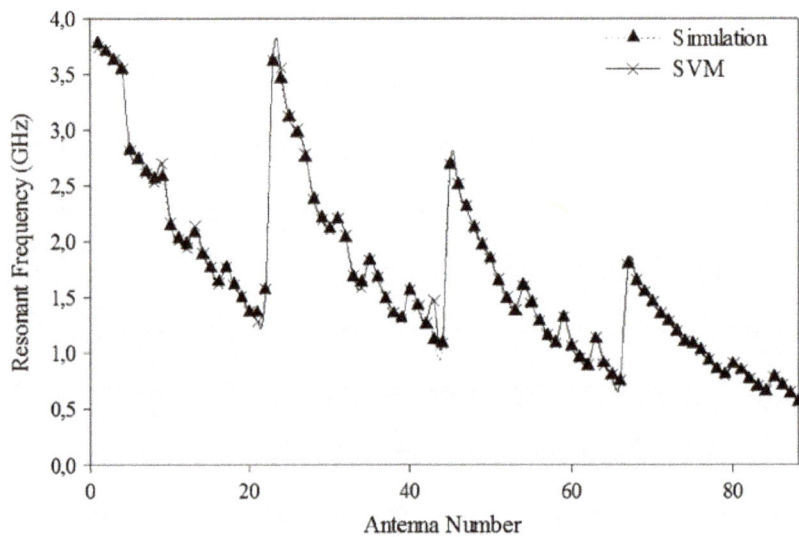

Figure 3. The comparative results of simulation and SVM for training phase.

Antenna number	Antenna parameters				
	Patch dimensions (mm)			ε_r	h/λ_d
	a_o	a_i	h		
1	15	4	2.5	9.8	0.045
2	15	6	1.57	2.33	0.027
3	15	10	3.175	2.2	0.056
4	20	3	0.64	4.5	0.009
5	20	6	1.57	2.33	0.021
6	20	15	2.5	9.8	0.027
7	25	4	3.175	2.2	0.035
8	25	12	1.57	2.33	0.015
9	30	10	0.64	4.5	0.005
10	30	20	3.175	2.2	0.024
11	35	12	0.64	4.5	0.005
12	35	24	2.5	9.8	0.015

Table 3. Physical and electrical parameters of 10 simulated ARCMAs for test process

Table 4. The resonant frequencies and APE values for test process

Antenna number	Resonant frequencies (GHz)						Percentage errors (%)				
	f_{IE3D}	SVM	Calculated by				SVM	Pintzos and Pregla (1978)	Wu and Rosenbaum (1973)	Bahl and Stuchly (1992)	Kumar and Dhubkarya (2011)
			Pintzos and Pregla (1978)	Wu and Rosenbaum (1973)	Bahl and Stuchly (1992)	Kumar and Dhubkarya (2011)					
1	1.734	1.733	1.735	1.814	1.813	2.888	0.029	0.058	4.614	4.556	66.551
2	3.323	3.285	3.265	3.181	3.181	3.749	1.149	1.745	4.273	4.273	12.820
3	3.563	3.496	2.917	2.843	2.844	2.396	1.882	18.131	20.208	20.180	32.753
4	2.006	2.023	1.995	2.025	2.025	5.234	0.825	0.548	0.947	0.947	160.917
5	2.594	2.573	2.576	2.534	2.535	3.700	0.829	0.694	2.313	2.274	42.637
6	1.033	1.030	1.041	1.027	1.026	0.804	0.339	0.774	0.581	0.678	22.168
7	2.258	2.249	2.135	2.354	2.355	5.748	0.400	5.447	4.252	4.296	154.562
8	1.833	1.830	1.832	1.785	1.785	1.854	0.179	0.055	2.619	2.619	1.146
9	1.189	1.197	1.193	1.160	1.159	1.564	0.697	0.336	2.439	2.523	31.539
10	1.547	1.616	1.420	1.395	1.395	1.175	4.461	8.209	9.825	9.825	24.047
11	1.021	1.016	0.960	0.920	0.936	0.901	0.490	5.975	9.892	8.325	11.753
12	0.582	0.571	0.542	0.689	0.495	0.786	1.890	6.873	18.385	14.948	35.052
APE							1.098	4.070	6.696	6.287	49.662

frequency results and corresponding percentage errors of the SVM model are tabulated in Table 4. For further comparison, the numerical results of several methods previously published in the literature (Bahl & Stuchly, 1992; Kumar & Dhubkarya, 2011; Pintzos & Pregla, 1978; Wu & Rosenbaum, 1973) are also listed in Table 4. It is apparent from Table 4 that all the SVM model give the remarkable results in comparison with those calculated by the methods presented in the literature (Bahl & Stuchly, 1992; Kumar & Dhubkarya, 2011; Pintzos & Pregla, 1978; Wu & Rosenbaum, 1973).

3. Numerical results and fabrication of ARCMA

To verify the validity of the proposed models, the resonant frequency results computed in this study were also compared with those of several suggestions reported elsewhere (Bahl & Stuchly, 1992; Kumar & Dhubkarya, 2011; Pintzos & Pregla, 1978; Wu & Rosenbaum, 1973) over several measurement data of ARCMAs published earlier in the literature (Bahl et al., 1980; Dahele & Lee, 1982; Dahele

Table 5. Physical and electrical parameters of of ARCMAs published earlier in literature

	Patch dimensions (mm)			ε_r	h/λ_d
	a_o	a_i	h		
Dahele, Lee, and Wong (1987)	50	25	1.59	2.32	0.007
Bahl et al. (1980)	20	10	3.18	2.32	0.040
Fan and Lee (1991)	50	25	1.59	2.32	0.007
Liu and Hu (1996b)	14.2	7.1	0.355	2.65	0.006
Kumar and Dhubkarya (2011)	17.2	8.6	1.6	4.2	0.022
Dahele and Lee (1982)	70	35	1.59	2.32	0.005
Lee et al. (1983)	70	35	1.59	2.3	0.005
Row (2004)	30	10	0.8	4.4	0.007
Shinde et al. (2010)	35	17.5	1.53	4.3	0.010
Shinde et al. (2010)	17.5	8.75	1.53	4.3	0.021
This study	13	2	2.54	4.5	0.054

Table 6. The measured and calculated resonant frequencies for ARCMAs								
					Resonant frequencies (GHz)			
	$f_{measured}$	**Simulated**		f_{SVM}	**Calculated by**			
		IE3D™	HFSS™		Pintzos and Pregla (1978)	Wu and Rosenbaum (1973)	Bahl and Stuchly (1992)	Kumar and Dhubkarya (2011)
Dahele, Lee, and Wong (1987)	0.878	0.880	0.880	0.880	0.886	0.850	0.854	0.877
Bahl et al. (1980)	2.450	2.500	2.410	2.539	2.337	2.171	2.149	2.297
Fan and Lee (1991)	0.891	0.880	0.880	0.880	0.886	0.850	0.854	0.877
Liu and Hu (1996b)	2.880	2.880	2.890	2.878	2.907	2.790	2.811	2.882
Kumar and Dhubkarya (2011)	1.989	2.030	1.980	2.046	2.035	1.850	1.855	1.997
Dahele and Lee (1982)	0.625	0.620	0.620	0.617	0.627	0.600	0.609	0.622
Lee et al. (1983)	0.626	0.630	0.620	0.623	0.629	0.610	0.612	0.625
Row (2004)	1.243	1.210	1.200	1.210	1.213	1.150	1.171	1.590
Shinde et al. (2010)	0.940	0.950	0.940	0.947	0.954	0.890	0.897	0.941
Shinde et al. (2010)	1.960	1.970	1.910	1.970	1.972	1.790	1.801	1.936
This study	3.000	3.030	2.970	3.050	2.800	3.070	2.518	8.392
APE (%)				1.393	1.940	5.530	6.150	19.820

et al., 1987; Fan & Lee, 1991; Lee et al., 1983; Liu & Hu, 1996a; Kumar & Dhubkarya, 2011; Row, 2004; Shinde et al., 2010), and over an ARCMA fabricated with the material of Rogers™ TMM 4, as well. Table 5 shows physical and electrical parameters of ARCMAs published earlier in the literature (Bahl et al., 1980; Dahele & Lee, 1982; Dahele et al., 1987; Fan & Lee, 1991; Lee et al., 1983; Liu & Hu, 1996a; Kumar & Dhubkarya, 2011; Row, 2004; Shinde et al., 2010).

The computed resonant frequency values and corresponding APEs have been given in Table 6. The resonant frequency results simulated by using IE3D™ and HFSS™ are also given in Table 6 so as to confirm the simulations performed in this study. It is shown that our simulated results are well agreed with measured ones. It is clearly seen from Table 6that our resonant frequency results are generally in very good agreement with the measured results as compared to those calculated by other suggestions (Bahl & Stuchly, 1992; Kumar & Dhubkarya, 2011; Pintzos & Pregla, 1978; Wu & Rosenbaum, 1973). Moreover, the proposed model here not only provides the better agreement but also allows us to compute the resonant frequency of ARCMA in a very simple manner without dealing with any other complicated functions and transformations.

The accuracy and validity of the proposed models were also tested on the measurement data of ARCMA, which was fabricated in this work. The simulated and measured return loss plots obtained by means of Agilent E5071B ENA Series RF network analyzer were illustrated in Figure 4. Notice that the measurement results may include some tolerances because of material production, geometry

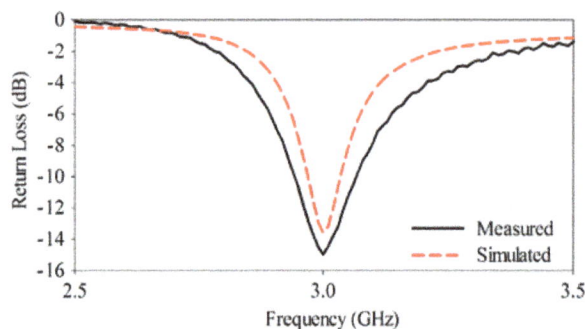

Figure 4. The simulated and measured return loss plots.

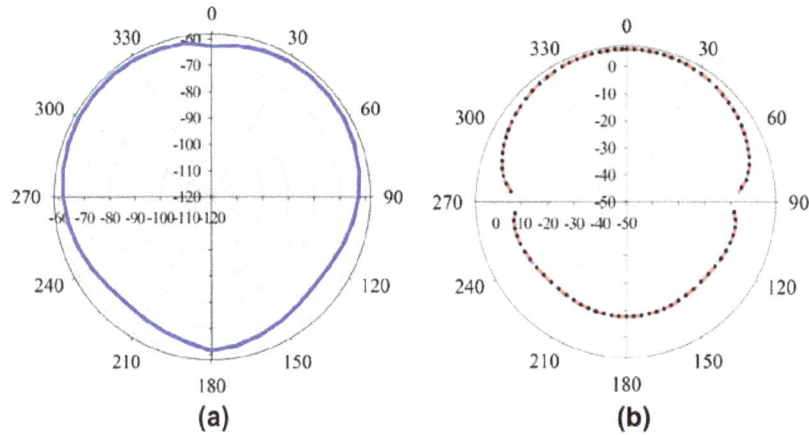

Figure 5. The simulated radiation patterns of fabricated ARCMA at 3 GHz for (a) ▬ E_ϕ for θ = 90°, (b) ▬ E_θ ϕ = 0° and ··· ϕ = 90°.

etching, and feed connector misalignment in the fabrication process. The measured and computed resonant frequency results of the antenna are also given in Table 6, and it can be seen that the proposed model for resonant frequency of the ARCMA provides the best fit with the measurement.

Figure 5 shows the measured two-dimensional (2D) radiation patterns E_θ and E_ϕ fields for the fabricated ARMA at 3 GHz. It is seen that the radiation patterns have good performance and approach omnidirectional radiation characteristics. The measured gain and half-power beam width (HPBW) achieved are 6.545 dBi and 110.5°, respectively.

4. Conclusion

In this paper, an application of SVM model is successfully implemented for the prediction of accurate resonant frequency of ARCMAs. IE3D™ simulation software was used to define resonant frequency of 100 ARCMAs. SVM model, physically, and electrical parameters of 88 ARCMAs were utilized training data; 12 ARCMAs were utilized for the test. It was seen that computed results with SVM for training and test data are in a good agreement with the simulation results. This method achieves the more accurate results as compared with those of the methods proposed in the literature. Additionally for validation study of our model, ARCMA is fabricated and measurement result is found in good agreement with the SVM result. The SVM approach is simple and fast modeling which produces more accurate results for the resonant frequency of the ARCMAs with less computational time and least errors. The most important advantages of this model are accuracy and easy to implement for the engineering problems which include the high nonlinearity.

Funding
This work is supported by the Scientific Research Fund Department of Mersin University [grant number BAP-FBE EEMB (AK) 2014-2 DR].

Author details
Ahmet Kayabasi[1]
E-mail: ahmetkayabasi@selcuk.edu.tr
ORCID ID: http://orcid.org/0000-0003-3047-0743
Ali Akdagli[2]
E-mail: aliakdagli@gmail.com
[1] Department of Electronic & Automation, Silifke-Tasucu Vocational School, Selcuk University, 33900, Silifke, Mersin, Turkey.
[2] Faculty of Engineering, Department of Electrical & Electronics Engineering, Mersin University, 33343, Ciftlikkoy, Mersin, Turkey.

References
Akdagli, A., Toktas, A., Kayabasi, A., & Develi, I. (2013). An application of artificial neural network to compute the resonant frequency of E-shaped compact microstrip antennas. *Journal of Electrical Engineering - Elektrotechnicky Casopis, 64*, 317–322.
Ali, S. M., Weng, C., & Kong, J. (1982). Vector Hankel transform analysis of annular-ring microstrip antenna. *IEEE Transactions on Antennas and Propagation, 30*, 637–644. http://dx.doi.org/10.1109/TAP.1982.1142870
Bahl, I. J., & Stuchly, S. S. (1992). Closed-form expressions for computer-aided design of microstrip ring antennas. *International Journal of Microwave and Millimeter-Wave Computer-Aided Engineering, 2*, 144–154. http://dx.doi.org/10.1002/(ISSN)1522-6301
Bahl, I. J., Stuchly, S. S., & Stuchly, M. A. (1980). A new microstrip radiator for medical applications. *IEEE Transactions on Microwave Theory and Techniques, 28*,

1464–1469.
http://dx.doi.org/10.1109/TMTT.1980.1130268

Bertsekas, D. P. (1995). *Nonlinear programming*. Belmont, MA: Athena Scientific.

Bhattacharyya, A. K., & Garg, R. (1985). Input impedance of annular ring microstrip antenna using circuit theory approach. *IEEE Transactions on Antennas and Propagation, 33*, 369–374. http://dx.doi.org/10.1109/TAP.1985.1143584

Chen, Z. N. (2000). Radiation pattern of a probe fed L-shaped plate antenna. *Microwave and Optical Technology Letters, 27*, 410–413. http://dx.doi.org/10.1002/(ISSN)1098-2760

Chew, W. (1982). A broad-band annular-ring microstrip antenna. *IEEE Transactions on Antennas and Propagation, 30*, 918–922. http://dx.doi.org/10.1109/TAP.1982.1142913

Christodoulou, C., Martinez-Ramon, M., & Balanis, C. (2006). *Support vector machines for antenna array processing and electromagnetics*. San Rafael, CA: Morgan & Claypool Publishers.

Cristianini, N., & Shawe-Taylor, J. (2000). *An introduction to support vector machines and other kernel-based learning methods*. Cambridge: Cambridge University Press. http://dx.doi.org/10.1017/CBO9780511801389

Dadgarnia, A., & Heidari, A. A. (2010). A fast systematic approach for microstrip antenna design and optimization using ANFIS and GA. *Journal of Electromagnetic Waves and Applications, 24*, 2207–2221. http://dx.doi.org/10.1163/156939310793699037

Dahele, J. S., & Lee, K. F. (1982). Characteristics of annular-ring microstrip antenna. *Electronics Letters, 18*, 1051–1052. http://dx.doi.org/10.1049/el:19820718

Dahele, J. S., Lee, K. F., & Wong, D. (1987). Dual-frequency stacked annular-ring microstrip antenna. *IEEE Transactions on Antennas and Propagation, 35*, 1281–1285. http://dx.doi.org/10.1109/TAP.1987.1143997

Deshmukh, A. A., & Kumar, G. (2007). Formulation of resonant frequency for compact rectangular microstrip antennas. *Microwave and Optical Technology Letters, 49*, 498–501. http://dx.doi.org/10.1002/(ISSN)1098-2760

Deshmukh, A. A., Phatak, N. V., Nagarbovdi, S., & Ahuja, R. (2013). Analysis of broadband E-shaped microstrip antennas. *International Journal of Computer Applications, 80*, 0975–8887.

El-khamy, S. E., El-Awadi, R. M., & El-Sharrawy, E. B. A. (1986). Simple analysis and design of annular ring microstrip antennas. *IEE Proceedings H Microwaves, Antennas and Propagation, 133*, 198–202. http://dx.doi.org/10.1049/ip-h-2.1986.0035

Fan, Z., & Lee, K. F. (1991). Hankel transform domain analysis of dual-frequency stacked circular-disk and annular-ring microstrip antennas. *IEEE Transactions on Antennas and Propagation, 29*, 867–870.

Gomez-Tagle, J., & Christodoulou, C. G. (1997). Extended cavity model analysis of stacked microstrip ring antennas. *IEEE Transactions on Antennas and Propagation, 45*, 1626–1635. http://dx.doi.org/10.1109/8.650074

Gurel, C. S., & Yazgan, E. (2010). Analysis of annular ring microstrip patch on uniaxial medium via Hankel transform domain immittance approach. *Progress in Electromagnetics Research M, 11*, 37–52. http://dx.doi.org/10.2528/PIERM09071404

Harrington, R. F. (1993). *Field computation by moment methods*. Piscataway, NJ: IEEE Press. http://dx.doi.org/10.1109/9780470544631

Kayabasi, A., Bicer, M. B., Akdagli, A., & Toktas, A. (2011). Computing resonant frequency of H-shaped compact microstrip antennas operating at UHF band by using artificial neural networks. *Journal of the Faculty of Engineering and Architecture of Gazi University, 26*, 833–840.

Kumar, R., & Dhubkarya, D. C. (2011). Design and analysis of circular ring microstrip antenna. *Global Journal of Researches in Engineering, 11*, 10–14.

Kumar, G., & Ray, K. P. (2003). *Broadband microstrip antennas*. Norwood, MA: Artech House.

Kumar, A., & Shukla, C. K. (2012). Artificial neural network employed to design annular ring microstrip antenna. *International Journal on Computer Science and Engineering, 4*, 556–564.

Lee, K.F., Dahele, J.S., & Ho, K.Y. (1983). Annular-ring and circular-disc microstrip antennas with and without air gaps. In *13th European Microwave Conference* (pp. 389–394). Nurnberg, Germany.

Liu, H., & Hu, X. F. (1996a). An improved method to analyze the input impedance of microstrip annular-ring antennas. *Journal of Electromagnetic Waves and Applications, 10*, 827–833. http://dx.doi.org/10.1163/156939396X00801

Liu, H., & Hu, X. F. (1996b). Input impedance analysis of microstrip annular ring antenna with thick substrate. *Progress in Electromagnetic Research, 12*, 177–204.

Motevasselian, A. (2011). Specteral domain analysis of resonant characteristics and radiation patterns of a circular disk and annular ring microstrip antenna on uniaxial substrate. *Progress in Electromagnetics Research, 21*, 237–251. http://dx.doi.org/10.2528/PIERM11091002

Pintzos, S. G., & Pregla, R. (1978). A simple method for computing the resonant frequencies of microstrip ring resonators. *IEEE Transactions on Microwave Theory and Techniques, 26*, 809–813. http://dx.doi.org/10.1109/TMTT.1978.1129491

Richards, W. F., Jai-Dong, O., & Long, S. (1984). A theoretical and experimental investigation of annular, annular sector, and circular sector microstrip antennas. *IEEE Transactions on Antennas and Propagation, 32*, 864–867. http://dx.doi.org/10.1109/TAP.1984.1143432

Row, J. S. (2004). Dual-frequency circularly polarised annular-ring microstrip antenna. *Electronics Letters, 40*, 153–154. http://dx.doi.org/10.1049/el:20040123

Sathi, V., Ghobadi, C. H., & Nourinia, J. (2008). Optimization of circular ring microstrip antenna using genetic algorithm. *International Journal of Infrared and Millimeter Waves, 29*, 897–905. http://dx.doi.org/10.1007/s10762-008-9382-5

Shinde, J., Shinde, P., Kumar, R., Uplane, M. D., & Mishra, B. K. (2010). Resonant frequencies of a circularly polarized nearly circular annular ring microstrip antenna with superstrate loading and airgaps. In *Kaleidoscope: Innovations for future networks and services* (pp. 1–7). Pune, India.

Tokan, N. T. (2008, April). Support vector design of the microstrip antenna. *Signal Processing, Communications and Applications, SIU 2008, IEEE 16th* (pp. 1–4). Aydin. http://dx.doi.org/10.1109/SIU.2008.4632716

Tokan, N. T., & Gunes, F. (2008). Support vector characterization of the microstrip antennas based on measurements. *Progress in Electromagnetics Research B, 5*, 49–61. http://dx.doi.org/10.2528/PIERB08013006

Vapnik, V. N. (1998). *Statistical learning theory*. New York, NY: Wiley.

Wolff, I., & Knoppik, N. (1971). Microstrip ring resonator and dispersion measurement on microstrip lines. *Electronics Letters, 7*, 779–781. http://dx.doi.org/10.1049/el:19710532

Wong, K. (2002). *Compact and broadband microstrip antennas*. New York, NY: Wiley. http://dx.doi.org/10.1002/0471221112

Wu, Y. S., & Rosenbaum, F. J. (1973). Mode chart for microstrip ring resonators (Short papers). *IEEE Transactions on Microwave Theory and Techniques, 21*, 487–489. http://dx.doi.org/10.1109/TMTT.1973.1128039

3

A review on future planar transmission line

Ashok Kumar[1], Garima Saini[1] and Shailendra Singh[2]*

*Corresponding author: Shailendra Singh, SGIT, Ghaziabad, India
E-mail: ershailendra1987@gmail.com
Reviewing editor: Wei Meng, Wuhan University of Technology, China

Abstract: Substrate Integrated Waveguide (SIW) is an emerging trend in planar technology. The SIW structure consists of two rows of conducting cylinders in a dielectric substrate. The average power-handling capacity of an SIW structure is primarily determined by its substrate materials and its geometric topology. The power-handling capability depends on the nature of SIW circuits. Microwave and millimeter wave components can be integrated on SIW. It acts as a bridge between planar and non-planar technology. By using innovative fabrication techniques, SIW technology combines the classical planar circuits and metallic waveguide. The animus of the study is to provide the idea on design and fabrication of SIW structures and comparison between different topologies used for size reduction and dominant mode bandwidth enhancement.

Subject: Telecommunication

Keywords: SIW; millimeter waves; SIW cavity; Q-factor

1. Introduction

The Substrate Integrated Waveguide (SIW) technology represents an emerging approach for the implementation and integration of microwave, millimeter components. SIW permits to realize traditional rectangular waveguide components in planar form. It is compatible with planar processing techniques such as standard printed circuit board (PCB) or low-temperature co-fired ceramic (LTCC) technology (Bozzi, Georgiadis, & Wu, 2011). SIW structures are used for many applications in microwave, millimeter wave and broadband wireless communications. SIW are high-density integration technique structures implemented by using two rows of conducting cylinders or slots in a dielectric substrate that electrically connect two parallel metal plates.

ABOUT THE AUTHOR

Ashok Kumar received his BTech degree in Electronics and Communication Engineering from Uttar Pradesh Technical University, Lucknow, India in 2008.Currently, he pursuing his M.E. (Master of Engineering) from NITTTR, Chandigarh, India. He has more than 4 year teaching experience. His research area of interest includes Waveguide, Antenna & Microwave and millimeter-wave passive components design and measurement.

PUBLIC INTEREST STATEMENT

Cell phones, Television, Radio etc. all these components are important part of our daily life. In all these we are using antenna, filter, couplers etc. as an integrating components. Day by day we are trying to miniaturize the things by using small size integrated components but these components can operate at very low power and fabrication of all these components on a chip is very difficult. So to reduce this small size and low power handling problem we are using a different type of medium which can provide a facility to design small size components on a single chip which are capable to handle moderate power also.

SIW technology allows integration of passive components; active components and antennas in a single substrate, due to this it reduce losses and parasitic effects (Bozzi, Perregrini, & Wu, 2008). Basically, the waveguide are the best transmission line for high power and high-frequency application. Planar transmission line has the advantage of operation at low frequency but suffers from power handling capacity. SIW is basically a combination of waveguide and planar transmission lines. It has moderate power-handling capacity and it is planar also. It is also referred as the "laminated waveguide". In laminated waveguide, the electromagnetic wave leakage is prevented by using conductor side wall via holes smaller than a quarter wavelengths ($\lambda/4$) (Hiroshi, Takeshi, & Funii, 1998).

Rectangular waveguides have been mostly used in the implementation of microwave and millimeter-wave components and systems with their features such as low insertion loss, high-quality factor (Q-factor) with high power-handling capability. The waveguide have some disadvantages such as bulky size, complex manufacturing and non-planar geometry (Tarck & Ke, 2013).

2. SIW characteristics

There are three types of losses present in the SIW structure i.e. conductor loss, dielectric loss and radiation loss (Bozzi et al., 2008). The conductor loss (α_c) is due to the finite conductivity of metal.

Conductor loss can be decreased by using highly conductive metal with increased substrate height. Dielectric loss (α_d) occurs due to finite conductivity of dielectric substrate. The radiation loss occurs in structure due to the energy leakage through the gaps. The conductor loss (α_c) and dielectric loss (α_d) is explained as (Deslandes & Wu, 2006)

$$\alpha_c = \frac{R_s \left(2h\pi^2 + l^3 k^2 \right)}{l^3 h \beta \eta} \tag{1}$$

$$\alpha_d = \frac{k^2 \tan \delta}{2\beta} \tag{2}$$

$$\tag{3}$$
$$\tan \delta = \frac{\sigma}{\omega \varepsilon}$$

where, α_c is the conductor loss, α_d is the dielectric loss; β is the phase constant, k is the free space wave number; h is the height of substrate, $\tan \delta$ is the dielectric loss tangent; η is the intrinsic impedance of the medium; R_s is the surface resistivity of the conductors; σ is the conductivity; ε is the permittivity.

Dielectric loss is not affected by the geometry of SIW structure, so it can be minimized by using good dielectric substrate. Radiation losses can be minimized by keeping space p is small and diameter d of the vias large. The radiation losses can be overcome by using condition as $p/d < 2.5$ (Bozzi, Pasian, Perregrini, & Wu, 2009). SIW modes practically coincide with a subset of guided modes of the rectangular waveguide, such as TE_{n0} modes with $n = 1, 2, ...$ and so on. TM modes are not supported by SIW due to the gaps between metal vias. In fact transverse magnetic fields determine longitudinal surface currents, which are subject to strong radiation due to presence of the gap (Daniels & Heath, 2007). If slots cut the surface current, a large amount of radiation will appear and if the slots are cut along the direction of flow of surface current then there is only small amount of radiation takes place as shown in Figure 2 (Xu & Wu, 2005).

$$\vec{n} * \vec{H} = \vec{J}_s \tag{4}$$

where, $\vec{n}$ = unit vector normal to the surface; $\vec{H}$ = magnetic field intensity; $\vec{J}_s$ = electric surface current density.

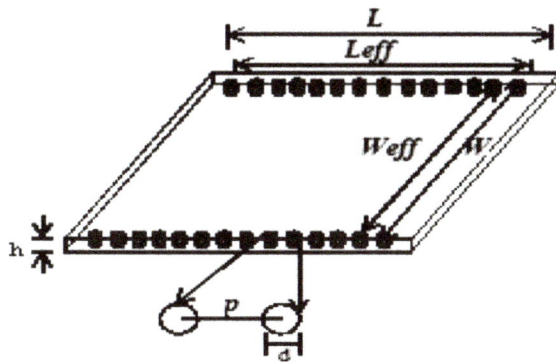

Figure 1. Basic SIW structure.

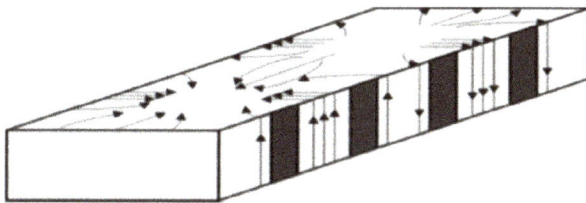

Figure 2. Surface current distribution on rectangular waveguide.

The design analysis is explained in terms of size and bandwidth. Cut-off frequency of the fundamental mode is determined by the width of the SIW. In recent years, there are different waveguide topologies have been proposed to improve the compactness and bandwidth of the SIW structures (Bozzi et al., 2011). The different topologies have been applied for the size reduction of SIW i.e. Substrate Integrated folded Waveguide (SIFW), Half-mode Substrate Integrated waveguide (HMSIW), Folded Half-mode Substrate Integrated Waveguide (FHMSIW), and Quadri-Folded Substrate Integrated Waveguide (QFSIW). In the SIFW, a metal septum permits folding of the waveguide, thus reducing the waveguide width by a factor of $\left(9\varepsilon_r\right)^{-1/2}$ which is depicted in Figure 3 (Grigoropoulos, Sanz-Izquierdo, & Young, 2005)

In Figure 4, the HMSIW permits the compactness early 50% which is based on the approximation of vertical cut of the Waveguide as virtual magnetic wall (Lai, Fumeaux, Wei, & Vahldieck, 2009). The FHMSIW is the combination of SIFW and HMSIW which help in size reduction (Zhai, Hong, & Wu, 2008).

The Quadri-folded Substrate Integrated Waveguide is designed using a special C-type slot in the middle conductor layer which is shown in Figure 5 (Zhang, Cheng, & Fan, 2011).

The comparison in topologies for size reduction is shown in Table 1.

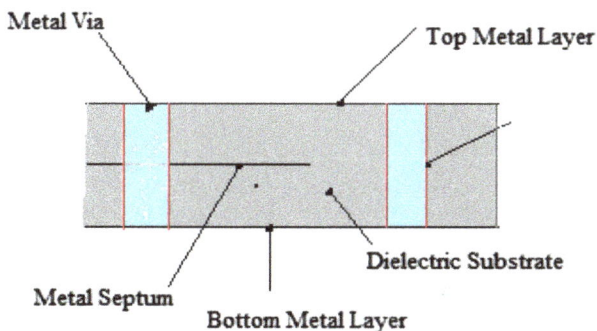

Figure 3. Substrate integrated folded waveguide (SIFW).

Table 1. The comparison in topologies for size reduction

SIW Topology	Size reduction approx. (%)
HMSIW (Lai et al., 2009)	50
FHMSIW (Zhai et al., 2008)	75
QFSIW (Zhang et al., 2011)	89

Table 2. The comparison in topologies for bandwidth enhancement

SIW Topology	Enhancement of bandwidth approx.(%)
SISW (Bozzi et al., 2005)	40
RSIW (Che et al., 2008)	73

For the bandwidth improvement, the topologies used are Substrate Integrated Slab Waveguide (SISW) and Ridge SIW.

In SISW, the dielectric medium is periodically pricked with air-filled holes, which are located in the lateral portion of the waveguide (Bozzi et al., 2005). In this approach, bandwidth was enhanced shown as in (Figure 6).

Ridge was applied through a row of thin, partial-height metal posts located at the center of the longer side of the waveguide. This modification is shown in Figure 7 (Che, Li, Russer, & Chow, 2008).

The comparison in topologies for bandwidth enhancement is shown in Table 2.

Low-loss material is the foundation for developing high-performance-integrated circuits and systems. The material selection is very critical for antenna development. The most used materials for

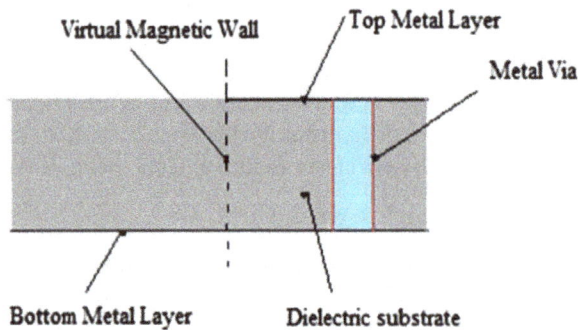

Figure 4. Half-mode substrate integrated waveguide (HMSIW).

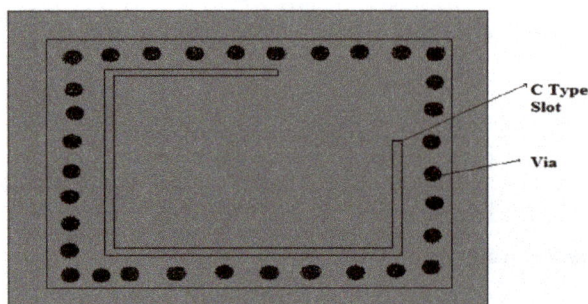

Figure 5. Quadri-folded substrate integrated waveguide (QFSIW).

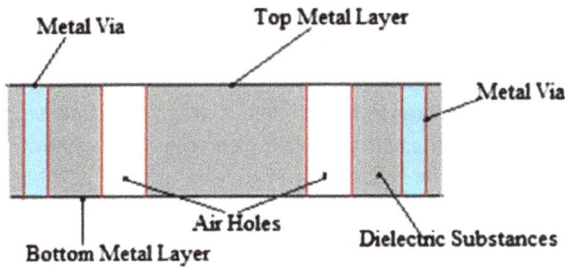

Figure 6. Substrate integrated slab waveguide (SISW).

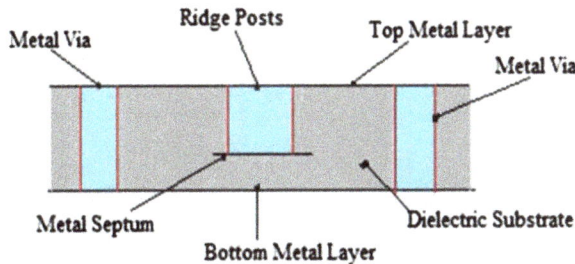

Figure 7. Ridge substrate integrated waveguide (RSIW).

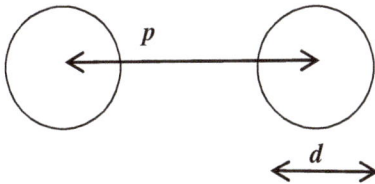

Figure 8. The diameter d and pitch p of via.

above-mentioned topologies are Rogers RT/duroid® 5880 glass microfiber-reinforced PTFE composite and RT/ duroid® 6002 for conventional PCB processing, which are easily sheared with laser and machined to the required shape. The holes can be easily drilled mechanically into these materials compared to ceramics (Tarck & Ke, 2013). The dimensional stability and a good thermal stability of the material is the important consideration in the SIW design. The novel materials as plastic, paper and textile are helpful to design cost-effective components in the field of academic and industrial research for various microwave applications. SIW antennas were fabricated using polyethylene terephthalate (PET) substrate (Moro, Collado, Via, Georgiadis, & Bozzi, 2012). It is a flexible and cheap substrate material for the development of microwave components. Paper is another easily available, eco-friendly and cheap material for SIW fabrication. The electronic circuit can be implemented on paper by ink-jet printing and there is no need of chemical etching or use of acids. The SIW interconnects and components on paper substrate by ink-jet printing were implemented (Moro, Kim, Bozzi, & Tentzeris, 2012). The paper-based implementation is well suited for the design of SIW components and used for conformal shape, arbitrary geometry and multilayered configuration. The textile material is used for the fabrication of a wearable SIW antenna (Agneessens, Bozzi, Moro, & Rogier, 2012).

The average power-handling capacity of an SIW structure is primarily determined by its substrate materials and its geometric topology. The power-handling capability depends on the nature of SIW circuits. Usually well-matched traveling-wave circuits can handle much more power than those counterparts with mismatch conditions and resonances (Tarck & Ke, 2013). In the case of filter designs, the SIW cavity resonators are fundamentally responsible for power handling capability in addition to the Microstrip-to-SIW transition (Chen & Ke, 2010; Ke & Chen, 2010).

3. Design analysis of siw component

The SIW parameters must be taken carefully in order to get desired result. The most significant advantage of SIW technology is to integrate all the components on the same substrate with high power-handling capacity as compared to other planar transmission lines i.e. Micro-strip line, strip line, Co-planar line, so there is the possibility to mount one or more chip-sets on the same substrate. SoS (system on substrate) represents the ideal system for cost effective, easy to fabricate and high performance mm-wave systems (Wu, 2006).

The size of the SIW cavity is determined by the corresponding Cut-off frequency (Deslands & Wu, 2002).

$$f_{cm,n} = \frac{c}{2\pi \sqrt{\mu_r \varepsilon_r}} \sqrt{\left(\frac{m\pi}{W_{eff}}\right)^2 + \left(\frac{n\pi}{L_{eff}}\right)^2} \tag{5}$$

$$f_{c1,0} = \frac{c}{2W_{eff} \sqrt{\mu_r \varepsilon_r}} \tag{6}$$

where, TE_{10} is the dominant mode. The W_{eff} and L_{eff} are the effective width and length of the SIW cavity respectively as shown in Figure 1.

$$W_{eff} = W - \frac{d^2}{0.95s} \tag{7}$$

$$L_{eff} = L - \frac{d^2}{0.95s} \tag{8}$$

where, W and L are the actual width and length of the SIW cavity; d is the diameter of the metal vias; and p is the distance between adjacent via holes as shown in Figure 8 (Sabri, Ahmad, & Bin Othman, 2012).

For SIW designs, following conditions need to be satisfied

$$d < \frac{\lambda_g}{5} \tag{9}$$

$$p \leq 4d \tag{10}$$

where,

$$\lambda_g = \frac{2\pi}{\sqrt{\frac{\varepsilon_r (2\pi f)^2}{c^2} - \left(\frac{\pi}{w}\right)^2}} \tag{11}$$

f is the operating frequency

c is the speed of light

The pitch p must keep small to reduce the leakage loss between adjacent posts. The post diameter may significantly affect the return loss of the waveguide.

4. SIW fabrication techniques

In the frequency range of 60–90 GHz and even at higher frequencies, PCB techniques have been widely used to implement the SIW structures. The PCB fabrication techniques have low manufacturing cost and great design flexibility. In this process, the metal holes are created either by micro-drilling or by laser cutting and their metallization are performed by using a conductive paste or metal plating (Deslandes, 2003).

At higher frequencies, radiation losses can occurs due to some technological limitations. The solution of this problem is that the holes are replaced by metallized slots in the circuit operation (Moldovan, Bosisio, & Wu, 2006). The recent development of LTCC substrate material extends the applicable frequency of the technique up to 100 GHz. The LTCC technology has the advantages of low conductor loss and low dielectric loss. It is very attractive for various integrated packaging.

LTCC provide a harmonic bed for embedded microwave and mm-wave passive components including antennas. LTCC

substrate material has a wide tunable range of thermal expansion coefficient (Wu & Huang, 2003). The SIW components were fabricated above 100 GHz by using Photo imageable thick-film materials. In this technique, there are best dimensional tolerances and low dielectric loss (Stephens, Young, & Robertson, 2005).

5. Applications of SIW

The applications of SIW are explained on the basis of passive and active components. The passive components of SIW are filters, circulators and couplers, etc. SIW filters provide good selectivity as compared to other planar filters. The inductive post filter is the simple form of SIW filters as shown in Figure 9 (Deslendes & Wu, 2003).

The zigzag filter topology as shown in Figure 10 (Mira, Mateu, & Cogollos, 2009) includes controllable cross-coupling to provide sharper response.

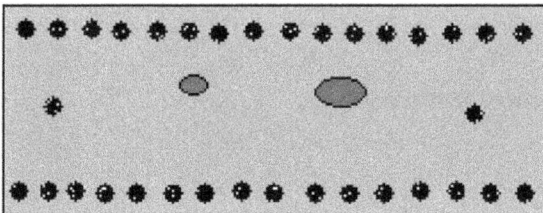

Figure 9. Inductive post filter.

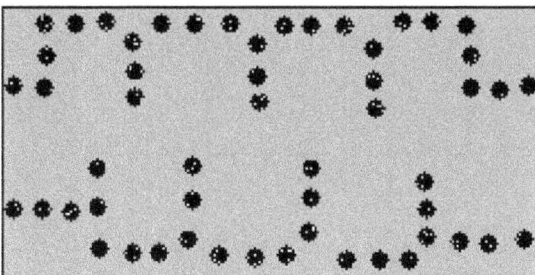

Figure 10. Zigzag meandered topology

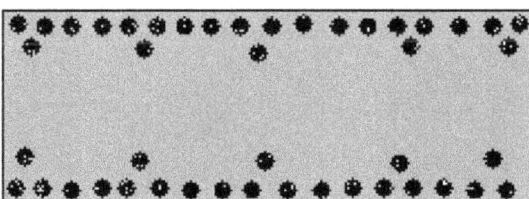

Figure 11. Iris-coupled filter.

The better electrical response provided by the post-wall iris coupled filter is discussed in (Hao, Hong, & Li, 2005; Mira & Bozzi, 2010) as shown in Figure 11.

SIW-based circulator used for high volume and medium power-level applications are shown in Figure 12 (D'Orazio & Wu, 2006; D'Orazio, Wu, & Helszajn, 2004).

Couplers find application in beam forming due to directional property and precision measurement. The most popular coupler is Riblet short-slot coupler which consists of two waveguides with coinciding H-planes and coupler outputs are in phase quadrature. The required output is obtained by the common wall is elimination as shown in Figure 13 (Cassivi, Deslendes, & Wu, 2002). The coupler geometry is based on even/odd mode analysis as TE_{10} mode is related to even mode and TE_{20} mode is related for odd mode.

The active SIW components are amplifier, oscillator and mixer, etc. The SIW technology is used in amplifiers for harmonic suppression. Harmonic suppression is achieved by lowering the frequency of second harmonic below cut-off frequency of SIW (He, Wu, & Hong, 2008; Wang & Park, 2012). The block diagram of the SIW amplifier is designed as shown in Figure 14 which consists of two iris-type inductive discontinuities and DC-decoupled transition.

SIW technology can be used to construct high Q resonant cavity. Low-phase noise oscillator could be designed by using high Q resonant cavity. The design concept of these SIW oscillators is shown in Figure 15 (Cassivi & Wu, 2003).

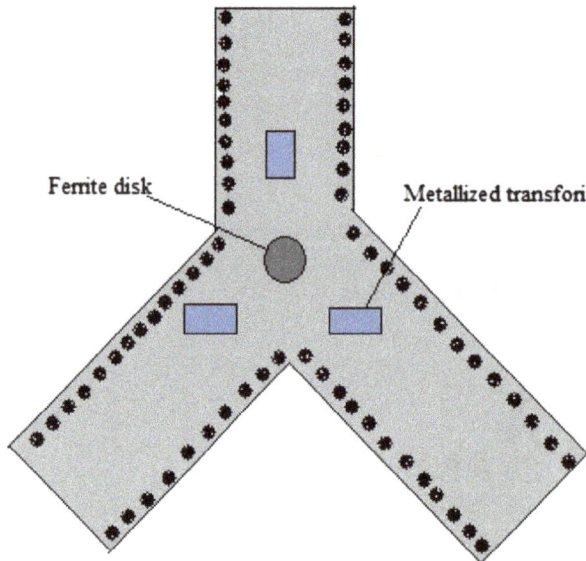

Ferrite disk

Metallized transfori

Figure 12. SIW-based circulator.

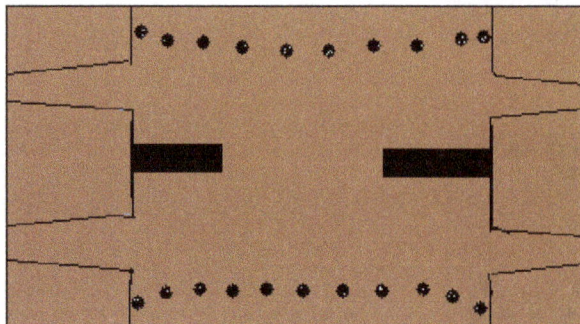

Figure 13. Riblet short-slot coupler.

Figure 14. SIW amplifier.

Figure 15. Positive feedback oscillator.

The positive feedback oscillator is the combination of an amplifier and SIW cavity which are formed on the same dielectric substrate. The SIW cavity acts as a frequency selector and the coupling level of the SIW cavity is maintained so that the gain of the loop is slightly greater than 1 dB. The total phase difference of the loop is 0 degree or an integer multiple of 2π (Barkhausen stability criterion).

6. Conclusion

Over the decades, different types of SIW has been evolved to overcome the bandwidth and size limitations of conventional SIW line. Different topologies have been applied for the size reduction of SIW i.e. Substrate Integrated folded Waveguide (SIFW), Half-mode Substrate Integrated waveguide(HMSIW), Folded Half-mode Substrate Integrated waveguide, and Quadri-folded Substrate Integrated Waveguide. Quadri-folded substrate integrated waveguide will give size reduction of approximate 90% but it will slightly increase the losses. For the bandwidth improvement, two different topologies have been used i.e. Substrate Integrated Slab Waveguide (SISW) and Ridge SIW. Ridged substrate integrated waveguide will give highest mono-modal bandwidth of operation as compared to conventional one. The use of novel material plays an important role for reducing the conductor and dielectric losses of SIW components.

Funding

The authors received no direct funding for this research.

Author details

Ashok Kumar[1]
E-mail: ashokmzn@gmail.com
ORCID ID: http://orcid.org/0000-0002-9090-9220
Garima Saini[1]
E-mail: garima@nitttrchd.ac.in
Shailendra Singh[2]
E-mail: ershailendra1987@gmail.com
ORCID ID: http://orcid.org/0000-0003-1862-5191
[1] Department of Electronics & Communication, NITTTR, Chandigarh, India.
[2] SGIT, Ghaziabad, India.

References

Agneessens, S., Bozzi, M., Moro, R., & Rogier, H. (2012). Wearable textile antenna in substrate integrated waveguide technology. *Electronics Letters, 48*, 985–987. http://dx.doi.org/10.1049/el.2012.2349

Bozzi, M., Deslandes, D., Arcioni, P., Perregrini, L., Wu, K., & Conciauro, G. (2005). Efficient analysis and experimental verification of substrate-integrated slab waveguides for wideband microwave applications. *International Journal of RF and Microwave Computer-Aided Engineering, 15*, 296–306. http://dx.doi.org/10.1002/(ISSN)1099-047X

Bozzi, M., Perregrini, L., & Wu, K. (2008). Modeling of conductor, dielectric, and radiation losses in substrate integrated waveguide by the boundary integral-resonant mode expansion method. *IEEE Transactions on Microwave Theory and Techniques, 56*, 3153–3161. http://dx.doi.org/10.1109/TMTT.2008.2007140

Bozzi, M., Pasian, M., Perregrini, L., & Wu, K. (2009). On the losses in substrate integrated waveguides and cavities. *International Journal of Microwave and Wireless Technologies, 1*, 395–401. http://dx.doi.org/10.1017/S1759078709990493

Bozzi, M., Georgiadis, A., & Wu, K. (2011). Review of substrate-integrated waveguide circuits and antennas. *IET Microwaves, Antennas & Propagation, 5,* 909–920. http://dx.doi.org/10.1049/iet-map.2010.0463

Cassivi, Y., & Wu, K. (2003). Low cost microwave oscillator using substrate integrated waveguide cavity. *IEEE Microwave and Wireless Components Letters, 13,* 48–50. http://dx.doi.org/10.1109/LMWC.2003.808720

Cassivi, Y., Deslendes, D., & Wu, K. (2002). Substrate integrated waveguide directional couplers. Asia Pacific Microwave Conference 2016, 1409–1412.

Che, W., Li, C., Russer, P., & Chow, Y. L. (2008). Propagation and band broadening effect of planar integrated ridged wave-guide in multilayer dielectric substrates. *IEEE MTT-S International Microwave Symposium Digest,* 217–220.

Chen, X. P., & Ke, W. (2010). Systematic overview of substrate integrated waveguide (SIW) filters: Design and performance tradeoffs. In *Asia-Pacific microwave conference workshop on recent progress in filters and couplers.*

D'Orazio, W., & Wu, K. (2006). Substrate-integrated-waveguide circulators suitable for millimeter-wave integration. *IEEE Transactions on Microwave Theory and Techniques, 54,* 3675–3680. http://dx.doi.org/10.1109/TMTT.2006.882897

Daniels, R. C., & Heath, R. W. (2007). 60 GHz wireless communications emerging requirements and design recommendations. *IEEE Vehicular Technology Magazine, 2,* 41–50. http://dx.doi.org/10.1109/MVT.2008.915320

Deslandes, D. (2003). Single-substrate integration technique of planar circuits and waveguide filters. *IEEE Transactions on Microwave Theory and Techniques, 51,* 593–596. http://dx.doi.org/10.1109/TMTT.2002.807820

Deslandes, D., & Wu, K. (2006). Accurate modeling, wave mechanisms, and design considerations of a substrate integrated waveguide. *IEEE Transactions on Microwave Theory and Techniques, 54,* 2516–2526. http://dx.doi.org/10.1109/TMTT.2006.875807

Deslands, D., & Wu, K. (2002). Design considerations and performance analysis of substrate integrated waveguide components. In *32nd European Microwave Conference* (pp. 1–4).

Deslendes, D., & Wu, K. (2003). Millimeter-wave substrate integrated waveguide filters. *IEEE Electrical Computer Engineering Canadian Conference, 3,* 1917–1920.

D'Orazio, W. D., Wu, K., & Helszajn, J. (2004). A substrate integrated waveguide degree-2 circulator. *IEEE Microwave and Wireless Components Letters, 14,* 207–209. http://dx.doi.org/10.1109/LMWC.2004.827867

Grigoropoulos, N., Sanz-Izquierdo, B., & Young, P. R. (2005). Substrate integrated folded waveguides (SIFW) and filters. *IEEE Microwave and Wireless Components Letters, 15,* 829–831. http://dx.doi.org/10.1109/LMWC.2005.860027

Hao, Z., Hong, W., & Li, H. (2005). A broadband substrate integrated waveguide (SIW) filter. *IEEE Antennas and Propagation Society International Symposium,* 598–601.

He, F. F., Wu, K., & Hong, W. (2008). Supression of second and third harmonics using lambda/4 low-impedance substrate integrated waveguide bias line in power amplifier. IEEE Xplore: Microwave and Wireless Components Letters, 18, 479–481.

Hiroshi, U., Takeshi, T., & Funii, M. (1998). Development of a laminated waveguide. *IEEE Transaction on Microwave Theory and Techniques, 46,* 2438–2443.

Ke, W., & Chen, X. P. (2010). Concept of substrate applied to filter design and reachable performances. In *European microwave week on recent advance in substrate integrated waveguide filters: Simulations, technologies and performances.*

Lai, Q., Fumeaux, Ch., Wei, H., & Vahldieck, R. (2009). Characterization of the propagation properties of the half-mode substrate integrated waveguide. *IEEE Transactions on Microwave Theory and Techniques, 57,* 1996–2004.

Mira, F., & Bozzi, M. (2010). Efficient design of SIW filters by using equivalent circuit models and calibrated space-mapping optimization. *International Journal of RF and Microwave Computer-Aided Engineering, 20,* 689–698. http://dx.doi.org/10.1002/mmce.v20:6

Mira, F., Mateu, J., & Cogollos, S. (2009). Design of ultra-wideband substrate integrated waveguide (SIW) filters in zigzag topology. *IEEE Microwave and Wireless Components Letters, 19,* 281–283. http://dx.doi.org/10.1109/LMWC.2009.2017589

Moldovan, E., Bosisio, R. G., & Wu, K. (2006). W-band multiport substrate-integrated waveguide circuits. *IEEE Transactions on Microwave Theory and Techniques, 54,* 625–632. http://dx.doi.org/10.1109/TMTT.2005.862670

Moro, R., Collado, A., Via, S., Georgiadis, A., & Bozzi, M. (2012). Plastic-based substrate integrated waveguide (SIW) components and antennas. *42th European Microwave Conference,* 1007–1010.

Moro, R., Kim, S., Bozzi, M., & Tentzeris, M. (2012). Novel inkjet printed substrate integrated waveguide (SIW) structures on low-cost Materials for wearable applications. *42th European Microwave Conference,* 72–75.

Sabri, S. S., Ahmad, B. H., & Bin Othman, A. R (2012). A review of substrate integrated waveguide (SIW) bandpass filter based on different method and design. *IEEE Asia-Pacific Conference on Applied Electromagnetics,* 210–215.

Stephens, D., Young, R., & Robertson, I. D. (2005). Millimeter-wave substrate integrated waveguides and filters in photoimageable thick-film technology *IEEE Transactions on Microwave Theory and Techniques, 53,* 3832–3838. http://dx.doi.org/10.1109/TMTT.2005.859862

Tarck, D., & Ke, W. (2013). Substrate integrated waveguide (SIW) techniques: The state-of-the- art developments and future trends. *Journal of University of Electronic Science & Technology of China, 42,* 171–192.

Wang, Z., & Park, C. W. (2012). Novel substrate integrated waveguide (SIW)-based power amplifier using SIW-based filter to suppress up to the fourth harmonic. *Asia-Pacific Microwave Conferences,* 830–832.

Wu, K. (2006). Towards system-on-substrate approach for future millimeter-wave and photonic wireless applications. In *Asia-Pacific Microwave Conference* (pp. 1895–1900).

Wu, K.-L., & Huang, Y. (2003). LTCC technology and its applications in high frequency front end modules. *International Symposium on Antennas Propagation and EM Theory,* 730–734.

Xu, F., & Wu, K. (2005). Guided-wave and leakage characteristics of substrate integrated waveguide. *IEEE Transaction on Microwave Theory and Techniques, 53,* 66–73.

Zhai, G. H., Hong, W., & Wu, K. (2008). Folded half mode substrate integrated waveguide 3 dB coupler. *IEEE Microwave and Wireless Components Letters, 18,* 512–514. http://dx.doi.org/10.1109/LMWC.2008.2001006

Zhang, C. A., Cheng, Y. J., & Fan, Y. (2011). Quadri-folded substrate integrated waveguide cavity and its miniaturized bandpass filter applications. *Progress In Electromagnetics Research C, 23,* 1–14. http://dx.doi.org/10.2528/PIERC11052401

4

Changes of transmission characteristics for different optic radiation incidence angles in filters protecting against hazardous infrared radiation

Grzegorz Gralewicz[1]*, Janusz Kubrak[2] and Grzegorz Owczarek[1]

*Corresponding author: Grzegorz Gralewicz, Department of Personal Protective Equipment, Central Institute for Labour Protection – National Research Institute, Czerniakowska 16, 00-701 Warsaw, Poland
E-mail: grgra@ciop.lodz.pl
Reviewing editor: Duc Pham, University of Birmingham, UK

Abstract: The paper presents the fundamental information concerning the types of protective optical filters used for protection against hazardous radiation within the visible and near-infrared spectrum range. The changes of transmission characteristics for different optic radiation angles of incidence with metallic reflective filters and interference filters have been analyzed. The results demonstrate that such changes exert no effect on the level of protection provided by the filters.

Subjects: Environmental & Health; Laser & Optical Engineering; Technology

Keywords: interference filters; infrared radiation; optical filters

1. Introduction

Protection against infrared radiation is provided by special filters mounted in glasses, goggles, or face protections. Such filters should block hazardous infrared radiation as well as reduce radiation within the visual spectrum range, which can cause dazzle, ensuring good visibility of the object/area of work at the same time.

Three basic types of protective optical filters that use protection against hazardous radiation within the visible (Vis) and near-infrared (NIR) spectrum range can be distinguished:

- Absorption filters,
- Metallic reflective filters, and
- Interference filters.

ABOUT THE AUTHORS

The subject of the Central Institute for Labour Protection – National Research Institute activity is conducting research and development works leading to new technical and organizational solutions in the field of labor protection, related to occupational safety, health, and ergonomics as well as other tasks essential for reaching the goals of the state's socioeconomic policy in this field.

VIGO SL Sp. z o.o. is highly specialized in production of high-quality interference optical coatings. Multilayer optical coatings designed and fabricated in our laboratories were applied in many different ranges of activities as ophthalmic, instrumental optics, spectroscopy, astronomy, medical science, and lightning technology.

PUBLIC INTEREST STATEMENT

This work discusses the information on the possibility of blocking the harmful infrared radiation in hot workstations with thin-film optics methods. The structure of the interference film and the results of laboratory tests on model filter solutions are presented. A new approach to the design of infrared radiation filters consists in developing thin-film coatings blocking the harmful infrared radiation with use of interferences occurring in the multilayered dielectric or metallic–dielectric coatings. The appropriate selection of thin-film structure, materials, their thickness, and sequence allows to produce a filter of highly effective infrared suppression and to control the values of transmittance in the visible range. The primary advantages of such solution are the extended filter life, high suppression of harmful radiation, and low absorption levels.

The absorption filters have been practically withdrawn from use due to an increase of filter temperature as a result of absorption of a large proportion of optic radiation against which they were intended to protect. Consequently, the comfort of their use was considerably deteriorated. In view of low level of use, the absorption filters have been excluded from the study.

The filters currently used in eye and face protection are made of glass or organic materials (mainly polycarbonate) and coated with a single metallic layer reflecting infrared radiation. They take advantage of characteristic properties of the metals used in their production, i.e. high radiation reflection coefficient within the IR spectrum and significant transmission coefficient within the visible spectrum range.

Interference filters (patent number P-401213) developed by the authors (Gralewicz et al., 2012a) making use of the principles of interference, i.e. overlapping of the waves leading to the reduction of the resultant wave amplitude (Feng, Elson, & Overfelt, 2005; Fuentes-Hernandez et al., 2011; Wang & Chen, 2005; Yaremchuk, Fitio, & Bobitski, 2006) are a novel solution designed for use in protective eyewear. In the case of interference filter, based on the theory of interference, there is a need to investigate the effect of changes of optic radiation incidence angle on the changes of filter properties.

In the present study, the selected features of transmission characteristics were tested for the effect of changes of optic radiation incidence angle in metallic reflective and interference filters. The analysis of this study achieved information, or changing the angle of incidence of a beam of optical radiation incident on the tested filters to protect against infrared radiation affects the change in the degree of protection filter.

2. Test samples—filters protecting against hazardous infrared radiation

For comparative analysis of the filters available in the market versus the interference filters developed by the authors, the following samples were prepared for the tests:

- Available on the market—metallic reflective filters on polycarbonate substrates—wafer of 50 mm diameter with a metallic layer of copper (Cu)—protection levels: 4–3, 4–5, 4–7 (the higher the protection level, the higher level of blocked IR radiation and appropriately lower transmission coefficient for visible spectrum radiation),
- Developed by the authors—interference filters on polycarbonate substrates—wafer of 50 mm diameter with interference coating made up of the layers of the following materials: aluminum, substance H4—LaTiO3 and silicon dioxide.

Filter samples used in the study are presented in Table 1.

A metallic reflective filter (Table 1: Numbers 4, 5, and 6) consists of the base: polycarbonate or mineral glass, and metallic coating layer of copper (Cu) or gold (Au). The metals used in their production are characterized by high radiation reflection coefficient within the IR spectrum and significant transmission coefficient within the visible spectrum range.

Table 1. Filter samples used in the study			
No	**Protection level**	**Filter type**	**Comments**
1	4–3	Interference filter	Developed by the authors
2	4–5	Interference filter	Developed by the authors
3	4–7	Interference filter	Developed by the authors
4	4–3	Metallic reflective filter	Commercially available
5	4–5	Metallic reflective filter	Commercially available
6	4–7	Metallic reflective filter	Commercially available

An interference optic radiation filter (Table 1: Numbers 1, 2, and 3) consists of the base: polycarbonate or mineral glass, and an appropriate sequence of layers made of dielectric materials with high refractive index and low refractive index, as well as metallic reflective layers (most frequently: silver for the visible portion of the spectrum, aluminum for UV, aluminum for IR), deposited with physical techniques involving evaporation under high vacuum conditions (Macleod, 2001; Sytchkova, 2011).

3. Testing methodology

The tests of filters protecting against hazardous IR radiation were performed using a Cary 5000-type spectrophotometer. The filter samples were positioned, so as to obtain incidence of the beam radiation perpendicular to the filter surface or parallel to the line of vision. Then, the angle of filter positioning was changed (30°, 45°) in relation to the beam radiation (Figure 1).

The measurement data used as the basis for determination of transmission characteristics of the filter were recorded. The characteristics of the tested filters were analyzed within the visible spectrum: 380–780 nm and infrared spectrum range 780–3,000 nm.

For the 380–780 nm range, the transmission coefficients which must meet the requirements specified in the relevant standards (EN 166, 2001; EN 171, 2002), i.e. ensure the required light transmission level, were determined.

For the 780–3,000 nm range, spectrum average IR transmission coefficients, which must meet the requirements specified in the relevant standards (EN 166, 2001; EN 171, 2002), i.e. ensure the required level of infrared radiation blockade were determined. The coefficients were calculated according to the following equations:

- Light transmission coefficient; light transmittance (Equation 1).

$$\tau_v = \frac{\int_{380\,nm}^{780\,nm} \tau_F(\lambda) \cdot V(\lambda) \cdot S_{D65\lambda}(\lambda) \cdot d\lambda}{\int_{380\,nm}^{780\,nm} V(\lambda) \cdot S_{D65\lambda}(\lambda) \cdot d\lambda} \tag{1}$$

- Transmission coefficient for the 780–1,400 nm spectrum range (Equation 2).

$$\tau_A = \frac{1}{63} \int_{780\,nm}^{1400\,nm} \tau(\lambda) \cdot d\lambda \tag{2}$$

- Transmission coefficient for the 780–2,000 nm spectrum range (Equation 3).

$$\tau_N = \frac{1}{123} \int_{780\,nm}^{2000\,nm} \tau(\lambda) \cdot d\lambda \tag{3}$$

- Transmission coefficient for the 780–3,000 nm spectrum range (Equation 4).

$$\tau_C = \frac{1}{223} \int_{780\,nm}^{3000\,nm} \tau(\lambda) \cdot d\lambda \tag{4}$$

where $S_{A\lambda}(\lambda)$ stands for spectral power distribution of standard illuminant CIE A (or a 3200 K light source for blue signal light). Cf. (ISO/CIE 10526, 1999); $S_{D65\lambda}(\lambda)$ stands for spectral power distribution of standard illuminant CIE D65. Cf. (ISO/CIE 10526, 1999); $V(\lambda)$ stands for the function of relative

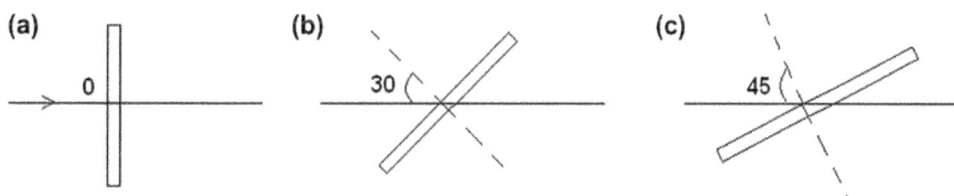

Figure 1. Changes in incidence angle of beam radiation: (a) 0° angle, (b) 30° angle, (c) 45° angle.

spectral luminous efficiency for daytime vision. Cf. (ISO/CIE 10526, 1999); and $\tau_F(\lambda)$ stands for spectral transmission coefficient of the filter.

The spectral values of spectral power distribution products $(S_{A\lambda}(\lambda), S_{D65\lambda}(\lambda)$ illuminants, relative spectral luminous efficiency $V(\lambda)$ of the eye, and spectral transmission coefficient $\tau_s(\lambda)$ of street signal light glass are specified in the European standards concerning personal eye protections (EN 166, 2001; EN 171, 2002).

The maximum transmittance (Equation 5) of the filter was also considered in the analysis of transmission characteristics.

$$T_{f\,max} = \frac{T^2}{(1-R)^2} = \frac{1}{(1+A/T)} \tag{5}$$

where R is the refraction index; T is the transmission coefficient; and A is the absorption coefficient.

Another value taken into account in the analysis of transmission characteristics of the filters is filter half-width $\Delta\lambda_{1/2}$, for which the transmittance is equal to 1/2 of the maximum transmittance to the following Equation 6 (Fuentes-Hernandez et al., 2011; Macleod, 2001).

$$\Delta\lambda_{1/2} = \frac{\lambda_{max} \cdot (1-R)}{m \cdot \pi \cdot \sqrt{R}} \tag{6}$$

where λ_{max} is the maximum transmittance; m is the row of interference; and R is the refraction index.

The filter half-width is dependent on the refraction index R and in the case of interference filters— on the order of interference. The higher R and the higher order of interference correlates with the higher filter half-width value.

The changes of transmission characteristics for different optic radiation angles of incidence with filters to the following Equation 7 (Macleod, 2001).

$$\lambda_{max}(\phi_1) = \frac{2 \cdot n_1 \cdot d_1}{m + \frac{\varepsilon}{\pi}} \cos\phi_1 \tag{7}$$

where n_1 is the transmission coefficient; m is the row of interference; and d is the thickness of the separating layer.

Figure 2 presents a sample of transmission characteristics $T(\lambda)$ of a filter within the 380–780 nm spectrum range, with $T_{f\,max}$ and $\Delta\lambda_{1/2}$ indicated.

4. Analysis of filters protecting against hazardous infrared radiation

Figures 3 and 4 present examples of transmission characteristics of a metallic reflective filter and the developed interference filter. The interference filters developed by the authors characterized by a steeper depth of characteristics are more significant as far as IR blocking is concerned in comparison with metallic reflective filters. The spectral transmittance values for ranges (780–1,400 and 780–2,000 nm) are in an order of magnitude of the value of the currently manufactured filter. Interference filters have a higher mechanical strength (according to EN 166, 2001; EN 171, 2002). This affects the reduction of costs in the long-term use (Gralewicz et al., 2012b).

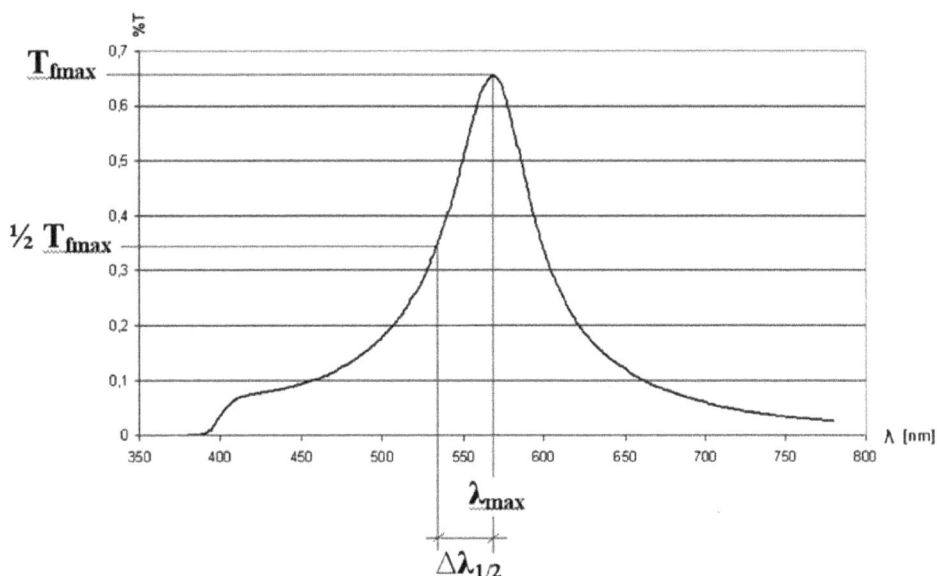

Figure 2. Sample transmission characteristics $T(\lambda)$ of a filter within the 380–780 nm spectrum range, with $T_{f\,max}$ and $\Delta\lambda_{1/2}$ indicated.

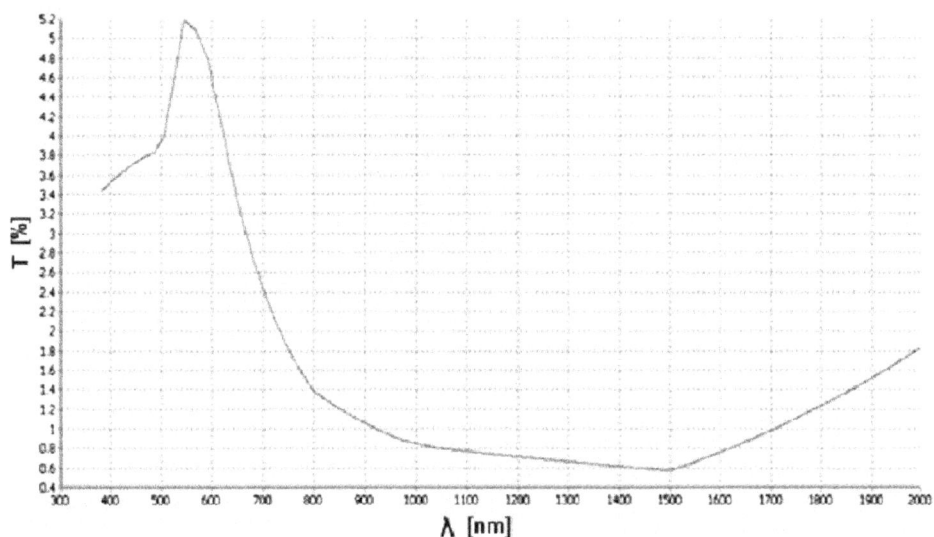

Figure 3. Transmission characteristics $T(\lambda)$ of a metallic reflective filter (protection level 4–5).

4.1. Results

A change of optic radiation incidence angle with the filter results in the change of $T_{f\,max}$ and $\Delta\lambda_{1/2}$. A shift of λ_{max} towards shorter wavelengths is observed. Figure 5 presents sample transmission characteristics of an interference filter, providing protection level of 4–5 for incidence angles: (a) 0°, (b) 30°, (c) 45°. Besides the shift of λ_{max} towards shorter wavelengths, broadening of the transmitted radiation beam corresponding to an increase of $\Delta\lambda_{1/2}$ occurs.

In Tables 2 and 3, maximum transmittance of the filter—$T_{f\,max}$ and filter half-width $\Delta\lambda_{1/2}$ corresponding with the change of optic radiation incidence angle with metallic reflective and interference filters are presented.

It is noteworthy that irrespectively of the angle at which the tested filter is positioned in relation to the beam radiation on spectrophotometer, the filter properties with respect to protection against IR radiation within the 780–2,000 nm spectrum range remain unaffected.

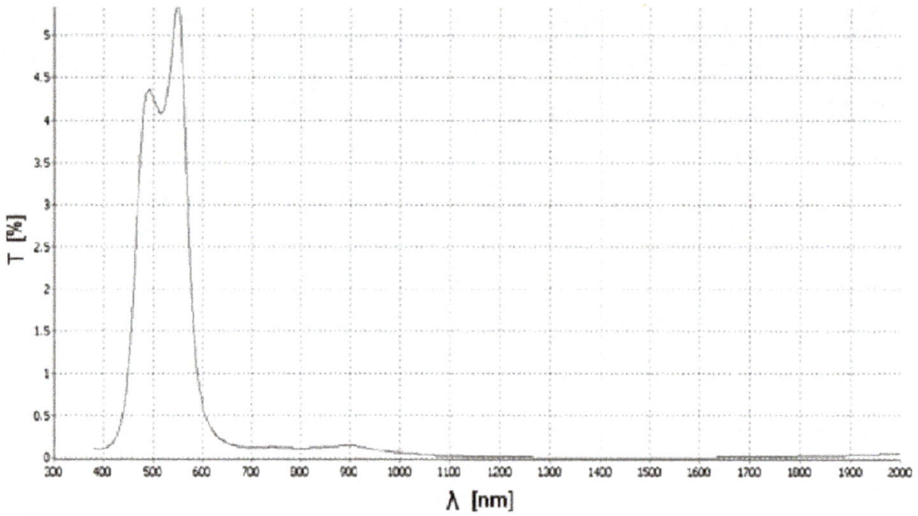

Figure 4. Transmission characteristics $T(\lambda)$ of an interference filter (protection level 4–5).

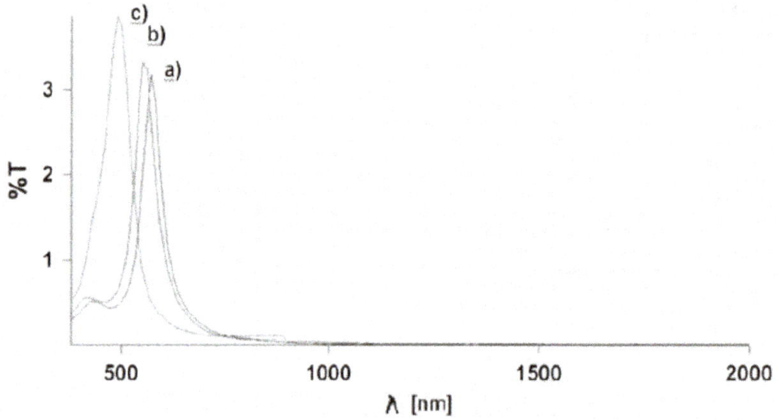

Figure 5. Sample transmission characteristics of an interference filter providing protection level of 4–5 for incidence angles: (a) 0°, (b) 30°, (c) 45°.

Table 2. Maximum filter transmittance values $T_{f\,max}$ correlated with the change of optic radiation incidence angle

$T_{f\,max}$ 380–780 nm	Optic radiation incidence angle		
	0°	30°	45°
Protection level 4–3			
Metallic reflective	18.784	18.409	17.341
Interference	13.654	15.493	15.401
Protection level 4–5			
Metallic reflective	2.937	2.622	2.055
Interference	3.944	4.617	5.898
Protection level 4–7			
Metallic reflective	0.600	0.473	0.312
Interference	0.319	0.323	0.439

Tables 3–5 present the results concerning light transmission coefficients and mean IR transmission coefficients within the 780–1,400 nm and 780–2,000 nm spectrum ranges for optic radiation incidence angles: 0°, 30°, 45°.

Table 3. Light transmission coefficients for optic radiation incidence angles: 0°, 30°, 45°—metallic reflective and interference filters

τ_v 380–780 nm	Optic radiation incidence angle		
	0°	30°	45°
Protection level 4–3			
Metallic reflective	14.133	14.333	14.750
Interference	9.082	10.546	12.319
Protection level 4–5			
Metallic reflective	2.021	1.857	1.594
Interference	1.470	2.887	3.199
Protection level 4–7			
Metallic reflective	0.353	0.292	0.203
Interference	0.160	0.212	0.219

Table 4. Mean IR transmission coefficients within the 780–1,400 nm range for optic radiation incidence angles: 0°, 30°, 45°—metallic reflective and interference filters

τ_A 780–1,400 nm	Optic radiation incidence angle		
	0°	30°	45°
Protection level 4–3			
Metallic reflective	1.486	1.428	1.658
Interference	1.003	0.575	0.681
Protection level 4–5			
Metallic reflective	0.216	0.190	0.190
Interference	0.038	0.008	0.003
Protection level 4–7			
Metallic reflective	0.043	0.032	0.033
Interference	0.037	0.039	0.039

Table 5. Mean IR transmission coefficients within the 780–2,000 nm range for optic radiation incidence angles: 0°, 30°, 45°—metallic reflective and interference filters

τ_N 780–2,000 nm	Optic radiation incidence angle		
	0°	30°	45°
Protection level 4–3			
Metallic reflective	0.942	0.935	1.167
Interference	0.563	0.367	0.528
Protection level 4–5			
Metallic reflective	0.134	0.118	0.131
Interference	0.006	0.021	0.013
Protection level 4–7			
Metallic reflective	0.023	0.0037	0.0038
Interference	0.039	0.042	0.042

Mean IR transmission coefficients for filters protecting against hazardous infrared radiation are determined for 780–1,400 nm and 780–2,000 nm spectrum range, according to the standard requirements (EN 166, 2001; EN 171, 2002), whereas the hazard levels at worksites exposed to infrared radiation are assessed for up to 3,000 nm range. In view of the above, the mean IR transmission coefficients were determined for the 780–3,000 nm spectrum range (Table 6).

Table 6. Mean IR transmission coefficients within the 780–3,000 nm range for optic radiation incidence angles: 0°, 30°, 45°—metallic reflective and interference filters

τ_C 780–3,000 nm	Optic radiation incidence angle		
	0°	**30°**	**45°**
Protection level 4–3			
Metallic reflective	0.524	0.527	0.678
Interference	0.362	0.258	0.414
Protection level 4–5			
Metallic reflective	0.055	0.043	0.053
Interference	0.013	0.037	0.008
Protection level 4–7			
Metallic reflective	0.018	0.052	0.053
Interference	0.050	0.054	0.055

5. Discussion

In the present research of transmission characteristics changes at different optic radiation incidence angles for metallic reflective and interference filters, the analysis was divided into two spectrum ranges: visible spectrum (380–780 nm) and IR spectrum (780–3,000 nm).

Within the visible 380–780 nm spectrum, on the basis of transmission characteristics, the light transmission coefficients τ_v and maximum transmittance $T_{f\,max}$ of the analyzed filters were determined for optic radiation incidence angles of 0°, 30°, 45°. For metallic reflective filters (protection levels: 4–3, 4–5, 4–7), an increase of the optic radiation incidence angle results in an increase of light transmission coefficients τ_v and a decrease of maximum transmittance $T_{f\,max}$ of the filter. In the case of interference filters (protection levels: 4–3, 4–5, 4–7), an increase of the optic radiation incidence angle results in an increase of light transmission coefficients τ_v and an increase of maximum transmittance $T_{f\,max}$ of the filter. A shift λ_{max} towards shorter wavelengths occurs, as well as broadening of the transmitted radiation band—an increase of $\Delta\lambda_{1/2}$. Higher transmittance of the visible spectrum in comparison with metallic reflective filters was observed for all the analyzed cases of interference filters.

For the 780–3,000 nm IR spectrum, IR transmission coefficients τ_A, τ_N, τ_C for optic radiation incidence angles: 0°, 30°, 45° were determined on the basis of transmission characteristics. The values of transmission coefficients obtained for the interference filters developed by the authors are lower by an order of magnitude in comparison with the values demonstrated for metallic reflective filters. The above finding provides evidence for more effective blocking of hazardous IR radiation by the developed interference filters.

6. Conclusions

- The investigated metallic reflective and interference filters comply with the requirements specified in the EN 166 and EN 171 standards.
- The values of transmission coefficients obtained for interference filters within the IR spectrum range from 780 to 2,000 nm are lower by an order of magnitude in comparison with the values demonstrated for metallic reflective filters.
- When the beam radiation is incident obliquely on the tested filters, $T_{f\,max}$ and $\Delta\lambda_{1/2}$ are changed. A shift of λ_{max} towards shorter wavelengths, as well as broadening of the transmitted beam (increase of $\Delta\lambda_{1/2}$) is observed.

- Irrespectively of the angle at which the tested filters are positioned in relation to beam radiation on spectrophotometer, the filter properties with respect to protection against IR radiation within the spectrum range up to 3,000 nm remained unaffected.

- Changes in optic radiation incidence angle in relation to the tested filters protecting against IR radiation cause no alterations in the protection level.

Funding

The publication has been based on the results of Phase II of the National Program "Safety and working conditions improvement," funded in the years 2011–2013 in the area of research and development works by the Ministry of Science and Higher Education. The Program coordinator: Central Institute for Labour Protection – National Research Institute.

Author details

Grzegorz Gralewicz[1]
E-mail: grgra@ciop.lodz.pl
Janusz Kubrak[2]
E-mail: jkubrak@vigo.com.pl
Grzegorz Owczarek[1]
E-mail: growc@ciop.lodz.pl

[1] Department of Personal Protective Equipment, Central Institute for Labour Protection – National Research Institute, Czerniakowska 16, 00-701 Warsaw, Poland.

[2] Vigo SL Sp. z o.o., Poznańska 129/133, 05-850 Ożarów Mazowiecki, Poland.

References

EN 166. (2001). Personal eye-protection—Specifications.

EN 171. (2002). Personal eye-protection—Infrared filters—Transmittance requirements and recommended use.

Feng, S., Elson, J., & Overfelt, P. (2005). Optical properties of multilayer metal-dielectric nanofilms with all-evanescent modes. *Optics Express, 13*, 4113–4124. http://dx.doi.org/10.1364/OPEX.13.004113

Fuentes-Hernandez, C., Owens, D., Hsu, J., Ernst, A. R., Hales, J. M., Perry, J. W., & Kippelen, B. (2011, May 1–6). The ultrafast nonlinear optical properties of induced transmission filters. In *Lasers and Electro-Optics (CLEO)* (pp. 1–2). Baltimore, MD: IEEE. ISBN: 978-1-4577-1223-4; INSPEC Accession Number: 12142085. Retrieved from http://ieeexplore.ieee.org/stamp/stamp.jsp?tp=&arnumber=5951283

Gralewicz, G., Owczarek, G., & Kubrak, J. (2012a, May). Interference filters protect against harmful infrared radiation for hot workplaces. Work safety. *Science and Practice, 5*, 12–15.

Gralewicz, G., Owczarek, G., & Kubrak, J. (2012b). Interference filters blocking harmful infrared radiation for hot workplaces. *Papers of the Institute of Electrical Engineering, 256*, 23–35.

ISO/CIE 10526. (1999). CIE standard illuminants for colorimetry.

Macleod, H. (2001, January 1). *Thin film optical filters* (3rd ed.). Taylor & Francis. http://dx.doi.org/10.1201/TFOPTICSOPT

Sytchkova, A. (2011). Reliable deposition of induced transmission filters with a single metal layer. *Applied Optics, 50*, C90–C94.

Wang, Q.-H., & Chen, R.-G. (2005). Interference filters in optically written display based on up-conversion of near infrared light. *Electronics Letters, 41*, 1217–1219. http://dx.doi.org/10.1049/el:20052509

Yaremchuk, I. Y., Fitio, V. M., & Bobitski, Y. V. (2006). Optical properties of multilayer thin-film interference filters. *8th International Conference on Laser and Fiber-Optical Networks Modeling, 8*, 117–120.

5

Decision about criticality of power transformers using whitenization weight functions on DGA caution levels

Vikal R. Ingle[1]* and V.T. Ingole[2]

*Corresponding author: Vikal R. Ingle, B.D. College of Engineering, Sevagram, Wardha, Maharashtra, India
E-mail: Chin_vikal@yahoo.com

Reviewing editor: Duc Pham, University of Birmingham, UK

Abstract: Power transformers are the most significant as well as the major asset of any power system network. The condition monitoring and assessment is the main concern in transformer management activities. As a first information source, dissolved gases-in-oil analysis (DGA) is universally accepted. The assessment of dissolved gases is characteristically observed analogous to grey system analysis. Grey system theory is supportive to the cases, when less information about the system is available. The cluster of grey incidences and whitenization weight functions classifies the factors of same type, in order to simplify a complex system. Three caution levels of key gases specified in IEEE standards are utilized in this study, to whiten the weight functions. The whitenization weight function with lower measure is selected for caution level-1. However, whitenization weight functions with middle measure are preferred for level-2 and level-3. Several key gas samples of the equal rating transformers are collected from gas analyzer section and utilized in condition assessment computations. The test samples are verified with variable and equal weight clustering criteria. The criticality judgment of transformer with variable weight clustering successfully identifies the crucial elements amongst samples.

ABOUT THE AUTHOR

Vikal R. Ingle was born on 1 January 1969. He received his BE (Electronics) and ME (Digital Electronics) degrees in 1996 and 2009, respectively. At present, he is an associate professor in Electronics engineering department of BDCE, Sevagram. He is a PhD researcher at PRMIT&R, Badnera. His major area of research includes electrical machines, Grey theory, and AI applications. He is a fellow of Institution of Engineers (India).

PUBLIC INTEREST STATEMENT

Power transformers are the most significant and the very expensive equipment of any electrical power supplying network. Reliable operation of this equipment is needed for uninterrupted power supply. Transformer outages can be catastrophic, and cause both direct and indirect costs to be incurred by industrial, commercial, and residential sectors. Direct costs include but are not limited to loss of production, idle facilities and labor, damaged or spoiled product, and damage to equipment. For commercial customers, the effects may include damage to electrical and electronic equipment, and in some cases, damage to goods. For residential customers, outages may cause food spoilage or damage to electrical equipment. In addition to direct costs, there are several types of indirect costs that may occur, such as accidental injuries, damage, legal costs, and increases in insurance rates. Therefore, vigilant management is the concern to extend their life, and to obtain the services for longer periods.

Subjects: Technology; Engineering & Technology; Electrical & Electronic Engineering; Fuzzy Systems

Keywords: DGA; key gas method; grey incidence analysis; whitenization weight functions; grey classes

1. Introduction

Power transformers play a key role in production and services, and in supplying the electricity to industrial and commercial sectors as well as to the domestic consumers. Maintaining the strength and reliability of the transformer has been a concern to avoid the power failure. There are several techniques for the maintenance, lifespan assessment, and condition evaluation of power system assets. Dealing with the problem of indicating and assessing the health of a transformer, several key measurements are available. The standards providing the guidance for use, analysis, and applications are included in ANSI/IEEE C57.104™ (2009) and IEC 60599 (1992). These standards are commonly known as the gas guides, which include the safety ranges of dissolved gases-in-oil. The dominating gases consist of hydrogen (H_2), methane (CH_4), acetylene (C_2H_2), ethylene (C_2H_4), ethane (C_2H_6), carbon monoxide (CO), and carbon dioxide (CO_2). These seven gases are referred as key gases in the literature. The IEC and IEEE specified that three caution levels of key gases are useful in condition judgments (Scatiggio & Pompili, 2013). In condition-based ranking, transformer's DGA data are evaluated against the established industry standards (Field, Cramer, & Antosz, 2002). Since the transformer condition was judged through health index or with criticality index, several groups assign scores as well (Jahromi, Piercy, Cress, Service, & Fan, 2009). These indices are vital in evaluating the state ranking of transformers (Abu-Siada, Arshad, & Islam, 2010). Researchers also attempt the different condition factors for every subsystem of transformer in preparing the concluding rank (Field et al., 2002; Hydroelectric Research and Technical Services Group, 2003; Toronto Hydro-Electric System Limited, 2010). However, condition-monitoring devices of power transformer are disseminated in nature and hardly interpret the inclusive and precise results for judgments about transformer health. As an immediate indicator, dissolved gas analysis is a simple and secured technique of power transformer testing. Significant weights are recommended in gas guide to main tank oil DGA. Furthermore, to draw a quantitative conclusion about the transformer reliability on numerical DGA data, appropriate assessment methods are desired.

Parametric data of transformer incorporated with soft-computing techniques are another kind of decision-making, applied in condition assessment of transformer. The several soft-computing methods are proposed and implemented based on DGA data intended for fault detection, criticality judgment, fault classifications, and state ranking. ANN with expert system (Wang, Liu, & Griffin, 1998), neuro-fuzzy inference system (Sun, Au, & Choi, 2007), fuzzy logic (Abu-Siada et al., 2010; Nemeth, Laboncz, & Kiss, 2009), and genetic algorithm (Zheng, Zhoa, & Wu, 2009) deduce the results effectively. However, these model-free methods necessitate massive data for precise analysis. Similarly, the existing parametric methods need large or reasonable samples with typical probability distributions. However, the conclusions drawn from quantitative analysis differ from that of qualitative results. In contrast, non-parametric test is competent in treating the distribution-free samples but applicable only to a continuous population distributions. Although in reality, the sample size may be prohibited from being large, either due to physical limitations or due to practical difficulties. Therefore, applying the statistical methods or model-free methods can hardly achieve useful solutions, when the system information becomes partially available. The solution to such problems with incomplete or non-deterministic information is always not unique. Whereas, the non-uniqueness is a basic law of the application of grey system theory and one can feel free to look at the problem with flexibility (Kuo, Yang, & Huang, 2008).

A system with partially known and partially unknown information is recognized as grey system. Grey system theory is useful in the condition, when less information about the system is available. It assists in determining the system's key factors and in identifying the factors' correlations. Grey numbers, grey relations, grey decision, grey predictions, and controls are the main subjects of grey system theory (Yang, 2008). The grey incidence analysis of grey theory is applied to the cases of different sample sizes and distributions. Relatively small computations are required and the conclusions drawn from quantitative analysis differ from that of qualitative results. The grey clusterings based on matrices of grey incidences or whitenization weight functions on grey numbers are useful in classifying the observational objects into predefined classes (Deng, 2005; Liu & Lin, 2006). Deng presented a grey whitenization weight function clustering method, wherein the weight of each index was calculated with the critical value of whitenization function (Deng, 2002). Zhang investigated the greyness of cluster result by establishing grey cluster on grey hazy set and combined the cluster result with cluster weight sequence (Zhang, 2002). Xiao et al. put forward grey optimal clustering, whitenization weight function constructed with the standard values of each class, and clustering performed with generalized weighted distance method (Xiao & Xiao, 1997). Liu et al. offered a grey fixed weight cluster decision analysis. The weight of each index has been determined by qualitative or quantitative analysis through Delphi method or analytic hierarchy process, and clustering carried out by whitenization function (Liu, Shen, Tan, & Guo, 2012). Liu and Xie proposed a grey cluster evaluation method based on triangle whitenization weight function; the method divides the values into a range of index "s" clusters so to fulfill the evaluation requirements. The calculation was conducted on grey fixed weight clustering (Liu & Xie, 2011). Qiu projected a grey correlation cluster analysis method (Qiu, 1995). Grey similarity matrices were calculated on the computation of grey correlation degree and clustering performed with maximal tree method or coding method.

The whitenization weight functions mainly classify the factors of same type in order to simplify the complex systems or phenomenon like DGA. This paper demonstrated the synthetic evaluation of DGA test samples, on both fixed and variable weight grey clustering decision. Three caution levels of key gases are utilized in whitening the three weight functions. The whitenization weight function with lower measure is selected for caution level-1. However, whitenization weight functions with middle measure are preferred for caution level-2 and level-3. While identifying the criticality of transformers, representative DGA samples are divided into three grey classes by means of grey clustering.

2. Grey clustering method

In classification of clustering method, grey clusters are divided into grey correlation cluster and grey whitenization weight function cluster. Among which, grey correlation cluster is mainly employed to incorporate the factors of same class for the simplification of a complex systems. However, grey whitenization weight function is majorly applied to inspect the presence of observational objectives in a predefined class. Grey clustering is also known as grey evaluation. The variable and fixed weight clustering is offered in whitenization weight functions. The fundamentals of both the methods are presented in next section, which covers a small part of grey incidence analysis but the conceptual framework is believed to be enough to realize the clustering methodology.

2.1. Grey clusters with variable weights

The grey clusterings based on matrices of grey incidences or whitenization weight functions of grey numbers are used in classifying the observational objects into predefined classes. In general, the whitenization weight function of the j-criterion and k-subclasses is determined by considering the objects of clustering or looking at all the same type of objects as a complete system. Assume that, there exist "n" objects to be clustered according to "m" cluster criteria into different grey classes. The clustering method based on the observational value of the ith objects, $i = 1, 2, \ldots, n$ with j-criterion, where $j = 1, 2, \ldots, m$. Then the ith objects are classified into kth grey class, where, $1 \leq k \leq s$. This process of computation is commonly known as grey clustering. Some of the imperative definitions are presented as follows:

Definition 1: All the s grey classes formed by the n objects, defined by their observational values at criterion j, are called the j-criterion with subclasses of k. The whitenization weight function on k-subclass of the j-criterion is denoted as $f_j^k(\cdot)$.

Definition 2: Assuming that the whitenization weight function $f_j^k(\cdot)$ for k-subclass of the j-criterion is shown in Figure 1 and the points $x_j^k(1)$, $x_j^k(2)$, $x_j^k(3)$ and $x_j^k(4)$ are called turning points of $f_j^k(\cdot)$.

Definition 3: Whitenization weight functions:
  (a) If the whitenization weight function $f_j^k(\cdot)$ above does not have first $x_j^k(1)$ and second $x_j^k(2)$ turning points then $f_j^k(\cdot)$ is called whitenization weight function of lower measure as shown in Figure 2.
  (b) If the second $x_j^k(2)$ and third $x_j^k(3)$ turning points of whitenization weight function $f_j^k(\cdot)$ coincide as shown in Figure 1 then the function $f_j^k(\cdot)$ is called a whitenization weight function of middle measure, shown in Figure 3.
  (c) If the whitenization weight function $f_j^k(\cdot)$ as shown in Figure 1 does not have third $x_j^k(3)$ and fourth $x_j^k(4)$ turning points then $f_j^k(\cdot)$ is called whitenization weight function of upper measure, shown in Figure 4.

PROPOSITION 1: (a) The typical whitenization weight function as shown in Figure 1 is expressed with:

$$f_j^k(x) = \begin{cases} 0, & x \notin \left[x_j^k(1), x_j^k(4)\right] \\ \frac{x - x_j^k(1)}{x_j^k(2) - x_j^k(1)}, & x \in \left[x_j^k(1), x_j^k(2)\right] \\ 1, & x \in \left[x_j^k(2), x_j^k(3)\right] \\ \frac{x_j^k(4) - x}{x_j^k(4) - x_j^k(3)}, & x \in \left[x_j^k(3), x_j^k(4)\right] \end{cases}$$

(b) The whitenization weight function of lower measure as shown in Figure 2 is given as:

$$f_j^k(x) = \begin{cases} 0, & x \notin \left[0, x_j^k(4)\right] \\ 1, & x \in \left[0, x_j^k(3)\right] \\ \frac{x_j^k(4) - x}{x_j^k(4) - x_j^k(3)}, & x \in \left[x_j^k(3), x_j^k(4)\right] \end{cases}$$

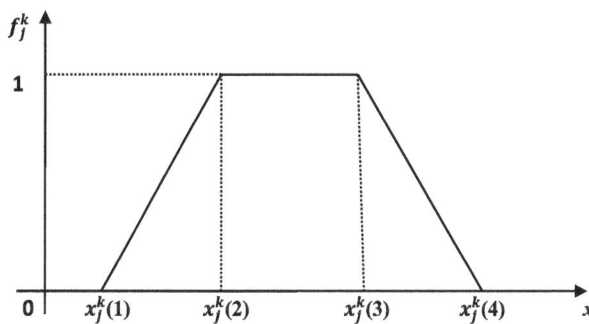

Figure 1. A typical whitenization function.

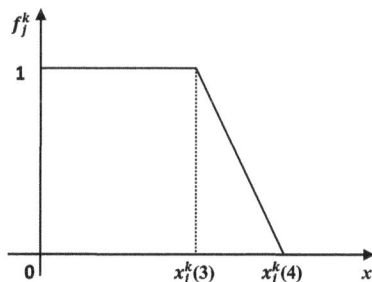

Figure 2. WW function of lower measures.

(c) The whitenization weight function of middle measure as shown in Figure 3 is given by:

$$
f_j^k(x) = \begin{cases}
0, & x \notin \left[x_j^k(1), x_j^k(4) \right] \\
\frac{x - x_j^k(1)}{x_j^k(2) - x_j^k(1)}, & x \in \left[x_j^k(1), x_j^k(2) \right] \\
1, & x = x_j^k(2) \\
\frac{x_j^k(4) - x}{x_j^k(4) - x_j^k(2)}, & x \in \left[x_j^k(2), x_j^k(4) \right]
\end{cases}
$$

(d) The whitenization weight function of upper measure as shown in Figure 4 is given as:

$$
f_j^k(x) = \begin{cases}
0, & x < x_j^k(1) \\
\frac{x - x_j^k(1)}{x_j^k(2) - x_j^k(1)}, & x \in \left[x_j^k(1), x_j^k(2) \right] \\
1, & x \geq x_j^k(2)
\end{cases}
$$

Definition 4: Critical value for *k*-subclass of the *j*-criterion is defined as:

The whitenization weights function in Figure 1

$$ \lambda_j^k = \frac{1}{2} \left[x_j^k(2), x_j^k(3) \right] $$

The whitenization weights function in Figure 2

$$ \lambda_j^k = x_j^k(3) $$

The whitenization weights function in Figures 1 and 4

$$ \lambda_j^k = x_j^k(2) $$

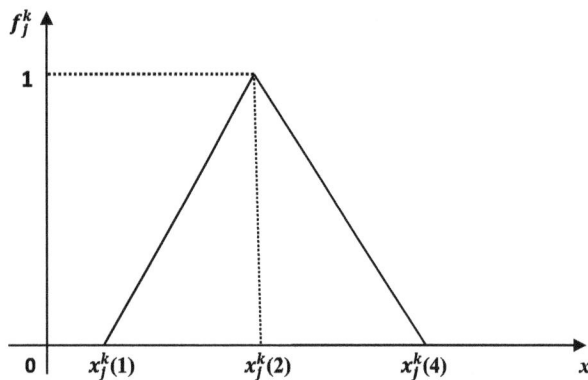

Figure 3. WW function of middle measures.

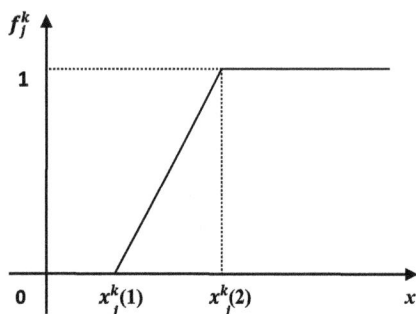

Figure 4. WW function of upper measures.

Definition 5: Assuming the critical value λ_j^k for k-subclass of the j-criterion, then the weight of the j-criterion with respect to k-subclass is:

$$\eta_j^k = \frac{\lambda_j^k}{\sum_{j=1}^m \lambda_j^k}$$

Definition 6: Assume that $\mathbf{X}ij$ is the observation values of object "i" and criterion-j, the whitenization weight function $f_j^k(\cdot)$ of k-subclass of the j-criterion and the η_j^k weight of the j-criterion with respect to k-subclass. Then

$$\sigma_i^k = \sum_{j=1}^m f_j^k(x_{ij}) \times \eta_j^k$$

is said to be the cluster coefficient of variable weight for object "i" that belongs to the kth grey class.

Definition 7: (a) The following

$$\sigma_i = \{\sigma_i^1, \sigma_i^2, \dots \sigma_i^s\}$$

is called the cluster coefficient vector of object "i".

(b) The matrix of such vector represented as:

$$\Sigma = \left[\sigma_i^k\right]_{n \times s}$$

and is called the cluster coefficient matrix.

Definition 8: If

$$\sigma_i^{k*} = \max_{1 \le k \le s} \{\sigma_i^k\}$$

Then object i belong to the grey class $k*$.

2.2. Fixed weights clustering
A fixed weight clustering equally weights all criteria under consideration and also applicable to the situations, where observational data or dimensions are different.

Definition 9: For any K_1 and $K_2 \in \{1, 2, \dots, s\}$ and if $\eta_j^{k1} = \eta_j^{k2}$ then η_j is applied instead of η_j^k. Therefore, fixed weight criteria coefficient is:

$$\sigma_i^k = \sum_{j=1}^m f_j^k\left(x_{ij}\right) \times \eta_j$$

where, $j = 1, 2, \dots, m$.

Definition 10: Assuming that x_{ij} ($i = 1, 2, \dots, n; j = 1, 2, \dots, m$) stands for the observational values of the object "i" with respect to criterion j, and $f_j^k(\cdot)$ is the whitenization weight function of the kth subclass of the j-criterion, then for any $j = 1, 2, \dots, m$, $\eta_j = \frac{1}{m}$ holds true and

$$\sigma_i^k = \sum_{j=1}^m f_j^k(x_{ij}) \times \eta_j = \frac{1}{m} \sum_{j=1}^m f_j^k(x_{ij}) \times \eta_j$$

is called the equal weight cluster coefficient for object "i" belongs to kth grey class.

3. Configuration of whitenization weight functions
Several dissolved gas analysis (IEC, IEEE, CIGRE, and MSZ National standard's ratio codes and graphical techniques) schemes are developed on empirical assumptions and experts' knowledge in the interpretations (IEC 60599, 1992; Scatiggio & Pompili, 2013). These standards provide the threshold limits for guidance, investigation, and analysis. The IEEE Std. C.57.104 specified key gas values of three evaluation levels (Table 1) are considered for three cluster criteria.

Applying the whitenization weight function $f_j^k(\cdot)$, synthetic clustering performed on three different caution levels, where $j = 1, 2, \ldots, 7$; for the criteria $k = 1, 2, 3$. The whitenization weight function of lower measure is used to figure out the caution level-1. However, whitenization weight functions with middle measure are preferred for caution level-2 and caution level-3 in experimentation. The chosen whitenization weight functions are configured with the equations to apprehend the three caution levels and displayed separately in the following in Tables 2–4.

The caution levels are used in tuning the preferred three whitenization weight functions. The critical values for k-subclass of the j-criterion with fixed weight clustering are given in Table 5. These critical values are used in computing the results of fixed weight clustering.

4. Identifying the criticality of transformers

This section presents execution of grey clustering on key gas data-set shown in Table 6. All seven key gases of every sample represent a characteristic of testing transformer. These key gas samples are the observational values represented as, x_{ij} ($i = 1, 2, \ldots, 21; j = 1, 2, \ldots, 7$) of different transformers. The object "i" with respect to criterion "j" is chosen for the specific key gas concentration. These 21 specimens are used to find the cluster coefficient vectors. The configured whitenization weight functions are employed for observation values with fixed weight criteria. In fixed or equal weight criteria, the considered key gases are treated with a weight of (1/7) for every elements and employed in computing the clustering coefficients.

Table 1. Key gas concentration in ppm (IEEE Std. C.57.104)

Key gases	Level-1	Level-2	Level-3
H_2	100	700	1,800
CH_4	120	400	1,000
CO	350	570	1,400
CO_2	2,500	4,000	10,000
C_2H_4	50	100	200
C_2H_6	65	100	150
C_2H_2	35	50	80

Table 2. Equations for lower measure WW function for caution level-1

Key gases	$f_j^1(x)$	$f_j^1(x)$	$f_j^1(x)$
H_2 ($j = 1$)	0	1	$(x-200)/100$
	$x \leq 0$ and $x \geq 200$	$x \leq 0$ and $x \leq 100$	$100 \leq x \leq 200$
CH_4 ($j = 2$)	0	1	$(200-x)/80$
	$x \leq 0$ and $x \geq 200$	$0 \leq x \leq 120$	$120 \leq x \leq 200$
CO ($j = 3$)	0	1	$(400-x)/50$
	$x \leq 0$ and $x \geq 400$	$0 \leq x \leq 350$	$350 \leq x \leq 400$
CO_2 ($j = 4$)	0	1	$(3,000-x)/500$
	$x \leq 0$ and $x \geq 3,000$	$0 \leq x \leq 2,500$	$2,500 \leq x \leq 3,000$
C_2H_4 ($j = 5$)	0	1	$(70-x)/20$
	$x \leq 0$ and $x \geq 70$	$0 \leq x \leq 50$	$50 \leq x \leq 70$
C_2H_6 ($j = 6$)	0	1	$(70-x)/5$
	$x \leq 0$ and $x \geq 70$	$0 \leq x \leq 65$	$65 \leq x \leq 70$
C_2H_2 ($j = 7$)	0	1	$(40-x)/5$
	$x \leq 0$ and $x \geq 40$	$0 \leq x \leq 35$	$35 \leq x \leq 40$

Table 3. Equations for middle measure WW function for caution level-2			
Key gases	$f_j^2(x)$	$f_j^2(x)$	$f_j^2(x)$	$f_j^2(x)$
H_2 (j = 1)	0	$(x-100)/600$	1, when x = 700	$(1,300-x)/600$
	$x \leq 100$ and $x \geq 1,300$	$100 \leq x \leq 700$		$700 \leq x \leq 1,300$
CH_4 (j = 2)	0	$(x-120)/280$	1, when x = 400	$(680-x)/280$
	$x \leq 350$ and $x \geq 790$	$120 \leq x \leq 400$		$400 \leq x \leq 680$
CO (j = 3)	0	$(x-350)/220$	1, when x = 570	$(790-x)/220$
	$x \leq 120$ and $x \geq 680$	$350 \leq x \leq 570$		$570 \leq x \leq 790$
CO_2 (j = 4)	0	$(x-2,500)/1,500$	1, when x = 4,000	$(5,500-x)/1,500$
	$x \leq 2,500$ and $x \geq 5,500$	$2,500 \leq x \leq 4,000$		$4,000 \leq x \leq 5,500$
C_2H_4 (j = 5)	0	$(x-50)/50$	1, when x = 100	$(150-x)/50$
	$x \leq 50$ and $x \geq 150$	$50 \leq x \leq 100$		$100 \leq x \leq 150$
C_2H_6 (j = 6)	0	$(x-65)/35$	1, when x = 100	$(135-x)/35$
	$x \leq 65$ and $x \geq 135$	$65 \leq x \leq 100$		$100 \leq x \leq 135$
C_2H_2 (j = 7)	0	$(x-35)/15$	1, when x = 50	$(65-x)/15$
	$x \leq 35$ and $x \geq 65$	$35 \leq x \leq 50$		$50 \leq x \leq 65$

Table 4. Equations for middle measure WW function for caution level-3			
Key gases	$f_j^3(x)$	$f_j^3(x)$	$f_j^3(x)$	$f_j^3(x)$
H_2 (j = 1)	0	$(x-700)/1,100$	1, when x = 1,800	$(2,900-x)/1,100$
	$x \leq 700$ and $x \geq 2,900$	$700 \leq x \leq 1,800$		$1,800 \leq x \leq 2,900$
CH_4 (j = 2)	0	$(x-400)/600$	1, when x = 1,000	$(1,600-x)/600$
	$x \leq 400$ and $x \geq 1,600$	$400 \leq x \leq 1,000$		$1,000 \leq x \leq 1,600$
CO (j = 3)	0	$(x-570)/830$	1, when x = 1,400	$(2,230-x)/830$
	$x \leq 570$ and $x \geq 2,230$	$570 \leq x \leq 1,400$		$1,400 \leq x \leq 2,230$
CO_2 (j = 4)	0	$(x-4,000)/6,000$	1, when x = 10,000	$(16,000-x)/6,000$
	$x \leq 4,000$ and $x \geq 16,000$	$4,000 \leq x \leq 10,000$		$10,000 \leq x \leq 16,000$
C_2H_4 (j = 5)	0	$(x-100)/100$	1, when x = 200	$(300-x)/100$
	$x \leq 100$ and $x \geq 300$	$100 \leq x \leq 200$		$200 \leq x \leq 300$
C_2H_6 (j = 6)	0	$(x-100)/50$	1, when x = 150	$(200-x)/50$
	$x \leq 100$ and $x \geq 200$	$100 \leq x \leq 150$		$150 \leq x \leq 200$
C_2H_2 (j = 7)	0	$(x-50)/30$	1, when x = 80	$(200-x)/50$
	$x \leq 50$ and $x \geq 110$	$50 \leq x \leq 80$		$80 \leq x \leq 110$

Table 5. Critical values of three whitenization weight functions						
λ_j^K	H_2 (j = 1)	CH4 (j = 2)	CO (j = 3)	CO_2 (j = 4)	C_2H_4 (j = 5)	C_2H_6 (j = 6)	C_2H_2 (j = 7)
K = 1	100	120	350	2,500	50	65	35
K = 2	700	400	570	4,000	100	100	50
K = 3	1,800	1,000	1,400	10,000	200	150	80

The cluster coefficient vectors of fixed weight clustering for all the observational objects are shown in Table 7. The result shows the classification of all observational objects which are divided into three desired grey classes.

Table 6. Dissolved gas-in-oil samples of testing transformers

Specimen	H_2	CH_4	CO	CO_2	C_2H_4	C_2H_6	C_2H_2
Tx_1	53	49	748	6,021	2,824	514	31
Tx_2	12	325	12	787	1	3	108
Tx_3	1	2	34	322	1	1	1
Tx_4	1	19	140	1,879	1	57	1
Tx_5	19	303	432	3,114	1	157	1
Tx_6	1	46	219	9,909	6	16	1
Tx_7	1	2	34	327	1	1	1
Tx_8	1	19	159	3,303	47	60	1
Tx_9	12	8,778	317	2,959	11,900	4,834	19
Tx_{10}	1	73	124	66,261	1	88	1
Tx_{11}	66	87	211	1,902	77	53	24
Tx_{12}	111	102	377	2,496	34	62	36
Tx_{13}	103	114	327	2,734	66	32	17
Tx_{14}	92	103	351	2,496	37	41	22
Tx_{15}	57	64	218	2,210	41	63	33
Tx_{16}	109	76	507	2,910	48	22	13
Tx_{17}	29	123	344	2,506	12	5	1
Tx_{18}	77	98	259	2,496	38	52	26
Tx_{19}	18	76	153	2,107	31	27	5
Tx_{20}	89	113	302	1,992	59	63	29
Tx_{21}	22	106	514	13,327	36	72	28

Table 7. Coefficient matrix of grey cluster with fixed weights

σ_i^k	K = 1	K = 2	K = 3	σ_i^k	K = 1	K = 2	K = 3
$i = 1$	0.4285	0.0272	0.0787	$i = 12$	0.8786	0.0296	0
$i = 2$	0.7143	0.1045	0.0095	$i = 13$	0.8146	0.0687	0
$i = 3$	1	0	0	$i = 14$	0.9971	0.0006	0
$i = 4$	1	0	0	$i = 15$	1	0	0
$i = 5$	0.4285	0.205	0.1228	$i = 16$	0.7271	0.1431	0
$i = 6$	0.8571	0	0.1406	$i = 17$	0.9929	0.0021	0
$i = 7$	1	0	0	$i = 18$	1	0	0
$i = 8$	0.8571	0.0764	0	$i = 19$	1	0	0
$i = 9$	0.4402	0.0437	0	$i = 20$	0.9357	0.0257	0
$i = 10$	0.7142	0.0938	0	$i = 21$	0.5714	0.135	0.0636
$i = 11$	0.8571	0.077	0				

In reference to fixed or equal weight criteria, maximum values are obtain as

$$\max_{1 \leq k \leq 3} \left\{ \sigma_1^1 \right\} = 0.4285; \quad \max_{1 \leq k \leq 3} \left\{ \sigma_2^1 \right\} = 0.7143; \quad \max_{1 \leq k \leq 3} \left\{ \sigma_3^1 \right\} = 1$$

$$\max_{1 \leq k \leq 3} \left\{ \sigma_4^1 \right\} = 1; \quad \max_{1 \leq k \leq 3} \left\{ \sigma_5^1 \right\} = 0.4285; \quad \max_{1 \leq k \leq 3} \left\{ \sigma_6^1 \right\} = 0.8571$$

$$\max_{1 \leq k \leq 3} \left\{ \sigma_7^1 \right\} = 1; \quad \max_{1 \leq k \leq 3} \left\{ \sigma_8^1 \right\} = 0.8571; \quad \max_{1 \leq k \leq 3} \left\{ \sigma_9^1 \right\} = 0.4402$$

$$\max_{1\leq k\leq 3}\left\{\sigma_{10}^{1}\right\}=0.7142; \quad \max_{1\leq k\leq 3}\left\{\sigma_{11}^{1}\right\}=0.8571; \quad \max_{1\leq k\leq 3}\left\{\sigma_{12}^{1}\right\}=0.8786$$

$$\max_{1\leq k\leq 3}\left\{\sigma_{13}^{1}\right\}=0.8146; \quad \max_{1\leq k\leq 3}\left\{\sigma_{14}^{1}\right\}=0.9971; \quad \max_{1\leq k\leq 3}\left\{\sigma_{15}^{1}\right\}=1$$

$$\max_{1\leq k\leq 3}\left\{\sigma_{16}^{1}\right\}=0.7271; \quad \max_{1\leq k\leq 3}\left\{\sigma_{17}^{1}\right\}=0.9929; \quad \max_{1\leq k\leq 3}\left\{\sigma_{18}^{1}\right\}=1$$

$$\max_{1\leq k\leq 3}\left\{\sigma_{19}^{1}\right\}=1; \quad \max_{1\leq k\leq 3}\left\{\sigma_{20}^{1}\right\}=0.9357; \quad \max_{1\leq k\leq 3}\left\{\sigma_{21}^{1}\right\}=0.5714$$

Among the three grey classes, only one class (DGA level-1) has shown the effective response in classification. Observing the classification system as a whole, sample no. 1, 5, 9, and 21 are found in a critical level of maintenance. The judgment about criticality biased with one caution level i.e. level-1; reason is that the coefficients of other levels are contributed extremely imperceptibly in the classification. Therefore, it is evidently unreasonable to consider the effective weights of all gases equally. Hence, it is obvious that all the specifications have different weights on every caution levels, such as in case of variable weight clustering.

The critical values for variable weights clustering for three caution levels are obtained and displayed in Table 8. The ratio of specified safety concentration of a gas to the total concentration of all gases presented in a particular level is assigned for critical values. The observation values of object, whitenization weight function, and weights for variable clustering resulted into the cluster coefficient matrix as shown in Table 9.

The variable weight criteria of grey subclass with highest magnitude led to following results:

$$\max_{1\leq k\leq 3}\left\{\sigma_{1}^{3}\right\}=0.2507; \quad \max_{1\leq k\leq 3}\left\{\sigma_{2}^{1}\right\}=0.9518; \quad \max_{1\leq k\leq 3}\left\{\sigma_{3}^{1}\right\}=1$$

$$\max_{1\leq k\leq 3}\left\{\sigma_{4}^{1}\right\}=1; \quad \max_{1\leq k\leq 3}\left\{\sigma_{5}^{2}\right\}=0.3565; \quad \max_{1\leq k\leq 3}\left\{\sigma_{6}^{3}\right\}=0.6731$$

Table 8. Critical values for variable weight clustering

η_{j}^{K}	H_2 (j = 1)	CH_4 (j = 2)	CO (j = 3)	CO_2 (j = 4)	C_2H_4 (j = 5)	C_2H_6 (j = 6)	C_2H_2 (j = 7)
K = 1	0.0310	0.0372	0.1086	0.7763	0.0155	0.0201	0.0108
K = 2	0.1182	0.0675	0.0962	0.6756	0.0168	0.0168	0.0084
K = 3	0.1230	0.0683	0.0956	0.6835	0.0136	0.0102	0.0054

Table 9. Variable weights clustering coefficient matrix of grey subclass

σ_{i}^{k}	K = 1	K = 2	K = 3	σ_{i}^{k}	K = 1	K = 2	K = 3
i = 1	0.0791	0.0183	0.2507	i = 12	0.9203	0.0145	0
i = 2	0.9518	0.0494	0.0003	i = 13	0.6232	0.1114	0
i = 3	1	0	0	i = 14	0.9978	0.0004	0
i = 4	1	0	0	i = 15	1	0	0
i = 5	0.0574	0.3565	0.0088	i = 16	0.2518	0.2555	0
i = 6	0.2236	0	0.6731	i = 17	0.9892	0.0034	0
i = 7	1	0	0	i = 18	1	0	0
i = 8	0.2236	0.3617	0	i = 19	1	0	0
i = 9	0.2142	0.2067	0	i = 20	0.9930	0.0030	0
i = 10	0.2114	0.0111	0	i = 21	0.0947	0.0751	0.3045
i = 11	0.9844	0.0091	0				

$$\max_{1 \leq k \leq 3} \left\{ \sigma_7^1 \right\} = 1; \qquad \max_{1 \leq k \leq 3} \left\{ \sigma_8^2 \right\} = 0.3617; \qquad \max_{1 \leq k \leq 3} \left\{ \sigma_9^1 \right\} = 0.2142$$

$$\max_{1 \leq k \leq 3} \left\{ \sigma_{10}^1 \right\} = 0.2114; \qquad \max_{1 \leq k \leq 3} \left\{ \sigma_{11}^1 \right\} = 0.9844; \qquad \max_{1 \leq k \leq 3} \left\{ \sigma_{12}^1 \right\} = 0.9203$$

$$\max_{1 \leq k \leq 3} \left\{ \sigma_{13}^1 \right\} = 0.6232; \qquad \max_{1 \leq k \leq 3} \left\{ \sigma_{14}^1 \right\} = 0.9978; \qquad \max_{1 \leq k \leq 3} \left\{ \sigma_{15}^1 \right\} = 1$$

$$\max_{1 \leq k \leq 3} \left\{ \sigma_{16}^2 \right\} = 0.2555; \qquad \max_{1 \leq k \leq 3} \left\{ \sigma_{17}^1 \right\} = 0.9892; \qquad \max_{1 \leq k \leq 3} \left\{ \sigma_{18}^1 \right\} = 0.9203$$

$$\max_{1 \leq k \leq 3} \left\{ \sigma_{19}^1 \right\} = 1; \qquad \max_{1 \leq k \leq 3} \left\{ \sigma_{20}^1 \right\} = 0.9930; \qquad \max_{1 \leq k \leq 3} \left\{ \sigma_{21}^3 \right\} = 0.3045$$

It follows that the DGA sample no. 2, 3, 4, 7, 9, 10, 11, 12, 13, 14, 15, 17, 18, 19, and 20 are classified in grey level-1. If the magnitudes of cluster coefficients are considered as score of the transformers, and then sample no. 3, 4, 7, 15, 18, and 19 are referred as absolutely healthy transformers. Whereas sample no. 2, 11, 12, 14, 17, and 20 are observed to be in the normal condition, except for sample no. 9, 10, and 13. Therefore, these three samples are observed as the critical elements in grey class-1. The sample no. 5, 8, and 16 are measured in criticality level-2 cluster. Sample no. 1, 6, and 21 are found in grey class-3 which implies that these samples are the most critical elements among the considered test samples and need immediate attention. The effect of variable weight clustering shows the criticality judgments on different levels and useful in setting the priorities about maintenance. The variable weight clustering is effective to the cases, when whitenization weight functions are selected based on experience.

5. Conclusions

Information from the analysis of gasses dissolved in insulating oil of transformer is a primary source of state assessment. The simple and reliable process, similar to variable weight grey clustering, facilitates the categorization of objects, which identify the criticality of transformers at three caution levels. The results of grey clustering certainly helped in setting the priorities about preventive maintenance and recommended the straight action for critical cases. The results obtained in this experimentation are limited to dissolved gas in oil samples and the three caution levels refer to IEEE standard. However, the variable weight clustering method will be effectively implemented, if additional monitoring parameters and their specified safety values are used for comprehensive analysis.

Funding
The authors received no direct funding for this research.

Author details
Vikal R. Ingle[1]
E-mail: Chin_vikal@yahoo.com
V.T. Ingole[2]
E-mail: vtingole@gmail.com
[1] B.D. College of Engineering, Sevagram, Wardha, Maharashtra, India.
[2] Prof. Ram Meghe Institute Technology & Research, Badnera, Amravati, Maharashtra, India.

References
Abu-Siada, A., Arshad, M., & Islam, S. (2010, July 25–29). Fuzzy logic approach to identify transformer criticality using dissolved gas analysis. *Power and Energy Society General Meeting 2010 IEEE*, 1–5. http://dx.doi.org/10.1109/PES.2010.5589789

Deng, J. L. (2002). *Elements of grey theory*. Wuchang: Press of Huazhong University of Science and Technology.

Deng, J. L. (2005). *The primary methods of grey system theory*. Wuhan: Huazhong University of Science and Technology Press.

Field, N., Cramer, S., & Antosz, S. (2002, November 8). Condition-based ranking of power transformers. In *Weidmann-ACTI Conference*. Las Vegas, NV.

Hydroelectric Research and Technical Services Group. (2003, June). *Transformer diagnostics-facilities instructions, standards, and techniques* (Vol. 3–31). Denver, CO: US Department of the Interior Bureau of Reclamation.

IEC 60599. (1992). *Mineral oil-impregnated electrical equipment in service-guide to the interpretation of dissolved and free gases analysis* (2nd ed.). IS 10593: 2006. Bureau of Indian Standards.

IEEE guide for the interpretation of gases generated in oil immersed transformers. (2009). IEEE Engineering Society,

ANSI/IEEE std.C57.104, 2008 (Revision of IEEE Standard. C57.104-1991, pp. CI–27).

Jahromi, A. N., Piercy, R., Cress, S., Service, J. R. R., & Fan, W. (2009). An approach to power transformer asset management using health index. *IEEE Electrical Insulation Magazine, 25*, 20–34. http://dx.doi.org/10.1109/MEI.2009.4802595

Kuo, Y., Yang, T., & Huang, G. W. (2008). The use of grey relational analysis in solving multiple attributes decision-making problems. *Computer & Industrial Engineering, 55*, 80–93.

Liu, K., Shen, X. L., Tan, Z. F., & Guo, W. Y. (2012). Grey clustering analysis method for overseas energy project investment risk decision. *Systems Engineering Procedia, 3*, 55–62.

Liu, N., & Xie, M. (2011). New grey evaluation method based on reformative triangular whitenization weight function. *Journal of Systems Engineering, 26*, 244–250.

Liu, S., & Lin, Y. (2006). *Grey information—Theory and practical applications*. London: Springer-Verlag.

Nemeth, B., Laboncz, S., & Kiss, I. (2009, May 31–June 3). Condition monitoring of power transformers using DGA and fuzzy logic. In *2009 IEEE Electrical Insulation Conference*. Montreal, Québec, Canada.

Qiu, X. J. (1995). A grey cluster relation analysis method and its application. *Systems engineering—Theory and practice, 15*, 15–21.

Scatiggio, F., & Pompili, M. (2013, June 2–5). Health index: The TERNA's practical approach for transformers fleet management. In *Electrical Insulation Conference* (pp. 178–182). Ottawa, Ontario, Canada.

Sun, Z.-L., Au, K.-F., & Choi, Tsan-Ming (2007). A neuro-fuzzy inference system through integration of fuzzy logic and extreme learning machines. *IEEE Transactions on Systems, Man and Cybernetics, Part B (Cybernetics), 37*, 1321–1331. http://dx.doi.org/10.1109/TSMCB.2007.901375

Toronto Hydro-Electric System Limited. (2010, July 23). *Asset condition assessment audit* (Report No. K-015466-RA-0001-R01). Toronto: Kinectrics.

Wang, Z., Liu, Y., & Griffin, P. J. (1998). A combined ANN and expert system tool for transformer fault diagnosis. *IEEE Transactions on Power Delivery, 13*, 1224–1229. http://dx.doi.org/10.1109/61.714488

Xiao, X. P., & Xiao, W. (1997). Grey optimal clustering theory model and its application. *Operations Research and Management, 6*, 21–26.

Yang, S. (2008, August 3–4). Application of grey target theory for handling multi-criteria vague decision making problems. *2008 ISECS International Colloquium on computing, communication, control and management*, 467–471.

Zhang, Q. S. (2002). Grey clustering result of greyness measurement. *Chinese Journal of management science, 10*, 54–56.

Zheng, R.-R., Zhao, J.-Y., & Wu, B.-C. (2009). Transformer oil dissolved gas concentration prediction based on genetic algorithm and improved gray verhulst model. *International Conference on artificial intelligence and computational intelligence AICI'09*, 575–579. http://dx.doi.org/10.1109/AICI.2009.100

6

Grey wolf optimizer based regulator design for automatic generation control of interconnected power system

Esha Gupta[1]* and Akash Saxena[1]

*Corresponding author: Esha Gupta, Department of Electrical Engineering, Swami Keshvanand Institute of Technology, Office no AC-201, Ramnagaria, Jagatpura, Jaipur 302017, Rajasthan, India
E-mail: esha.gupta@outlook.com
Reviewing editor: Siew Chong Tan, University of Hong Kong, Hong Kong

Abstract: This paper presents an application of grey wolf optimizer (GWO) in order to find the parameters of primary governor loop for successful Automatic Generation Control of two areas' interconnected power system. Two standard objective functions, Integral Square Error and Integral Time Absolute Error (ITAE), have been employed to carry out this parameter estimation process. Eigenvalues along with dynamic response analysis reveals that criterion of ITAE yields better performance. The comparison of the regulator performance obtained from GWO is carried out with Genetic Algorithm (GA), Particle Swarm Optimization, and Gravitational Search Algorithm. Different types of perturbations and load changes are incorporated in order to establish the efficacy of the obtained design. It is observed that GWO outperforms all three optimization methods. The optimization performance of GWO is compared with other algorithms on the basis of standard deviations in the values of parameters and objective functions.

Subjects: Algorithms & Complexity; Power Engineering; Systems & Controls

Keywords: automatic generation control (AGC); integral square error (ISE); grey wolf optimizer (GWO); gravitational search algorithm (GSA); genetic algorithm (GA)

1. Introduction
Ongoing electricity demand and exponential growth in population have laid a heavy burden on conventional generation, transmission, and distribution system. To match high load demands with exponential increment of utilities at transmission and distribution ends, the problem of the power system operation and control emerged as a challenging design problem. Automatic generation control (AGC) is a common denominator used to maintain the fair balance between the real power generation, system load demand, and associated system losses (Nanda & Kaul, 1978). IEEE defines

ABOUT THE AUTHORS
Esha Gupta is currently working as a lecturer in Department of Electrical Engineering at Swami Keshvanand Institute of Technology, Management and Gramothan, Jaipur, India. Her area of interest includes power system operation and control and optimization techniques.

Akash Saxena is an associate professor of Electrical Engineering at Swami Keshvanand Institute of Technology, Management and Gramothan, Jaipur, India. His research interests include power system operations and control, automatic generation control, and application of nature-inspired meta-heuristic algorithms.

PUBLIC INTEREST STATEMENT
With the ongoing demand for power, existing utilities are working at their operating limits. Successful operation and control of power system are a major thrust area of research. This manuscript is an effort to present a control paradigm to design an efficient automatic generation control regulator based on the hunting strategy of the grey wolfs. After reading this manuscript, the readers will be benefited with the application of this nature-based algorithm and various aspects of power system operation and control.

AGC as "The regulation of the power output of electric generators with in a prescribed area in response to changes in system frequency, tie-line loading, or the relation of these to each other, so as to maintain the scheduled system frequency and/or the establish interchange with other areas within predetermined limits" (*IEEE Standard Definitions of Terms*, 1970).

Figure 1 shows the schematic diagram of load frequency control (LFC) and automatic voltage regulator (AVR) of a turbo generator. The automatic control of turbo generator consists of two major loops, i.e. LFC loop and AVR loop. The LFC loop controls the real power output and frequency of the system and AVR loop regulates the magnitude of terminal voltage of all the generators and the reactive power output. The AGC works on two control modes. In primary control, speed governors are responsible for the operation of control valve of turbine power input. The secondary control is slow and maintains the tie line power interchange.

A rich literature survey on existing AGC techniques was reported in Ibraheem and Kumar (2005). In the paper, the authors explained AGC schemes, types of power system models, control techniques, control strategies, sensitivity features, etc. The critical issue was to obtain the modeling of an interconnected power system near any operating equilibrium, Elgerd (1983) and Sadat (1999). Over the past few years, many researchers have done significant researches on the subject of better AGC of large interconnected power systems. Some of the approaches were based on Pole Placement Technique (Sivaramaksishana, Hariharan, & Srisailam, 1984), coefficient diagram method (CDM) Bernard, Mohamed, Qudaih, and Mitani (2014) and Ali, Mohamed, Qudaih, and Mitani (2014), neural networks (NN) (Kanniah, Tripathy, Malik, & Hope, 1984; Kothari, Satsangi, & Nanda, 1981; Valk et al., 1985), fuzzy logic (FL) (Banerjee, Mukherjee, & Ghoshal, 2014; Sahu, Pati, Mohanty, & Panda, 2015; Sudha & Vijaya Santhi, 2012; Wu, Er, & Gao, 2001), Super Magnetic Energy Storage Device (Pandan, Sahu, & Panda, 2014), and evolutionary algorithms (EA) (Abdel-Magid & Abido, 2003; Abdel-Magid & Dawoud, 1996; Debbarma, Chandra Saikia, & Sinha, 2014; Emary, Zawbaa, Grosan, & Hassenian, 2015; Gozde, Cengiz Taplamacioglu, & Kocaarslan, 2012; Mirjalili, 2015; Mirjalili, Mirjalili, & Lewis, 2014; Muro, Escobedo, Spector, & Coppinger, 2011; Nanda & Mishra, 2009; Puja, Chandra, & Nidul, 2014; Rout, Sahu, & Panda, 2013; Sahu, Panda, & Padhan, 2014; Saxena, Gupta, & Gupta, 2012; Song, Sulaiman, & Mohamed, 2014; Song et al., 2015). In view of the literature survey, it has been observed that CDM is an algebraic way to solve these compensator design problems. The process of obtaining coefficients is time-consuming and not suitable for fast and online applications. Adaptive leaning paradigms like neural networks (Kanniah et al., 1984; Kothari et al., 1981; Valk et al., 1985) are based

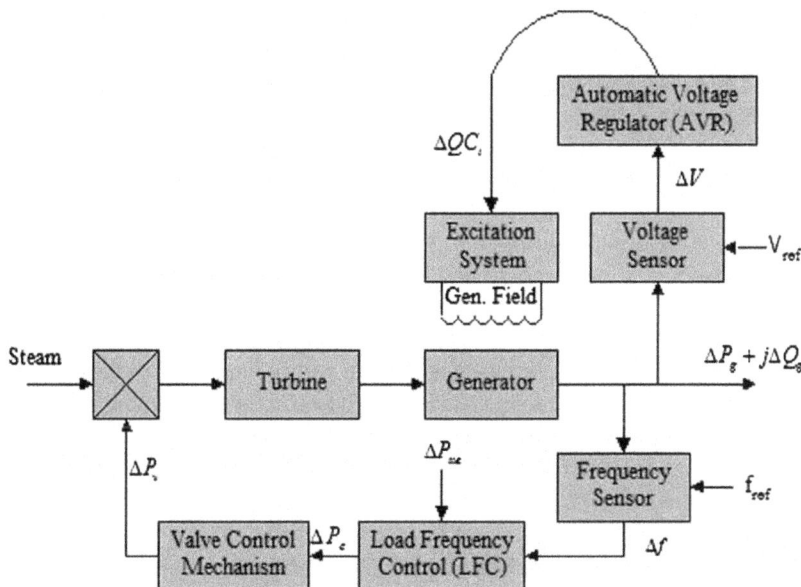

Figure 1. Schematic diagram of LFC and AVR of a turbo generator.

on data generation. Learning of the network is dependent of data-sets. This time-consuming activity makes the approach very lethargic and inappropriate. Fuzzy logic Wu et al. (2001) and Sahu et al. (2015) models are based on uncertainty modeling. Models are based on approximation often failed to maintain the accuracy on real-world problems as thousands of fuzzy models are based on conjunction, disjunction, implications, and defuzzification choices. In AGC problems, high degree of precision is required.

The traditional approach is to optimize the parameters of the secondary loop of governor; however, in this study, it has been shown by the authors that the effect of primary governor loop parameters on the controller setting is predominant. Now a days, meta-heuristic techniques are used to solve these problems due to their flexibility, avoidance of local optima, and derivation-free mechanism. Some of these approaches include gravitational search algorithm (GSA) (Sahu et al., 2014), particle swarm optimization (PSO) (Abdel-Magid & Abido, 2003), genetic algorithm (GA) (Abdel-Magid & Dawoud, 1996), bacterial foraging (BF) (Nanda & Mishra, 2009), differential evolution (DE) (Rout et al., 2013), artificial bee colony (ABC) (Gozde et al., 2012), firefly algorithm (FA) (Debbarma et al., 2014), and cuckoo search (CS) (Puja et al., 2014). Investigations have been carried out using PSO (Abdel-Magid & Abido, 2003) and GA (Abdel-Magid & Dawoud, 1996) and found that they are getting trapped at local optima. These difficulties were overcome by bacterial foraging (BF) technique. Nanda and Mishra (2009) implemented BF technique with integral controller. The resulting performance was better as compared to classical and GA methods. Some new algorithms like ABC (Gozde et al., 2012), FA (Debbarma et al., 2014), CS (Puja et al., 2014), etc. were also successfully applied in AGC. Debbarma et al. (2014) presented the FA to design fractional controller gains. A comparison was presented by authors with the conventional Proportional Integral and Differential Controllers. Shivaie, Kazemi, and Ameli (2015) presented a modified harmony search algorithm to solve the LFC.

In this work, a digital simulation is used with grey wolf optimizer (GWO) to optimize the parameter of AGC for two areas' system. Mirjalili et al. (2014) proposed a population-based algorithm known as GWO inspired by nature of grey wolf in 2014. It mimics the leadership hierarchy and the hunting behavior of grey wolf (Muro et al., 2011). This algorithm shows a very promising response to deal with the optimization process of uni-modal, multi-modal, fixed dimension multi-modal and composite functions. The algorithm outperforms other conventional population-based techniques. The comparison was based on ability of exploration, exploitation, local optima avoidance, and convergence (Muro et al., 2011). For population-based techniques, exploration and exploitation are the common features. However, there is no mathematical analogy found between these two features. In fact, these two features are contradicting in nature. Exploration of search space for potential solutions and the exploitation performance to converge on the global optima are the main features of any algorithm and moreover responsible for the performance of the algorithm. Exploitation process is controlled by control parameters. GWO maintains a fair balance between the exploration and exploitation phenomena. Recently, GWO has been applied on real optical engineering (Mirjalili et al., 2014), combined economic load dispatch problem (Song et al., 2014), and parameter estimation in surface waves (Song et al., 2015). Mirjalili (2015) employed GWO for the training of multi-layer perceptron; further, the performance of this perceptron is tested on three function approximation sets and five classification problems. The performance of GWO-trained MLP is superior to well-known evolutionary trainers like GA, PSO, and Evolution Strategy. Song et al. (2015) applied GWO as a powerful surface wave dispersion curve inversion scheme. The proposed scheme is benchmarked on noise-free, noisy, and field data. Further, the results of algorithms were compared with the conventional techniques. Emary et al. (2015) proposed a classification-based fitness function to eliminate the redundant, irrelevant, and noisy data. This fitness function is optimized through GWO. GWO performed feature selection and classification task efficiently. Salient features of the algorithm are described below:

(1) This algorithm is based on the leadership hierarchy and hunting behavior of the grey wolf. The simplicity of the algorithm allows scientists to simulate the natural concepts in a lucid manner.

(2) The noteworthy feature of this algorithm is that it has less control parameters and possesses a derivative-free mechanism. This mechanism is a boon and advocates the superiority of GWO to avoid local optima trap.

(3) Algorithm is flexible and can be applied to many real-world problems without changing the main structure.

(4) This algorithm can be applied to non-differentiable, stochastic, and discontinuous functions.

In view of the above literature survey, salient features of this algorithm become the primary motivation to apply GWO in AGC regulator design. In this paper, two standard objective functions, integral time-multiplied absolute error (ITAE) and integral squared error (ISE), are used for the analysis which are function of time and error.

In Section 2, a brief description of the modeling of two areas' interconnected system, its investigation, and the proposed approach is presented. Section 3 describes the grey wolf optimization technique. Section 4 exhibits the dynamic responses of frequency deviation in both the areas and the tie line power at different loading conditions. Section 6, the conclusion, shows the efficacy of the proposed approach by comparing it with GSA (Sahu et al., 2014), PSO (Abdel-Magid & Abido, 2003) and GA (Abdel-Magid & Dawoud, 1996).

2. System modeling

2.1. AGC model

The two areas' non-reheat thermal interconnected power system is shown in Figure 2. The main components of the power system include speed governor, turbine, rotating mass, and load. The inputs of the power system are controller output u, load disturbance ΔP_L, and tie line power ΔP_{tie}, while the outputs are frequency deviations Δf and area control error ACE. The ACE signal controls the steady-state errors of frequency deviation and tie power deviation. Mathematically, ACE can be defined as

$$ACE = B\Delta f + \Delta P_{tie} \qquad (1)$$

where B indicates the frequency bias parameter.

The operating behavior of the power system is dynamic, so it must be assumed that the parameters of the system are linear. For the mathematical modeling, transfer function is used.

The transfer function of a governor is represented by Elgerd (1983):

$$G_g(s) = \frac{1}{1 + sT_g} \qquad (2)$$

Turbine is represented by the transfer function as (Elgerd, 1983):

$$G_t(s) = \frac{1}{1 + sT_t} \qquad (3)$$

The transfer function of rotating mass and load (Elgerd, 1983):

$$G_L(s) = \frac{K_p}{1 + sT_p} \qquad (4)$$

where $T_p = \frac{2H}{fD}$ and $K_p = \frac{1}{D}$.

ΔP_G and ΔP_L are the two inputs of rotating mass and load with $\Delta f(s)$ being the output and represented by Elgerd (1983).

$$\Delta f(s) = G_L(s)[\Delta P_G(s) - \Delta P_L(s)] \qquad (5)$$

2.2. System investigated

The system is investigated on the two equal thermal areas connected by a weak tie line having the same generation capacity of 1,000 MVA. The parameters of the system are taken from (Sadat, 1999). A sudden step perturbation of 0.1875 p.u. occurs in area 1 and 0.1275 p.u. in area 2. The transfer function model of two areas' thermal system is shown in Figure 2. The system is implemented using MATLAB 2013 and run on a Pentium IV CPU, 2.69 GHz, and 1.84-GB RAM computer (MATLAB, http://www.mathworks.com).

2.3. The proposed approach

The controller used in AGC system is PI controller as it determines the difference between set point and reference point as well as removes the steady-state error. For the design of PI controller, parameters' proportional gain (K_p) and integral gain (K_I) are essential. However, in this work, for the ease and simplicity of optimization process, we consider proportional gain 1. Area control errors are the input of the controllers for area 1 and area 2 which are defined as

$$ACE_1 = B_1 \Delta f_1 + \Delta P_{tie} \qquad (6)$$

$$ACE_2 = B_2 \Delta f_2 + \Delta P_{tie} \qquad (7)$$

where $B_1 = \frac{1}{R_1} + D_1$ and $B_2 = \frac{1}{R_2} + D_2$.

The output of the controllers are u_1 and u_2 and are obtained as

$$u_1 = K_{P1} ACE_1 + K_{I1} \int ACE_1 \qquad (8)$$

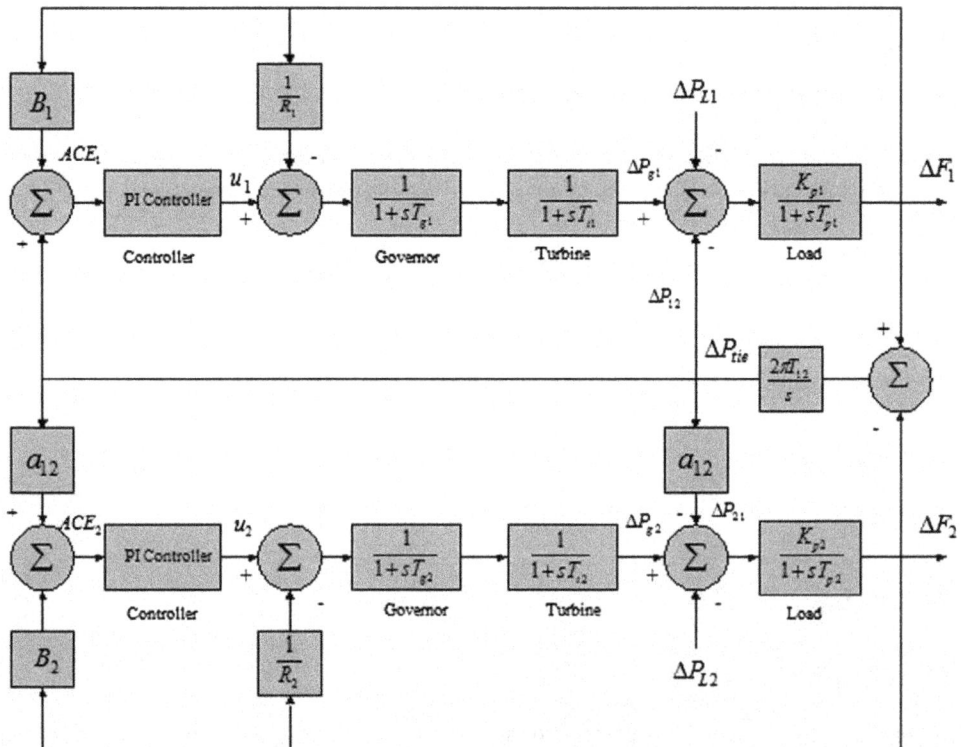

Figure 2. Transfer function model of two areas' non-reheat thermal interconnected system.

$$u_2 = K_{P2}ACE_2 + K_{I2} \int ACE_2 \tag{9}$$

In this paper, the estimation of integral gains and parameters of primary governor loop are based on two objective functions (ITAE and ISE) which are mentioned in Equations (10) and (11). It aims to reduce the steady-state error to zero and maximize the damping ratio of the system.

$$J1 = ITAE = \int_0^T \left(|\Delta f_1| + |\Delta f_2| + |\Delta P_{tie}| \right) \cdot t dt \tag{10}$$

$$J2 = ISE = \int_0^T \left(|\Delta f_1|^2 + |\Delta f_2|^2 + |\Delta P_{tie}|^2 \right) dt \tag{11}$$

The problematic constraints are the parameters of AGC regulator which contains integral gains, speed regulations, and the frequency sensitivity coefficients as they are bounded with the limits. These parameters are system specific. Hence, the design problem can be formulated as

   Minimize J

   Subjected to

$$K_{I_{min}} \leq K_I \leq K_{I_{max}} \tag{12}$$

$$R_{min} \leq R \leq R_{max} \tag{13}$$

$$D_{min} \leq D \leq D_{max} \tag{14}$$

where J is the objective function (J_1 and J_2).

3. Grey wolf optimizer

A recent population-based swarm intelligence technique, called GWO, inspired by the nature of grey wolf is discussed in this section. This technique was proposed by Mirjalili et al. (2014) in 2014. In GWO, the leadership hierarchy and the hunting behavior of grey wolf are mimicked. Grey wolves belong to Canidae family and prefer to live in a pack of 5–12 members on average. This pack is categorized into four groups, namely: alpha, beta, delta, and omega for the simulation of leadership hierarchy. They have very strict social-dominant hierarchy.

   Alphas are the first level and are the leaders of the pack. Alphas are the decision-makers regarding hunting, sleeping place, and time to wake up and that decision will be followed by the pack. Hence, the alpha wolf is also known as the dominant wolf. Alpha is not essentially the strongest member in the pack, but good in organizing and disciplining the pack.

   Beta comes in the second level on the hierarchy of grey wolves. Betas help alpha wolves in decision-making and the activities of the pack. Betas are the best candidates to get the position of alpha in case of the alpha wolves pass away or become very old. The beta supports alpha's command throughout the pack.

   Delta is the third level in the pack. Delta wolves have to submit alpha and beta, but they dominate omega. The scouts, elders, hunters, sentinel, and care takers belong to this group.

   Omega wolves have the lowest ranking in the pack. They always have to surrender to all other dominant wolves. Omega is not a main member, but everyone face fighting and problems in case of losing an omega.

As hunting is also an interesting behavior of grey wolves, the three important steps of hunting are employed to carry out the optimization, which are: searching for prey, encircling the prey, and attacking the prey. According to Muro et al. (2011), the main stages of grey wolf hunting are tracking, chasing, pursuing, encircling, and attacking the prey.

In the mathematical modeling of social hierachy of wolf, alpha (α) is considered as the fittest solution, beta (β) and delta (δ) are the second- and the third-best fittest solutions, respectively, in designing GWO. The rest of the candidate solutions are considered as omega (ω). The hunting is guided by α, β, and δ. The ω wolves follow α, β, and δ wolves.

For the modeling of encircling the prey, the following equations are proposed.

$$\vec{D} = |\vec{C} \cdot \vec{X}_p(t) - \vec{X}(t)| \tag{15}$$

$$\vec{X}(t+1) = \vec{X}_p(t) - \vec{A} \cdot \vec{D} \tag{16}$$

where t represents current iteration, $\vec{A}$ and $\vec{C}$ are coefficient vectors, $\vec{X}_p$ is the position vector of the prey, and $\vec{X}$ is the position vector of the grey wolf. The vectors $\vec{A}$ and $\vec{C}$ can be calculated as follows:

$$\vec{A} = 2\vec{a} \cdot \vec{r}_1 - \vec{a} \tag{17}$$

$$\vec{C} = 2 \cdot \vec{r}_2 \tag{18}$$

The components of $\vec{a}$ are decreased linearly from 2 to 0 over the course of iterations and r_1, r_2 are random vectors in [0, 1]. Figure 3 shows the 3D position of prey at (X^*, Y^*, and Z^*) and grey wolf at (X, Y, Z) with its possible next locations.

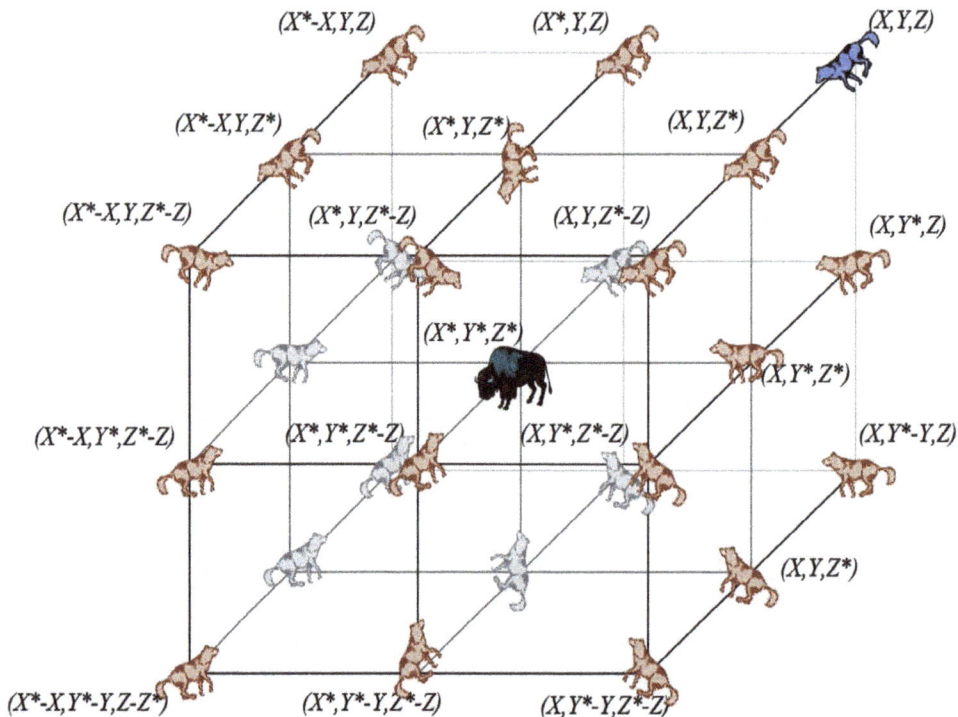

Figure 3. Three-dimensional position vector with their next possible locations.

During hunting, the first three best solutions (α, β, and δ) obtained are saved and coerce other search agents (including the omega) to update their positions according to the best search agent. The following are the proposed formula.

$$\vec{D}_\alpha = |\vec{C}_1 \cdot \vec{X}_\alpha - \vec{X}|, \vec{D}_\beta = |\vec{C}_2 \cdot \vec{X}_\beta - \vec{X}|, \vec{D}_\delta = |\vec{C}_3 \cdot \vec{X}_\delta - \vec{X}| \qquad (19)$$

$$\vec{X}_1 = \vec{X}_\alpha - \vec{A}_1 \cdot (\vec{D}_\alpha), \vec{X}_2 = \vec{X}_\beta - \vec{A}_2 \cdot (\vec{D}_\beta), \vec{X}_3 = \vec{X}_\delta - \vec{A}_3 \cdot (\vec{D}_\delta) \qquad (20)$$

$$\vec{X}(t+1) = \frac{\vec{X}_1 + \vec{X}_2 + \vec{X}_3}{3} \qquad (21)$$

Figure 4 shows the updating position of search agent according to the alpha, beta, and delta. It can be observed that alpha, beta, and delta estimate the position of the prey and other wolves update their position stochastically around the prey and the final position is randomly within the circle.

The searching of grey wolves depends on the position of the alpha, beta, and delta. For searching, they diverge from each other. Mathematically, $\vec{A}$ varies with random values greater than 1 or less than −1 to oblige the search agent to diverge from the prey. This brings out exploration and allows GWO algorithm to search globally. If $|A| > 1$, grey wolves diverges from the prey to find the fitter prey.

When the prey stops moving, the grey wolf finishes its hunt by attacking it. If $|A| < 1$, grey wolves converge toward the prey and attack it. The vector A is a random value in the interval $[-a, a]$. This process is known as exploitation.

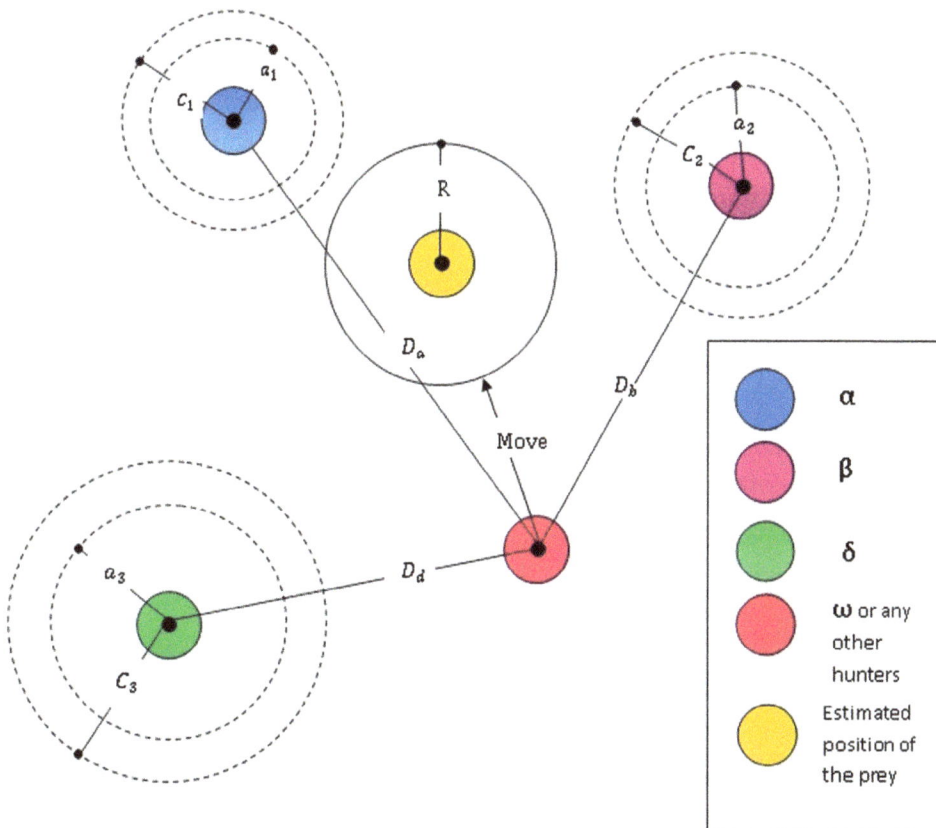

Figure 4. Position updating in GWO.

4. Results and analysis

This section presents the critical analysis of regulator performance on two areas' interconnected thermal units under different load perturbations in both areas. Table 1 shows the optimal parameters of integral regulator with primary governor loop constants, namely: frequency sensitivity coefficient 'D' and speed regulation 'R'. The values obtained from the optimization process of proposed GWO technique with GSA, PSO, and GA from objective functions J1 (ITAE) and J2 (ISE) are shown in this table. The lowest value of integral regulator gain and the highest value of R are obtained with GWO optimization process, which shows that controller setting obtained from this technique is more robust as compared to others (Nanda & Mishra, 2009). Table 2 shows the values of system modes (eigenvalues) and the minimum damping ratio obtained from the objective functions J1 and J2 by the proposed GWO-tuned AGC regulator. These values are compared with the regulator obtained by modern heuristic optimization algorithms GSA, PSO, and GA. For stability studies, eigenvalues are important, as they give the information about the system's behavior when the system is subjected to any physical disturbance. Both real and imaginary parts have their interpretation and physical significance. Oscillatory instability is due to the real positive part of the complex conjugate eigenvalues. These eigenvalues are also known as swing modes. It is observed from Table 2 that when the system is tested with GA-tuned AGC regulator with criteria J2, the swing modes possess positive real part (0.0361). Real positive part indicates the oscillation of growing amplitude. The real part of the complex conjugate eigenvalue shows the damping behavior which represents the damp oscillations, meaning: larger the magnitude, more the rate of decay. Imaginary components show the frequency of oscillations. With PSO setting J2, the response contains high frequency modes (2.69 Hz and 2.18 Hz). Higher frequency modes are not good for control equipment's health. On the other hand, GWO regulator not only possesses moderate values of frequency oscillation (1.49 Hz, 1.73 Hz), but also contains larger real negative parts (−0.4638, −0.2848, and −0.29, −0.05) as compared with other regulators. The eigenvalues after the employment of GWO AGC regulator possess bigger negative part as compared with any other algorithms which indicates that the system is comparatively stable. The values of minimum damping ratio, when the optimization process is carried out by objective function J1, are 0.1875 for GWO, 0.1601 for GSA, 0.1345 for PSO, and 0.1668 for GA. It is observed that the system's damping performance is comparatively improved with GWO technique. However, less value of damping ratios is obtained by GA as compared to any other algorithm. Prima facie design obtained by the J2 criteria is rejected due to the positive real part in GA regulator, appearance of higher frequencies of oscillations in all regulators, and smaller negative real part of the system's modes. To justify this: firstly, the responses of GWO-based regulator are taken into consideration with both the objective functions J1 and J2. It is then compared with the GSA (Sahu et al., 2014), PSO (Abdel-Magid & Abido, 2003), and GA (Abdel-Magid & Dawoud, 1996). Figure 5 shows three dynamic responses obtained from GWO. The frequency deviation of area 1 is shown in Figure 5(a–d). Deviation in the tie line power is shown in Figure 5(e) and the frequency deviation of area 2 is shown in Figure 5(f). Figure 5(a–d) shows the frequency deviation in area 1 with increase in load for the two objective functions J1 and J2 with changes in loads as:

Table 1. Optimized parameters of AGC regulator

Parameters	GWO		GSA (Sahu et al., 2014)		PSO (Abdel-Magid & Abido, 2003)		GA (Abdel-Magid & Dawoud, 1996)	
	J1	J2	J1	J2	J1	J2	J1	J2
KI1	0.2072	0.4000	0.3817	0.4171	0.3131	0.4498	0.3031	0.6525
KI2	0.2055	0.5000	0.2153	0.2028	0.1091	0.2158	0.3063	0.7960
R1	0.0555	0.0400	0.0401	0.0435	0.0581	0.0201	0.0794	0.0503
R2	0.0689	0.0500	0.0657	0.0635	0.0531	0.03	0.0737	0.0609
D1	0.5943	0.6000	0.5889	0.4778	0.4756	0.5910	0.7591	0.7216
D2	0.5507	0.8000	0.8946	0.8744	0.6097	0.8226	0.8950	0.8984

Table 2. Eigenvalues and minimum damping ratio								
Parameter	GWO		GSA (Sahu et al., 2014)		PSO (Abdel-Magid & Abido, 2003)		GA (Abdel-Magid & Dawoud, 1996)	
	J1	J2	J1	J2	J1	J2	J1	J2
System modes	−5.8597	−5.9843	−5.8468	−5.976	−5.846	−6.5657	−5.6586	−5.808
	−4.2274	−4.3812	−4.313	−4.4257	−4.4443	−4.8155	−4.2083	−4.2168
	−0.4638 ± 1.7329i	−0.2900 ± 1.9211i	−0.3994 ± 1.7029i	0.2511 ± 1.9124i	**−0.4010 ± 1.7004i**	−0.0030 ± 2.6953i	−0.4925 ± 1.3799i	−0.2024 ± 1.6817i
	−0.2848 ± 1.4917i	−0.0582 ± 1.7136i	−0.2606 ± 1.6066i	−0.1924 ± 1.7420i	**−0.2406 ± 1.7718i**	−0.0220 ± 2.1889i	−0.2491 ± 1.4729i	**0.0361 ± 1.5786i**
	−0.121	−0.0879	−0.3395	−0.5169	**−0.0983 ± 0.0157i**	−0.4666	−0.1353	−0.1058
	−0.2008	−0.4496	−0.1102	−0.0884	−0.3521	−0.0494	−0.3294	−0.7991
	−0.2217	−0.5606	−0.2061	−0.2416		−0.2144	−0.3712	−0.9209
Minimum damping ratio	**0.1875**	0.0339	0.1601	0.1098	0.1345	0.0011	0.1668	0.0229

Figure 5(a)—load is increased by 10% in area 1.

Figure 5(b)—load is increased by 20% in area 2.

Figure 5(c)—load is increased by 10% in area 1 and 20% in area 2.

Figure 5(d)—load is increased and decreased by ±25% and ±50% from their nominal values in area 1.

The critical analysis of these responses suggests that the overshoot and the settling time of ITAE (J1) are less than ISE (J2) in all conditions and achieve better dynamic performance. Due to the superior performance of J1, ITAE is considered in further cases. It can be observed in Figure 5(d) that as the load increases, the oscillations in the system are less with GWO (J1) regulator and dampen out quickly as compared with GWO (J2) regulator. Figure 5(e) and Figure 5(f) shows the response of tie line power and frequency deviation in area 2 for the range ±25%–±50% of the nominal load. In Figure 5(f) it is observed that there is negligible effect of variation of load on the frequency deviation in area 2 when increased in area 1. It is also empirical to judge with the responses obtained under different conditions that although both regulators achieve zero steady-state error, the regulator tuned with the criteria J1 obtained less settling time. The effect of the increment in the load in areas is not prominently seen in the response with J1 setting. It can be concluded from Figure 5 that overall robust regulator design can be obtained with criteria J1.

The comparisons of all the algorithms are examined by the four cases.

Case A: Load change in area 1 by 10%. The dynamic responses of Δf_1, Δf_2, and ΔP_{tie} are given in Figures 6–8 for all the algorithms. Figure 9 shows the representation of OS, ST, and FOD under all cases.

Case B: Load change in area 2 by 20%. Figures 10–12 show the dynamic responses of the system.

Case C: Load is increased in area 1 by 25%. In Figures 13–15, the system dynamic responses are shown.

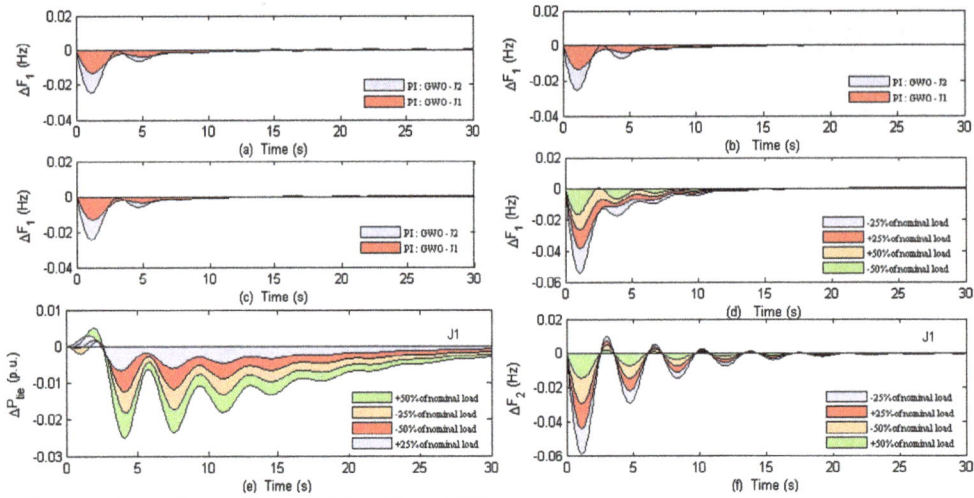

Figure 5. Dynamic responses obtained from GWO.

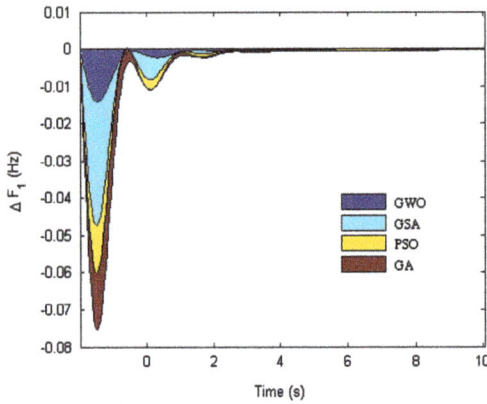

Figure 6. Change in frequency of area 1 by 10% load change in area 1.

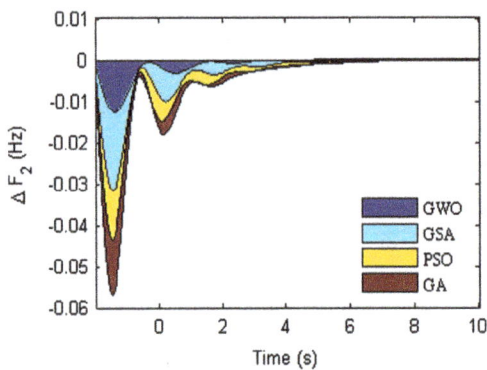

Figure 7. Change in frequency of area 2 by 10% load change in area 1.

Case D: Load is decreased in area 1 by 25% and its responses are given in Figures 16–18.

It is observed from Figures 6–8 that GWO-based controller exhibits the better dynamic performance as compared with others. Percentage overshoot and settling time are much less in these cases. The low oscillatory response exhibited by GWO is also good for the equipment's health. Figure9 shows the preliminary calculations for all the algorithms under all cases to show the effectiveness of proposed GWO over GSA (Sahu et al., 2014), PSO (Abdel-Magid & Abido, 2003) and GA (Abdel-Magid & Dawoud, 1996). It has been observed that minimum value of settling time is obtained from GWO-based regulator. This value is considered as a close replica of dynamic

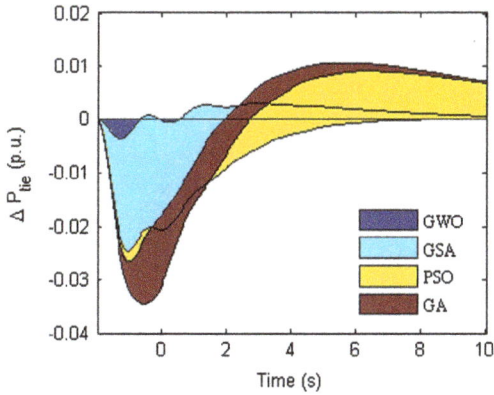

Figure 8. Change in tie line power by + 10% load change in area 1.

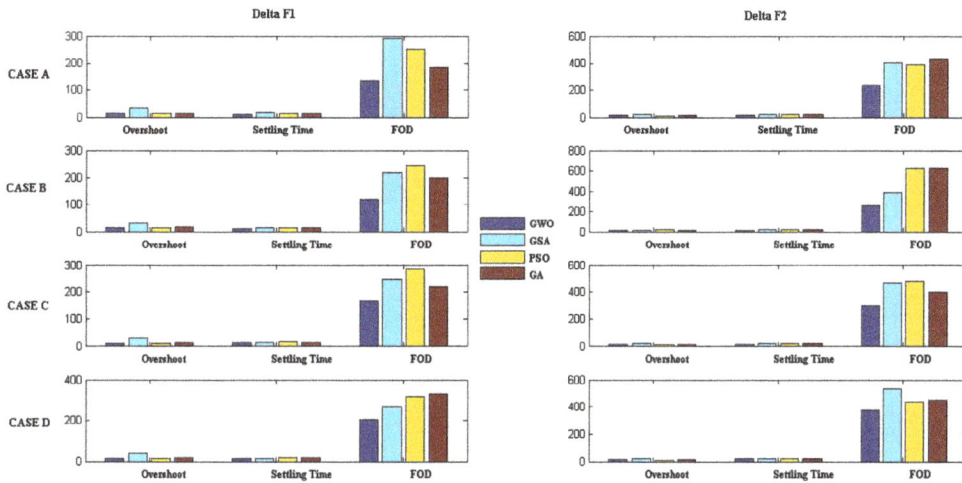

Figure 9. Representation of OS, ST, and FOD under all cases.

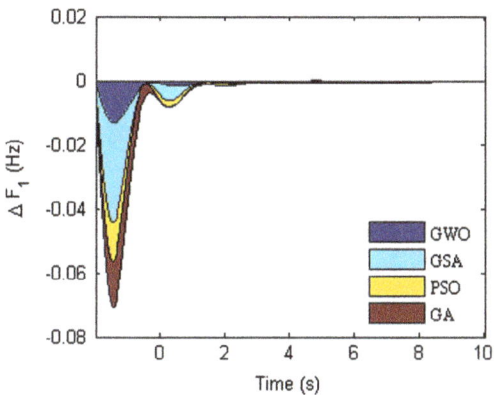

Figure 10. Change in frequency of area 1 by +20% load change in area 2.

performance of controller. It is also empirical to mention here that for frequency deviation in area 1, the settling time obtained from GWO is 4.7, whereas from GSA, PSO, and GA, settling times are 7.1, 7.8, and 6.6, respectively. The frequency deviation in area 2 also shows that the value of settling time is less when GWO is used. The value of settling time, when frequency deviation, in area 2 is 5.4 for GWO, 7.2 for GSA, 7.8 for PSO, and 8.7 for GA. It is also interesting to observe that with the 10% increase in the load, PSO gives erroneous results and the flow of tie line power behaves in a different manner. Hence, critical analysis of dynamic responses clearly reveals the better dynamic performance exhibited by GWO. By examining the responses in Figures 10–12, it is clearly seen that the

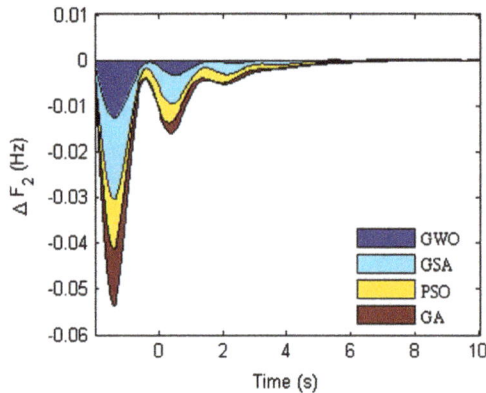

Figure 11. Change in frequency of area 2 by +20% load change in area 2.

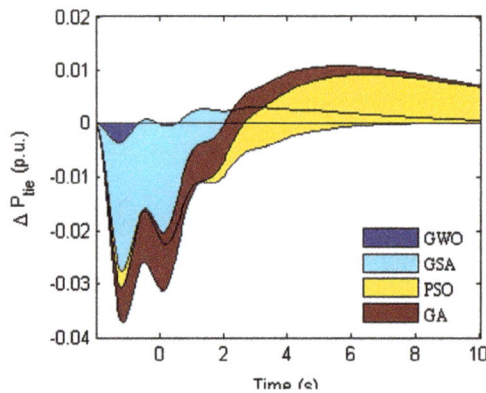

Figure 12. Change in tie line power by +20% load change in area 2.

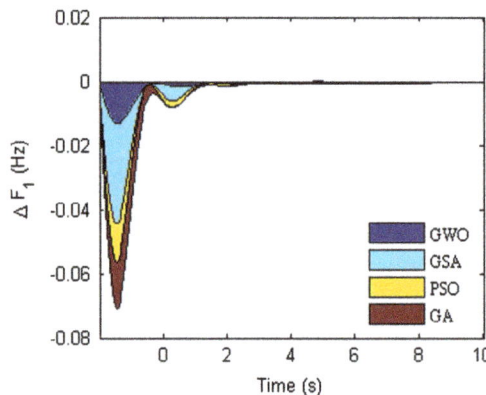

Figure 13. Change in frequency of area 1 with 25% increase in load change in area 1.

settling time and peak overshoot are less when load changes in area 2 are by 20%. It can be observed from the Figure 11 that when area 2 observes 20% increase, GA-based controller is not able to mitigate the frequency oscillations. This inculcates oscillatory instability in the system. However, GWO-based controller shows a better dynamic response and yields a satisfactory performance over a wide range of loading conditions. For case B, the settling time of GWO is 4.9, 6.4, 7.1, and 7.3 for GSA, PSO, and GA, respectively. This analysis can be seen in Figure 9. Figures 13 and 14 show the frequency deviations of areas 1 and 2. Figure 9 illustrates the overshoot, settling time, and FOD of case 3 when load is increased in area 1 by 25%. From dynamic response and graphical representation of overshoot, settling time, and FOD, it is clear that GWO provides competitive results as compared to all other algorithms. The dynamic responses for case D are shown in Figures 16–18 and it has been observed that GWO-tuned controller yields a better dynamic performance. The minimum

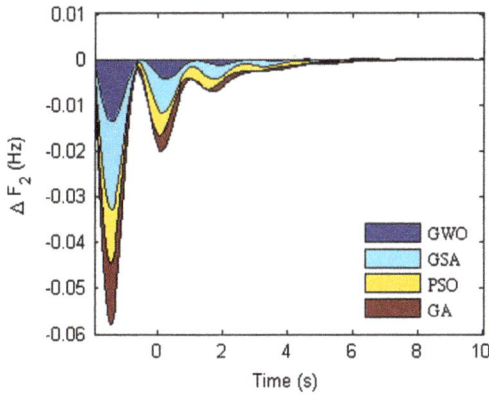

Figure 14. Change in frequency of area 2 on +25% load change in area 1.

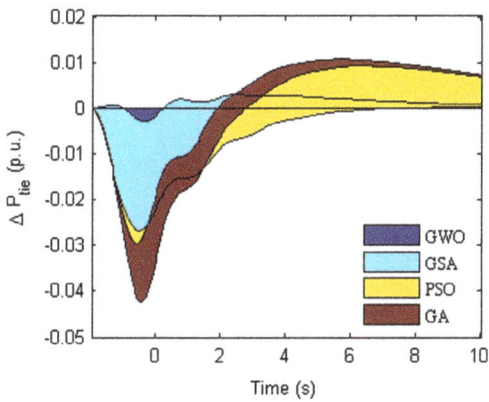

Figure 15. Change in tie line power by +25% load change in area 1.

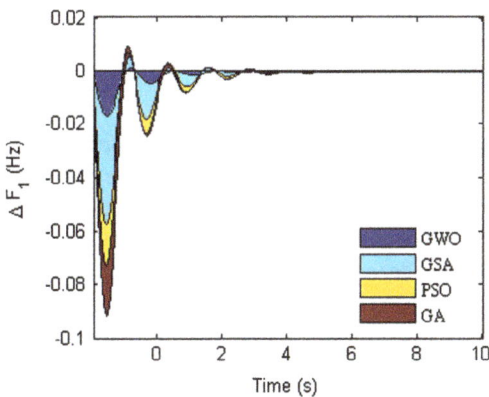

Figure 16. Change in frequency of area 1 on −25% load change in area 1.

settling time is obtained from GWO which is 4.1 for frequency deviation in area 1 and 4.5 for frequency deviation in area 2. However, in case of GSA, the settling time is 6.1. An oscillatory response is obtained by the GA-, GSA-, and PSO-tuned controllers.

The eigenvalues obtained from J1 and J2 for each case of all the algorithms are provided in Table 3. It has been observed that all modes which come from GWO technique lie in the left half of the s-plane and thus sustain the stability of the system. However, in case of GA, few modes lie in the right half of s-plane and make the system unstable. This phenomenon can be observed in Figures 11 and 12. It can also be observed that the modes obtained after the realization of controller through J1 criterion possess a bigger negative part, and in few operating cases, like B, C, and D, this

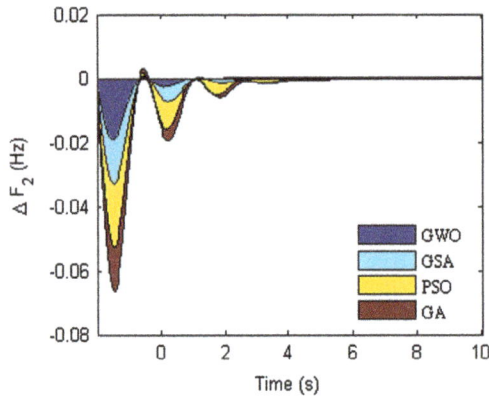

Figure 17. Change in frequency of area 2 on −25% load change in area 1.

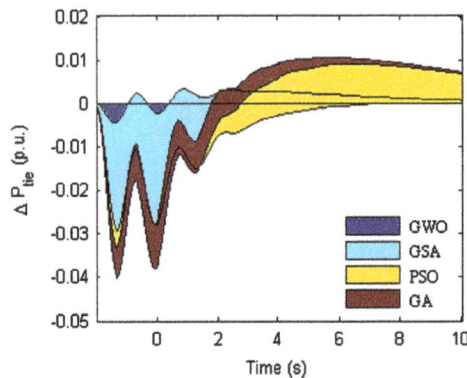

Figure 18. Change in tie line power by −25% load change in area 1.

phenomenon is more prominent. Eigenvalues are true replica of the system's behavior. In different operating cases, these calculations show different behaviors of the system.

5. Optimization performance

Around 100 trials of optimization are carried out to judge the efficiency of the optimization process carried out by all the above-said algorithms. To provide a fair comparison, population size (100) and maximum number of iterations (1,000) are kept the same. Stopping criterion for the optimization process is maximum run of the iteration. To observe the optimization process in a critical way, the standard deviations of optimized parameters of the regulator along with the values of objective functions are calculated and shown in Table 4. It is observed that high values of standard deviations are obtained in regulator parameters and values of objective functions when optimization process is handled by GSA. Comparatively large values of standard deviations are found in GA and PSO for base load conditions when it is compared with GWO. High values of standard deviation are observed in speed regulation parameters after each run of optimization obtained with GA regulators (J1 and J2). The impact of speed regulation parameters on the dynamic response is shown in Saxena et al. (2012). The lowest values of standard deviations are observed, when the parameters are optimized by GWO. It basically means that in each run of optimization, GWO exhibits precision in computing the parameters. The standard deviation in the values of integral gains for area1 by J1 and J2 are minimum for GWO (0.012654 and 0.0200); for GA, these values are 0.97 and 0.268; similarly, for GSA, 0.024 and 0.07 and for PSO, 0.02 and 0.08. It can be concluded that regulator setting integral gain observes least variation in numerical values when the parameter is optimized through GWO (J1). The values of standard deviation in objective functions J1 and J2 are lowest for GWO process and highest for GA. The values of standard deviation in the values of J1 for GA, PSO, GSA, and GWO are 1.76, 0.38, 0.0212, and 0.000978 and for the J2 are 0.013, 0.011, 0.000419, and 0.0000078. It has been observed that the values obtained by GWO are precise and the optimization processes are reliable

Parameters	GWO		GSA (Sahu et al., 2014)		PSO (Abdel-Magid & Abido, 2003)		GA (Abdel-Magid & Dawoud, 1996)	
	J1	J2	J1	J2	J1	J2	J1	J2
Case A	−5.8014	−5.8014	−5.7891	−5.9112	−5.7884	−6.4711	−5.5532	−5.752
	−4.2274	−4.2274	−4.313	−4.4257	−4.4443	−4.8155	−4.1792	−4.2168
	−0.4924 ± 1.6361i	−0.4924 ± 1.6361i	−0.4288 ± 1.6043i	−0.2773 ± 1.8079i	**−0.4277 ± 1.6059i**	−0.0430 ± 2.5784i	−0.2588 ± 1.4307i	−0.2211 ± 1.5866i
	−0.2842 ± 1.4933i	−0.2842 ± 1.4933i	−0.2570 ± 1.6085i	−0.1941 ± 1.7460i	**−0.2395 ± 1.7695i**	−0.0222 ± 2.1888i	−0.5271 ± 1.1657i	**0.0370 ± 1.5795i**
	−0.1208	−0.1208	−0.3454	−0.5259	**−0.0983 ± 0.0157i**	−0.4806	−0.1466	−0.1058
	−0.2021	−0.2021	−0.1101	−0.0884	−0.3584	−0.0494	−0.3344	−0.9221
	−0.2229	−0.2229	−0.2062	−0.2416		−0.2144	−0.401	−0.8182
Case B	−5.8597	−5.9843	−5.8468	−5.976	−5.846	−6.564	−5.5965	−5.808
	−4.1275	−4.2672	−4.2059	−4.3093	−4.3269	−4.6691	−4.0832	−4.1155
	−0.4646 ± 1.7341i	−0.2906 ± 1.9218i	−0.4002 ± 1.7063i	−0.2534 ± 1.9127i	**−0.3943 ± 1.7014i**	**0.0014 ± 2.6941i**	−0.5108 ± 1.2557i	−0.2029 ± 1.6828i
	−0.3315 ± 1.3466i	−0.1072 ± 1.5609i	−0.3117 ± 1.4571i	−0.2466 ± 1.5904i	**−0.3057 ± 1.6218i**	−0.0944 ± 2.0213i	−0.3032 ± 1.2918i	−0.0013 ± 1.4380i
	−0.1204	−0.0879	−0.3394	−0.5169	**−0.0986 ± 0.0155i**	−0.4774	−0.1467	−0.1059
	−0.2047	−0.4498	−0.1101	−0.0884	−0.3521	−0.0494	−0.344	−0.8003
	−0.2234	−0.5752	−0.2096	−0.245		−0.2155	−0.3879	−0.9453
Case C	−5.7282	−5.8373	−5.7167	−5.8312	−5.716	−6.353	−5.4991	−5.6819
	−4.2274	−4.3812	−4.313	−4.4278	−4.4443	−4.8155	−4.1792	−4.2168
	−0.5297 ± 1.5095i	−0.3560 ± 1.6872i	−0.4632 ± 1.4784i	−0.3257 ± 1.6774i	**−0.4587 ± 1.4794i**	−0.0990 ± 2.4269i	−0.2590 ± 1.4294i	−0.2413 ± 1.4636i
	−0.2816 ± 1.4943i	−0.0567 ± 1.7151i	−0.2541 ± 1.6063i	−0.1938 ± 1.7505i	**−0.2396 ± 1.7679i**	−0.0228 ± 2.1889i	−0.5411 ± 1.0386i	**0.0382 ± 1.5795i**
	−0.1204	−0.0878	−0.355	−0.5367	**−0.0982 ± 0.0157i**	−0.4855	−0.1462	−0.1058
	−0.2039	−0.4672	−0.11	−0.0886	−0.3686	−0.0494	−0.3357	−0.9251
	−0.2252	−0.5609	−0.2063	−0.2401		−0.2144	−0.4258	−0.8475
Case D	−6.0556	−6.2024	−6.041	−6.1949	−6.0401	−6.8711	−5.7445	−5.9969
	−4.2274	−4.3812	−4.313	−4.4278	−4.4443	−4.8155	−4.1792	−4.2168
	−0.3695 ± 2.0323i	−0.1897 ± 2.2365i	−0.3106 ± 2.0088i	−0.1627 ± 2.2305i	**−0.3055 ± 2.0076i**	**0.1510 ± 3.0616i**	−0.4440 ± 1.5429i	−0.1319 ± 1.9807i
	−0.2838 ± 1.4894i	−0.0573 ± 1.7122i	−0.2589 ± 1.6017i	−0.1958 ± 1.7442i	**−0.2459 ± 1.7656i**	−0.0217 ± 2.1890i	−0.2680 ± 1.4339i	**0.0368 ± 1.5765i**
	−0.1216	−0.0879	−0.3261	−0.4948	**−0.0983 ± 0.0158i**	−0.4698	−0.1478	−0.1059
	−0.1971	−0.434	−0.1105	−0.0886	−0.3379	−0.0494	−0.3276	−0.7544
	−0.2194	−0.5605	−0.2059	−0.24		−0.2144	−0.3632	−0.9192

Table 3. System modes for each case of all the algorithms

enough for obtaining the regulator design. However, high values of standard deviations in parameters of regulator and in the objective functions (1.76, Table 4) show that the optimization process loses its relevance when it is handled by GA. Figure 11 shows the convergence characteristics of all the optimization algorithms. It can be observed from the figure that GA converges prematurely at 235th iteration; it converges to local minima for the ITAE objective function. The value of the function at this instance is 11.035. In a similar manner, the value of objective function when treated with

Table 4. Standard deviation of optimized parameters of AGC regulator

Parameters	GWO		GSA		PSO		GA	
	J1	**J2**	**J1**	**J2**	**J1**	**J2**	**J1**	**J2**
KI1	0.01265	0.02000	0.02425	0.07720	0.02604	0.08461	0.09729	0.20687
D1	0.00666	0.00299	0.04475	0.09833	0.08416	0.04759	0.18872	0.21313
R1	0.00124	0.00602	0.00045	0.00168	0.00153	0.01403	0.00892	0.00717
KI2	0.00538	0.00564	0.10256	0.00769	0.09639	0.08158	0.05800	0.15550
D2	0.00765	0.05750	0.04916	0.11771	0.16726	0.08363	0.17955	0.07945
R2	0.00195	0.00252	0.00173	0.00385	0.00280	0.01080	0.00310	0.0105937
	0.000978	**0.0000078**	**0.0212**	**0.000419**	**0.38**	**0.011**	**1.76**	**0.013**

Table 5. Performance parameters of optimized parameters of AGC regulator

Performance parameters	GWO	GSA	PSO	GA
Best performance	0.0000023	0.0276	0.000014	0.000091
Average performance	0.000495	0.04353	0.003137	0.02222
Poor performance	0.0032	0.0907	0.03	0.06

PSO is 6.60 and shows the convergence at 256th iteration. Authors found that the optimization process handled by GSA is difficult to converge and time taking; on the other hand, GWO handled the optimization process in a pace and the convergence is faster. It is also worth mentioning here that for this optimization process and for similar run of iterations, the time taken by GA, PSO, and GSA is much more than GWO. This shows the faster convergence of GWO for ITAE functions. It has also been observed that the optimization process by J2 (ISE) is more time-consuming and less accurate. Table 5 shows the optimization performance of algorithms in terms of average, best, and poor values of objective function J1. For real problems like AGC, the time for evaluation of regulator parameter through optimization process is a critical issue to be addressed. Moreover, the efficacy of the obtained results through dynamic responses is also a major parameter for regulator design. The following section summarizes the major contributions and findings of this work in a conclusive manner.

6. Conclusion

This paper presents an application of recently introduced algorithm, GWO, to find optimal parameters of the AGC regulator. The GWO regulator is employed on a test system of two thermal units connected with a weak tie line of limited capacity for AGC. Following are the major findings of the work:

(1) Comparison of the application of two objective functions, namely ISE and ITAE, in optimization process for finding the regulator parameters, under different contingencies, is investigated. Results reveal that ITAE is a better choice to optimize the regulator parameters.

(2) Eigenvalue analysis is performed to test the effectiveness of the proposed approach and to compare the results of the proposed approach with the recently published approaches. It is observed that the damping obtained from GWO regulator is more positive as compared with other algorithms.

(3) Convergence characteristics of the algorithms are exhibited to monitor the health of algorithms and their flow. It is empirical to judge that GA has a major problem of premature convergence and the time taken by the optimization process is much more in comparison with PSO, GSA, and GWO. GWO shows promising results in terms of overshoot, settling time

obtained from the frequency responses of both areas under different loading cases, standard deviations in regulator's parameter values, ISE and ITAE values, and optimization performance.

(4) Damping performance is evaluated with different contingencies, load changes, and step disturbances in both areas. PI controller setting obtained through GWO exhibits better dynamic performance and overall low settling time.

Application of other new meta-heuristic algorithms in AGC regulator design on different models of power system considering different renewable energy power sources lays in future scope.

Nomenclature

I	subscript referred to area i (1,2)
Δf_i	frequency deviation in area i (Hz)
ΔP_{Gi}	incremental generation of area i (p.u.)
ΔP_{Li}	incremental load change in area i (p.u.)
ACE_i	area control error of area i
B_i	frequency bias parameter of area i
R_i	speed regulation of the governor of area i (Hz/p.u.MW)
T_{gi}	time constant of governor of area i (s)
T_{ti}	time constant of turbine of area i (s)
K_{pi}	gain of generator and load of area i
T_{pi}	time constant of generator and load of area i (s)
ΔP_{tie}	incremental change in tie line (p.u.)
T_{12}	synchronizing coefficient
T	simulation time (s)
α	alpha wolf
β	beta wolf
δ	delta wolves
ω	omega wolves
t	current iteration
$\vec{X}_P$	position vector of the prey
$\vec{X}$	position vector grey wolf

Funding
The authors received no direct funding for this research.

Author details
Esha Gupta[1]
E-mail: esha.gupta@outlook.com
ORCID ID: http://orcid.org/0000-0001-5117-2329
Akash Saxena[1]
E-mail: aakash.saxena@hotmail.com
ORCID ID: http://orcid.org/0000-0002-1820-8024
[1] Department of Electrical Engineering, Swami Keshvanand Institute of Technology, Office no AC-201, Ramnagaria, Jagatpura, Jaipur 302017, Rajasthan, India.

References
Abdel-Magid, Y. L., & Abido, M. A. (2003). AGC tuning of interconnected reheat thermal systems with particle swarm optimization. *IEEE International Conference on Electronics, Circuits and Systems*, 1, 376–379.
Abdel-Magid, Y. L., & Dawoud, M. M. (1996). Optimal AGC tuning with genetic algorithms. *Electric Power Systems Research*, 38, 231–238. http://dx.doi.org/10.1016/S0378-7796(96)01091-7
Ali, R., Mohamed, T. H., Qudaih, Y. S., & Mitani, Y. (2014). A new load frequency control approach in an isolated

small power systems using coefficient diagram method. *International Journal of Electrical Power & Energy Systems, 56*, 110–116.

Banerjee, A., Mukherjee, V., & Ghoshal, S. P. (2014). Intelligent controller for load-tracking performance of an autonomous power system. *Ain Shams Engineering Journal, 5*, 1167–1176. http://dx.doi.org/10.1016/j.asej.2014.06.004

Bernard, M. Z., Mohamed, T. H., Qudaih, Y. S., & Mitani, Y. (2014). Decentralized load frequency control in an interconnected power system using coefficient diagram method. *International Journal of Electrical Power & Energy Systems, 63*, 165–172.

Debbarma, S., Chandra Saikia, L., & Sinha, N. (2014). Solution to automatic generation control problem using firefly algorithm optimized IλDμ controller. *ISA Transactions, 53*, 358–366. http://dx.doi.org/10.1016/j.isatra.2013.09.019

Elgerd, O. I. (1983). *Energy systems theory. An introduction.* New Delhi: McGraw-Hill.

Emary, E., Zawbaa, H. M., Grosan, C., & Hassenian, A. E. (2015). Feature subset selection approach by gray wolf optimizer. *Advances in Intelligent Systems and Computing, 334*, 1–13. http://dx.doi.org/10.1007/978-3-319-13572-4

Gozde, H., Cengiz Taplamacioglu, M., & Kocaarslan, İ. (2012). Comparative performance analysis of Artificial Bee Colony algorithm in automatic generation control for interconnected reheat thermal power system. *International Journal of Electrical Power & Energy Systems, 42*, 167–178. http://dx.doi.org/10.1016/j.ijepes.2012.03.039

Ibraheem, P. K., & Kothari, D. P. (2005). Recent philosophies of automatic generation control strategies in power systems. *IEEE Transactions on Power Systems, 20*, 346–357. http://dx.doi.org/10.1109/TPWRS.2004.840438

IEEE standard definitions of terms for automatic generation control on electric power systems. (1970). *IEEE Trans. Power Apparatus and Systems, 89*, 1356–1364.

Kanniah, J., Tripathy, S. C., Malik, O. P., & Hope, G. S. (1984). Microprocessor-based adaptive load-frequency control. *IEE Proceedings C Generation, Transmission and Distribution, 131*, 121–128. http://dx.doi.org/10.1049/ip-c.1984.0020

Kothari, M. L., Satsangi, P. S., & Nanda, J. (1981). Sampled-data automatic generation control of interconnected reheat thermal systems considering generation rate constraints. *IEEE Transactions on Power Apparatus and Systems, PAS-100*, 2334–2342. http://dx.doi.org/10.1109/TPAS.1981.316753

Mirjalili, S. (2015). How effective is the grey wolf optimizer in training multi-layer perceptrons. *Applied Intelligence.* doi:10.1007/s10489-014-0645-7

Mirjalili, S., Mirjalili, S. M., & Lewis, A. (2014). Grey wolf optimizer. *Advances in Engineering Software, 69*, 46–61. http://dx.doi.org/10.1016/j.advengsoft.2013.12.007

Muro, C., Escobedo, R., Spector, L., & Coppinger, R. (2011). Wolf-pack (*Canis lupus*) hunting strategies emerge from simple rules in computational simulations. *Behavioural Processes, 88*, 192–197. http://dx.doi.org/10.1016/j.beproc.2011.09.006

Nanda, J., & Kaul, B. (1978). Automatic generation control of an interconnected power system. *IEEE Proceedings, 125*, 385–390.

Nanda, J., & Mishra, L. C. (2009). Maiden application of bacterial foraging-based optimization technique in multiarea automatic generation control. *IEEE Transactions on Power Systems, 24*, 602–609. http://dx.doi.org/10.1109/TPWRS.2009.2016588

Pandan, S., Sahu, R. K., & Panda, S. (2014). Automatic generation control with thyristor controlled series compensator including superconducting magnetic energy storage units. *Ain Shams Engineering Journal, 5*, 759–774.

Puja, D., Chandra, S. L., & Nidul, S. (2014). Comparison of performances of several cuckoo search algorithm based 2DOF controllers in AGC of multi-area thermal system. *International Journal of Electrical Power & Energy Systems, 55*, 429–436.

Rout, U. K., Sahu, R. K., & Panda, S. (2013). Design and analysis of differential evolution algorithm based automatic generation control for interconnected power system. *Ain Shams Engineering Journal, 4*, 409–421. http://dx.doi.org/10.1016/j.asej.2012.10.010

Sadat, H. (1999). *Power system analysis.* Boston, MA: McGraw-Hill.

Sahu, R. K., Panda, S., & Padhan, S. (2014). Optimal gravitational search algorithm for automatic generation control of interconnected power systems. *Ain Shams Engineering Journal, 5*, 721–733. http://dx.doi.org/10.1016/j.asej.2014.02.004

Sahu, B. K., Pati, S., Mohanty, P. K., & Panda, S. (2015). Teaching-learning based optimization algorithm based fuzzy-PID controller for automatic generation control of multi-area power system. *Applied Soft Computing, 27*, 240–249. http://dx.doi.org/10.1016/j.asoc.2014.11.027

Saxena, A., Gupta, M., & Gupta, V. (2012). Automatic generation control of two area interconnected power system using Genetic algorithm. *2012 IEEE International Conference on Computational Intelligence & Computing Research (ICCIC)*, 1–5.

Shivaie, M., Kazemi, M. G., & Ameli, M. T. (2015). A modified harmony search algorithm for solving load-frequency control of non-linear interconnected hydrothermal power systems. *Sustainable Energy Technologies and Assessments, 10*, 53–62. http://dx.doi.org/10.1016/j.seta.2015.02.001

Sivaramaksishana, A. Y., Hariharan, M. V., & Srisailam, M. C. (1984). Design of variable structure load-frequency controller using pole assignment techniques. *International Journal of Control, 40*, 437–498.

Song, H. M., Sulaiman, M. H., & Mohamed, M. R. (2014). An application of grey wolf optimizer for solving combined economic emission dispatch problems. *International Review on Modelling and Simulations, 7*, 838–844.

Song, X., Tang, L., Zhao, S., Zhang, X., Li, L., Huang, J., & Cai, W. (2015). Grey wolf optimizer for parameter estimation in surface waves. *Soil Dynamics and Earthquake Engineering, 75*, 147–157.

Sudha, K. R., & Vijaya Santhi, R. (2012). Load frequency control of an interconnected reheat thermal system using type-2 fuzzy system including SMES units. *International Journal of Electrical Power & Energy Systems, 43*, 1383–1392.

Valk, I., Vajta, M., Keviczky, L., Haber, R., Hetthessy, J., & Kovacs, K. (1985). Adaptive load-frequency control of hungarian power system. *Automatica, 21*, 129–137.

Wu, S., Er, M. J., & Gao, Y. (2001). A fast approach for automatic generation of fuzzy rules by generalized dynamic fuzzy neural networks. *IEEE Transactions on Fuzzy Systems, 9*, 578–594. http://dx.doi.org/10.1109/91.940970

7

Hybrid evolutionary algorithm based fuzzy logic controller for automatic generation control of power systems with governor dead band non-linearity

Omveer Singh[1]* and Ibraheem Nasiruddin[2]

*Corresponding author: Omveer Singh, Electrical Engineering Department, Maharishi Markandeshwar University, Ambala, Haryana, India
E-mail: omveers@gmail.com

Reviewing editor: Yunhe Hou, University of Hong Kong, Hong Kong

Abstract: A new intelligent Automatic Generation Control (AGC) scheme based on Evolutionary Algorithms (EAs) and Fuzzy Logic concept is developed for a multi-area power system. EAs i.e. Genetic Algorithm–Simulated Annealing (GA–SA) are used to optimize the gains of Fuzzy Logic Algorithm (FLA)-based AGC regulators for interconnected power systems. The multi-area power system model has three different types of plants i.e. reheat, non-reheat and hydro, and are interconnected via Extra High Voltage Alternate Current transmission links. The dynamic model of the system is developed considering one of the most important Governor Dead Band (GDB) non-linearity. The designed AGC regulators are implemented in the wake of 1% load perturbation in one of the control areas and the dynamic response plots are obtained for various system states. The investigations carried out in the study reveal that the system dynamic performance with hybrid GA–SA-tuned Fuzzy technique (GASATF)-based AGC controller is appreciably superior as compared to that of integral and FLA-based AGC controllers. It is also observed that the incorporation

ABOUT THE AUTHORS

Omveer Singh is working as an associate professor with Electrical Engineering Department, Maharishi Markandeshwar University, Ambala, India. He did PhD from Jamia Millia Islamia University, New Delhi, India. He has published over 42 articles in international journals and conferences. Currently, he is supervising 2 research scholars. He has diversified research interests in the areas of automatic generation control, advanced power systems, soft computing techniques and renewable power generation.

Ibraheem Nasiruddin is presently with the Department of Electrical Engineering, Qassim Engineering College, Qassim University, Kingdom of Saudi Arabia, where he is working as Professor of Electrical Engineering. He has published over 55 research papers in journal of national and international repute, also over 100 research articles in conferences and workshops. He has guided 14 PhD thesis and many graduate and postgraduate students for their project and dissertation work. His research interests include power system operation and control, automatic generation control of power systems.

PUBLIC INTEREST STATEMENT

The AGC problem has been one of the major concerns for power system engineers and is becoming more significant these days owing to the increasing size and complexity of interconnected power systems. For efficient and successful operation of an interconnected power system, the main requirement is the retention of an electrical power system characterized by nominal frequency, voltage profile and load flow configuration. Designing an efficient AGC strategy is necessary to ensure the fulfilment of requirements. An efficient AGC scheme based on GASA-tuned fuzzy approach is proposed in this manuscript. This hybrid GASATF technique exhibits its efficacy to great extent in power system which constitutes different types of turbines, and also has system load perturbations in one of the power system area.

of GDB non-linearity in the system dynamic model has resulted in degraded system dynamic performance.

Subjects: Computer & Software Engineering; Electrical & Electronic Engineering; Power Engineering

Keywords: automatic generation control; governor dead band; non-linearity; fuzzy logic algorithm; evolutionary algorithms

1. Introduction
The growth of large interconnected power system is to minimize the occurrence of the black outs and providing an increasing power interchange among distinct system under the huge interconnected electric networks. The power system operators enhance load-interchange-generation balance between control areas, and adjust the system frequency as close as possible nominal values. Automatic Generation Control (AGC) is necessary to keep the system frequency and the inter-area tie-line power as close as predefined nominal values. A reliable and committed power utility should cope with load variations and disturbance effectively. It should give permissible high quality of power while controlling frequency within acceptable limits. The problem of AGC has been extensively analysed during the last few decades. Elgerd and Fosha in 1970 were the first to propose the design of optimal AGC regulators using modern control theory for interconnected power systems (Elgerd & Fosha, 1970). The proposed scheme provided better control performance for a wide range of operating conditions than the performance of conventionally designed control schemes. Since the classical gain scheduling methods may be unsuitable in some operating conditions due to the complexity of the power systems such as non-linear load characteristics and variable operating points, the modern approaches were preferred for use. Following the work of Elgerd and Fosha, many AGC schemes based on modern control theory have been suggested in literature (Ibraheem & Kothari, 2005; Shayeghi, Shayanfar, & Jalili, 2009; Singh & Ibraheem, 2013). They were followed by AGC schemes based on intelligent control concepts after a very long time.

In Chown and Hartman (1998), a Fuzzy Logic Controller (FLC) as a part of the AGC system in Eskom's National Control Centre based on Area Control Error (ACE) as control signal for the plant is described. Moreover, Yousef proposed an adaptive fuzzy logic load frequency control of multi-area power system in Yousef (2015), Yousef et al. (2014). Anand and Jeyakumar (2009) incorporated the system with governor dead band, generation rate constraint and boiler dynamics non-linearities in the system models. Fuzzy Logic Algorithm (FLA) has been employed to design FLC for the system to overcome the drawback of conventional Proportional–Integral Controller. It has circumvented the controller gain problem to some extent but did not give more accurate (Albertos & Sala, 1998) and precise optimal gains for FLC in the AGC due to need of the exact system operating conditions. Therefore, Evolutionary Algorithms (EAs) are introduced to fight with the controller optimum gain problem (Boroujeni, 2012; Devi & Avtar, 2014; Ghoshal, 2004; Yousef, AL-Kharusi, & Mohammed, 2014; Pratyusha & Sekhar, 2014; Saini & Jain, 2014; Singhal & Bano, 2015). Authors discussed classical controller gains tuning through Non-Dominated Shorting Genetic Algorithm-II (NSGA-II) technique for AGC of an interconnected system. Integral Time multiply Absolute Error, minimum damping ratio of dominant eigenvalues and settling times in frequency and tie-line power deviations considered as multiple objectives and NSGA-II is employed to generate Pareto optimal set. Further, a fuzzy-based membership value assignment method was employed to choose the best compromise solution obtained from Pareto solution set. This method was also investigated with non-linear power system model (Yegireddi & Panda, 2013). EAs were found capable to give global optimum gains for FLCs to handling sensitive controlling issue of AGC. FLCs are characterized by a set of parameters, which are optimized using EAs i.e. GA, SA to improve their performance (Boroujeni, 2012; Ghoshal, 2004; Saini & Jain, 2014). EAs were efficiently applied to AGC of power systems and have shown its ameliorated performance without systematic and precise data of the power system model.

In this article, the design of optimal AGC scheme for a three-area interconnected power system is investigated with three diverse controller's i.e. classical integral control, FLA and GASA-tuned FLC controllers. Moreover, a hybrid GASATF technique constitutes GA and SA approach to determine output fitness function from the fuzzy Mamdani algorithm. This is used as the input for GA–SA technique to design the optimal gains for AGC scheme. The designed AGC scheme yielded ameliorated system dynamic performance under various operating conditions of a three-area interconnected hydro-thermal power systems with and without considering Governor Dead Band (GDB). In this article, studies have been carried out by considering 0.01 p.u.MW perturbation in one of the power system areas. The simulation study carried out using MATLAB/SIMULINK toolbox 2014a version platform. The investigations include the non-linear effect of GDB. Power systems dynamic performance has been studied by investigating the response plots of the disturbed areas (ΔF_1, ΔF_2, ΔF_3, ΔP_{tie12}, ΔP_{tie23} and ΔP_{tie31}). Also the investigations of response plots obtained for ΔX_g, ΔF_3, ΔP_{tie31}, ACE_3 and U_3 with inclusion of GDBs. The system nominal system parameters are presented in Appendix A.

2. Power system model for investigation

It is a three-area interconnected power system consisting of power plants with reheat, non-reheat thermal and hydro turbines, and is interconnected via EHV AC tie-line. The single-line diagram of the multi-area interconnected power system model is presented by Figure 1. The optimal AGC controllers are designed considering (i) linear model of the system and (ii) non-linear model of the system with GDBs. The transfer function model of both the models with and without GDB non-linearity is shown in Figure 2. This model exhibits linear and non-linear both characteristics of the power system behaviour.

3. Effect of governor dead band non-linearity

Most of the real-time AGC of power systems include non-linearity in itself through several ways. The sort of non-linearity may be materialistic affect, Generation Rate Constraints (GRCs), GDB and load, etc. The real-time simulation cannot be realized without incorporation of non-linearity and the demo of real-time AGC model may also be incomplete without non-linearity. Therefore, one of the most prominent non-linearity in the form of GDB is considered in the proposed power systems model. Movement of thermal or hydro GDB can not be permitted without specified tolerable limits for flow of steam/water.

The GDB effect indeed can be significant in AGC studies. In the model, a GDB impression is supplemented to all control areas to simulate non-linearity (Gozde & Taplamacioglu, 2011; Yegireddi & Panda, 2013). Explaining function method is utilized to exhibit the GDB in the control areas. The GDB non-linearity moves to create a continuous sinusoidal oscillation of natural time schedule near to $T_0 = 2$ s. GDB linearization is done by the method using means of deviation and rate of deviation in the movement. With these considerations, GDB is taken into account by adding limiters to the turbine input valve. In this work, the backlash of approximately 0.5% is selected. The GDB transfer function in the power systems model is presented as:

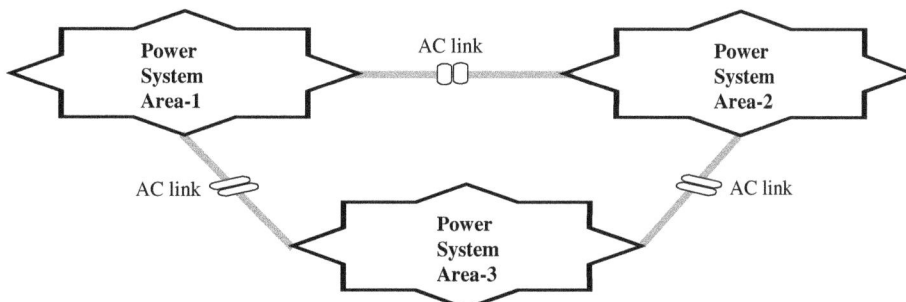

Figure 1. Block diagram of multi-area interconnected power system with AC link.

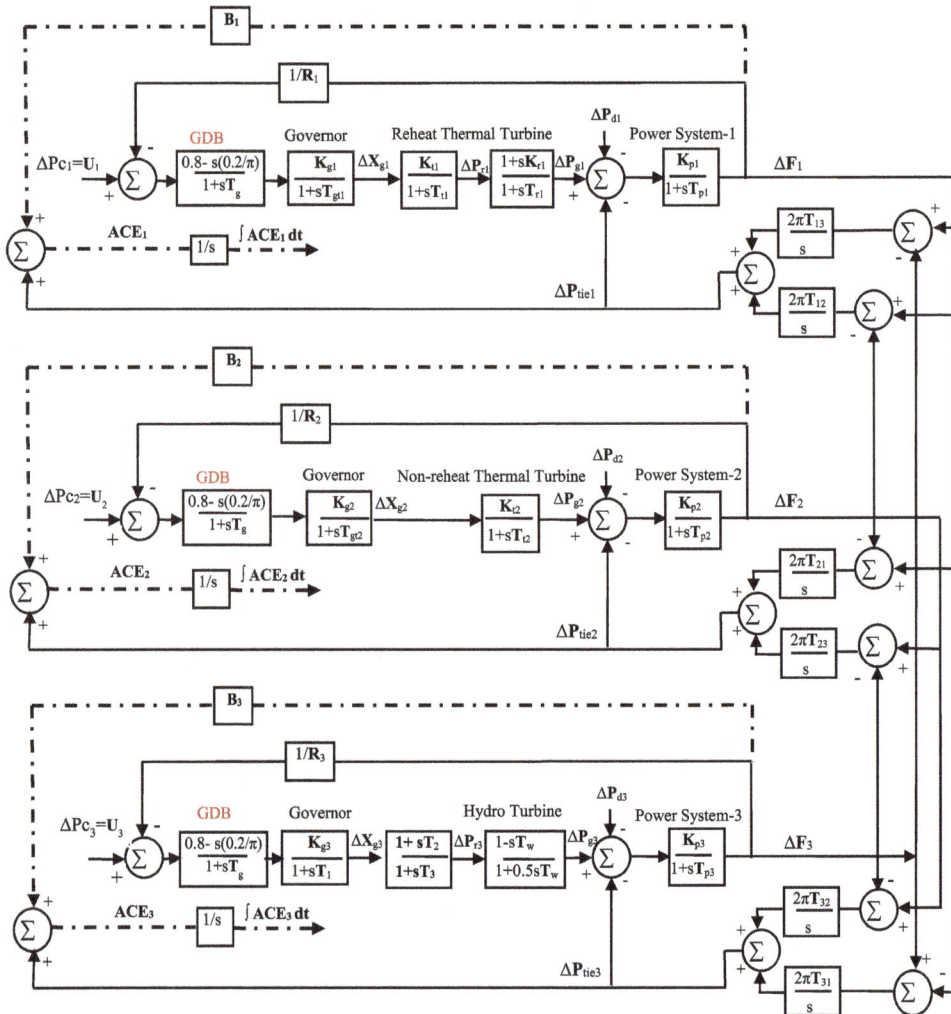

Figure 2. Transfer function model of multi-area interconnected power system without GDB and with GDB non-linearity.

$$G(s) = \frac{0.8 - \frac{0.2s}{\pi}}{1 + sT_g} \tag{1}$$

4. Hybrid EA-based fuzzy logic controller structure

Basically, ACE is the measure of short-term error between generation and electric consumer demand. Power system performance is termed as good if a control area closely matches generation with load demand. AGC of the power systems means minimizing the ACE to zero. Hence, the systems frequency and tie-line power flows are maintained at their scheduled values (Elgerd & Fosha, 1970), respectively:

$$ACE_i = \sum_i \left(B_i \Delta F_i + \Delta P_{tieij} \right) \tag{2}$$

where $i = 1, 2, 3$.

The input to ACE_i is rate of change for integral control action. It can be defined as:

$$U_i = - \int K_i \left(ACE_i \right) dt \tag{3}$$

The ACE_i is tuned in using of Integral Square Error (ISE) criterion which has the form of:

$$ISE = \int_0^{\infty} e^2(t)\, dt \qquad (4)$$

For the present investigation; this error is considered as ACE_i and evaluating ISE as a fitness function as:

$$\text{Fitness function } (ISE) = \int_0^{20} ACE_i^2\, dt \qquad (5)$$

Over the last few decades, FLCs have been developed for analysis and control of several types of systems (Anand & Jeyakumar, 2009; Chown & Hartman, 1998; Yousef, 2015; Yousef et al., 2014). But FLCs have not been exploited to its full capability to counter the problems associated with the systems having non-linear characteristics. It has been demonstrated that its performance is far from being stringent and time optimal (Albertos & Sala, 1998). Subsequently, this problem has been successfully solved using EAs for AGC problem of power systems (Yegireddi & Panda, 2013). Therefore, researchers have proposed the combination of FLA and EAs to design the controllers exploiting the salient features of both the algorithms. The hybrid structure of these techniques exhibited their adaptable qualities in non-linear systems.

FLC consists of FLA approach with integral control action in the closed loop control system. The FLC has input signal, namely ACE_i of the systems, and then its output signal (U_i) is the input signal for GASATF. FLC feedback gains (K_i) are optimized with the help of EAs i.e. combination of GA and SA technique. Finally, the output signal from the GASATF called the new U_i is used in the AGC of power systems.

These optimal gains are globally optimal for the system to be controlled. These global optimal gains give exact value of the fitness function for developing AGC scheme. These global optimal gains can also handle the AGC problem when system is operating with nominal system parameters. The gain scheduling of the power systems is a very important and effective process of its controlling (Saini & Jain, 2014). In view of this, ultimately GASA-tuned FLC is proposed in the present article. This scheme is shown in Figure 3. The presented GASA strategy enables to evaluate global optimal value of fuzzified K_i followed by the modification of fuzzified fitness function of the FLC.

The GASATF heuristic incorporates SA in the selection process of Evolutionary Programming (Boroujeni, 2012). The solution string comprises all feedback gains and is encoded as a string of real numbers. The population structure of GASATF is shown in Figure 4. The GASATF heuristic employs blend crossover and a mutation operator suitable for real number representation to provide it a better search capability. The main objective was to minimize the ACE_i augmented with penalty terms corresponding to transient response specifications in the system frequency and tie-line power flows. These two disturbances yield the error for the AGC systems. The heuristic is quite general and various aspects like non-linear, discontinuous functions and constraints are easily incorporated as per requirement. The fuzzy fitness function is taken as the summation of the absolute values of the three

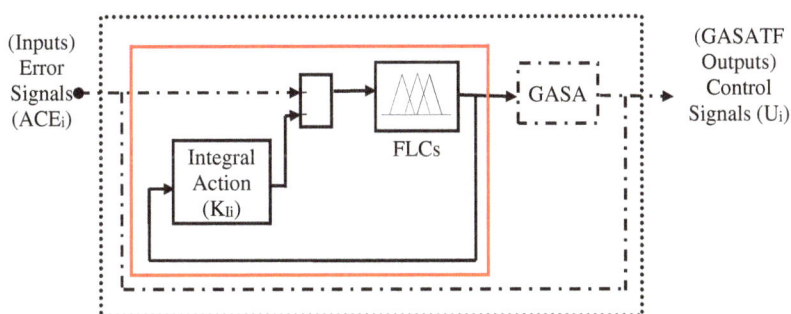

Figure 3. Internal architecture of GASATF-based controller within the "subsystem".

Family 1   Family 2   Family 3   ....   Family n

Parent 1   Parent 2   Parent 3   ....   Parent n
Child 11   Child 21   Child 31   ....   Child n1
Child 12   Child 22   Child 32   ....   Child n2
.   .   .   .
.   .   .   .
.   .   .   .
Child 1w   Child 2x   Child 3y   ....   Child nz

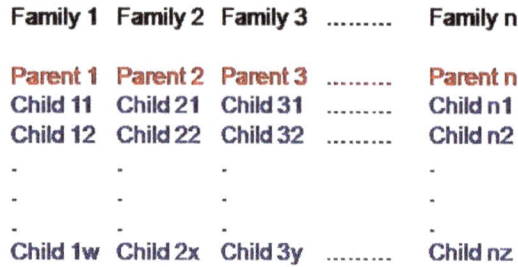

Figure 4. Population structure of GASATF.

at every discrete time instant in the simulation. An optional penalty term is added to take care of the transient response specifications.

The FLA has been investigated with seven triangular membership functions of the FLC. FLC has seven rules for the control of ACE_i. This controller has a Negative Big (NB), Negative Medium (NM), Negative Small (NS), Zero (Z), Positive Small (PS), Positive Medium (PM), Positive Big (PB) functions. Mamdani fuzzy rule list is shown in Table 1. In the FLCs, there is an input and output selected in the fuzzy Mamdani inference system which is described in Figure 5. The input and output membership functions of FLA for ACE_i in both test cases are shown in Figures 6 and 7. A triangular membership function shapes of the derivative error and gains according to integral controller are chosen to be

Table 1. Mamdani-based fuzzy rule	
Mamdani-based Rules for FLCs	
Rule1	If(ACE_i is NB) then (U_i is PB)
Rule2	If(ACE_i is NM) then (U_i is PM)
Rule3	If(ACE_i is NS) then (U_i is PS)
Rule4	If(ACE_i is Z) then (U_i is Z)
Rule5	If(ACE_i is PS) then (U_i is NS)
Rule6	If(ACE_i is PM) then (U_i is NM)
Rule7	If(ACE_i is PB) then (U_i is NB)

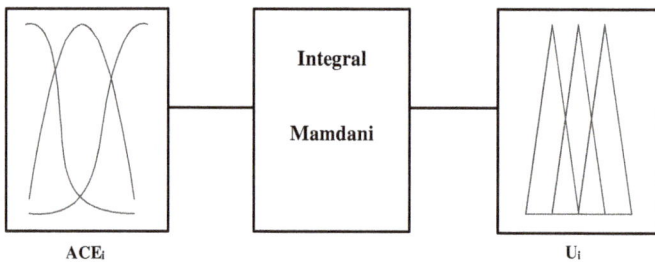

Figure 5. Fuzzy inference system.

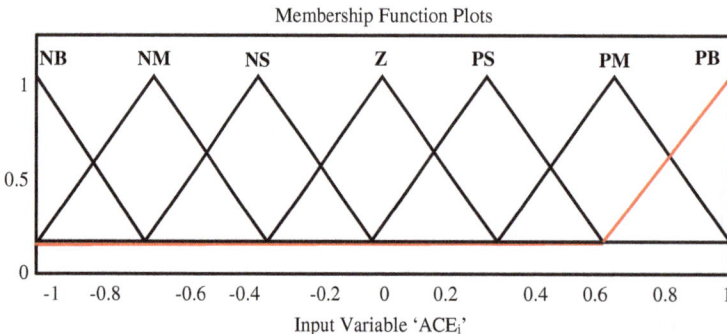

Figure 6. Membership function without GDB for FLC input.

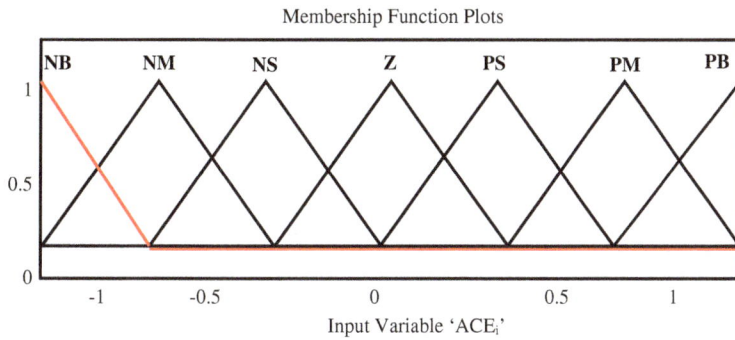

Figure 7. Membership function with GDB for FLC input.

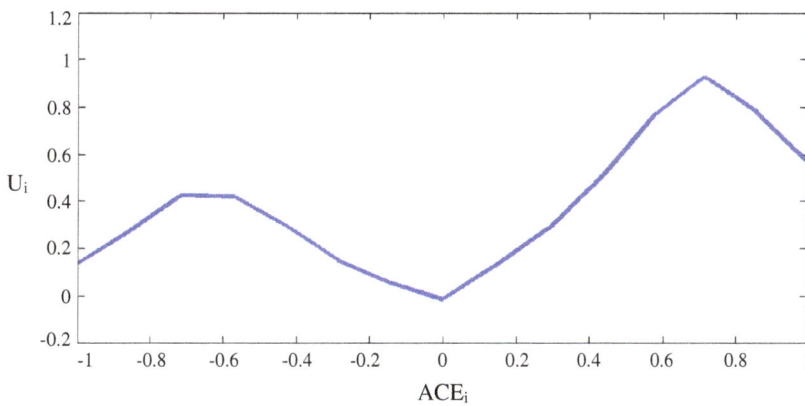

Figure 8. Surface view of input/output without GDB for FLC.

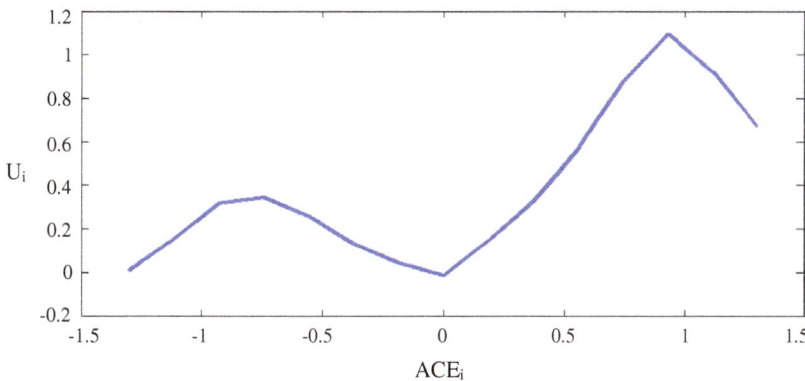

Figure 9. Surface view of input/output with GDB for FLC.

identical for the FLC. However, its horizontal axis limits are taken at different values for evaluating the controller output. The characteristics of the FLC present a conditional relationship between ACE_i and U_i as shown in Figures 8 and 9.

In this graph, x-level shows the controller's input and y-level presents output of the FLC. Rule base are defined in the scaling range of $[-1, 1]$ without GDB, and $[-1.3, 1.3]$ with GDB of the power system.

Initially, population size is 100 and number of parents are 10. Subsequentially, Boltzmann initial probability factor is 0.99 and final is 0.0001. Approximate parameters are utilized as initial cost (1,000,000), total iterations (120), crossover rate (0.6), probability: exp(generation/maximum generation) and mutation probability rate (0.002).

4.1. Pseudo code

The proposed technique can be understood from the following pseudo code which is given below in the steps of implementation;

Step 1: *Create random population.*

Step 2: *In the beginning, assume the current temperature (T), initial temperature (T^1) and final temperature (T^{MAXIT}), parent strings (N), children strings (M) as per FLC structure with scaling factor limits (fitness function limits) (Singh & Das, 2008).*

Step 3: *For each and every parent number is defined by i. Then, create m(i) children using crossover process.*

Step 4: *Modify the mutation operator with the probability margin P$_m$.*

Step 5: *Obtain the best child for every parent (initial competition level in that process).*

Step 6: *Select the best child as a parent for the future generation.*

For every family, accept the best child as the parent for future generation if objective quantity of the best child (Y_1) is less than objective quantity of its parent (Y_2) that is equal to $Y_1 < Y_2$ or $\frac{exp(Y_2 - Y_1)}{T} \geq \rho$ (a random number uniformly divided between 0 and 1).

Step 7: *Reschedule steps 8–11 for every family.*

Step 8: *Counting initiate from zero (count = 0).*

Step 9: *Reschedule step 10 for every child; go to step 11.*

Step 10: *Increase every count by 1, If ($Y_1 < Y_2$) or $\frac{exp(Y_{LOWEST} - Y_1)}{T_c} \geq \rho$ in which T_c is the current temperature at that time and Y_{LOWEST} is the lowest objective quantity ever got in the calculations.*

Step 11: *Acceptance number of the family is equal to count (A).*

Step 12: *Add up the acceptance numbers of all the families (S).*

Step 13: *For every family i, evaluate the number of children to be generated in the next generation as per the following formula: $m(i) = \frac{exp(TC \times A)}{S}$; in which total number of children generated by all the families (TC).*

Step 14: *Decrease the temperature after the each iteration.*

Step 15: *Repeat the steps number 3–14 until a defined number of iterations have been achieved or desired result has been completed.*

Table 2. New generation (best population)	
Best population	
GASATF (without GDB)	[1.2934 −0.2751 0.7940 0.5983 0.6839 −0.1796 0.5248 −0.0002 0.9710 0.1931 1.3231 0.6587 0.6180 0.8957]
GASATF (with GDB)	[1.4658 0.9298 −1.4513 −6.1149 0.6008 −0.1285 0.474 3 0.0085 0.1239 0.0053 1.1172 0.1076 −0.0378 −0.3649]

The GASATF presented for straight forward calculation of the fuzzy fitness function. After the fuzzy fitness function tuning, GASATF technique creates global optimal feedback gains for the AGC. For adopting the best generation the criteria is created by optimal GASATF approach which is reported in Table 2.

5. Results and discussion

The power system model under investigation is simulated on MATLAB/SIMULINK platform to carry out investigations with GASA-tuned FLC for the AGC scheme. The dynamic responses of power system models are obtained using GASATF-based AGC schemes by creating 1% load perturbation in area-3. The optimal feedback gains for all investigated controllers are presented in Table 3. In Table 4, frequency dynamic of area-3 is exhibited by numerical analysis. These dynamic responses are plotted in Figures 10–15. The proposed controller resulted in the response plots with less settling time, minimum peak overshoot and under shoot as compared to those obtained with IC and FLC-based AGC schemes. Further, the investigation of these dynamic responses reveals that the dynamic performance with GASATF controller is better than that obtained other compared controllers. However, an appreciable improvement in the dynamic performance of power system is visible while using proposed controller rather than using IC and FLC.

The dynamic performance of proposed AGC controller is also obtained by considering effect of GDB non-linearity. The plot of Figure 16 presents the effect of the GDB on dynamic response of speed governor systems. Dynamic response of the governor valve reveals that opening of valve is faster in non-reheat turbine as compared the reheat and hydro turbines. Figure 16 shows that governor valve movement in hydro turbine is slow. The dynamic responses of speed governor of non-reheat thermal turbine are found to be more oscillatory than speed governor of other turbines. The speed governor responses are found to be uniform and proportional to the area inertia, the control output U_3 remains unchanged from its value just following the disturbance until AGC scheme becomes effective after 1–4 s. The investigation of the dynamic response of Figures 17–20 reveals that the dynamic responses are more oscillatory due to presence of GDB. The realistic behaviour of GDB is shown by these dynamic responses which are taken different scenarios.

Further inspection of these system responses reveals that AGC controllers based on integral control action offer a very sluggish response trends associated with large number of oscillatory modes, large settling time and a considerable amount of steady state error. Whereas, the dynamic response

Table 3. Optimal feedback gains and biasing coefficients of the controllers

Controllers	K_{Ii}	B_1	B_2	B_3
IC	0.341	0.425	0.425	0.425
	0.132			
	0.129			
FLC	0.299	0.115	0.115	0.115
	0.275			
	0.264			
GASATF	0.349	0.287	0.287	0.287
	0.358			
	0.357			

Table 4. Power systems dynamic performance of area-3

Controllers	Peak overshoot (ΔF_3), p.u.	Peak undershoot (ΔF_3), p.u.	Settling time (ΔF_3), s
IC	0.0554	−0.1113	More than 12
FLC	0.1431	−0.1328	4.46
GASATF	0.0103	−0.1015	1.56

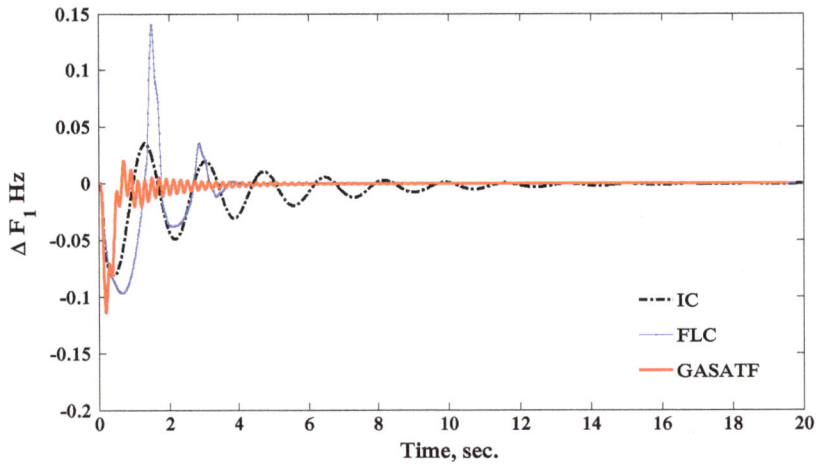

Figure 10. Dynamic response of ΔF_1 for power system model.

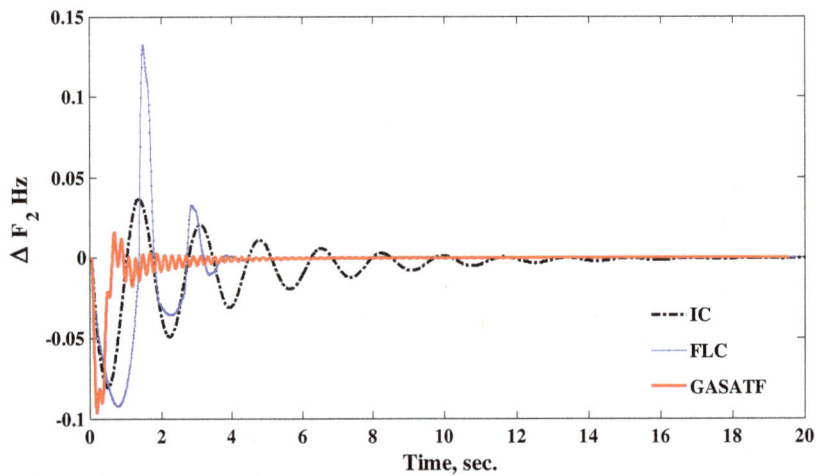

Figure 11. Dynamic response of ΔF_2 for power system model.

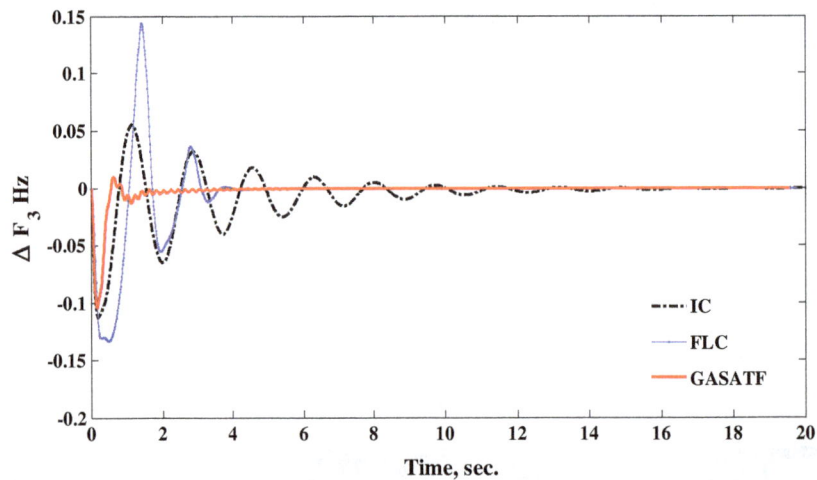

Figure 12. Dynamic response of ΔF_3 for power system model.

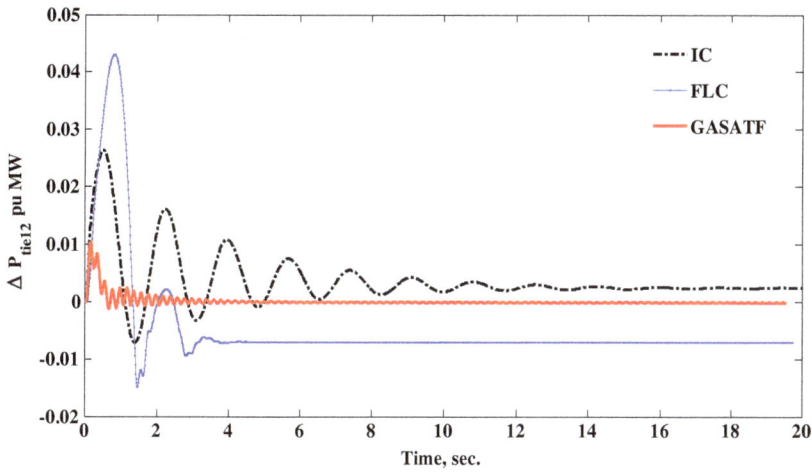

Figure 13. Dynamic response of ΔP_{tie12} for power system model.

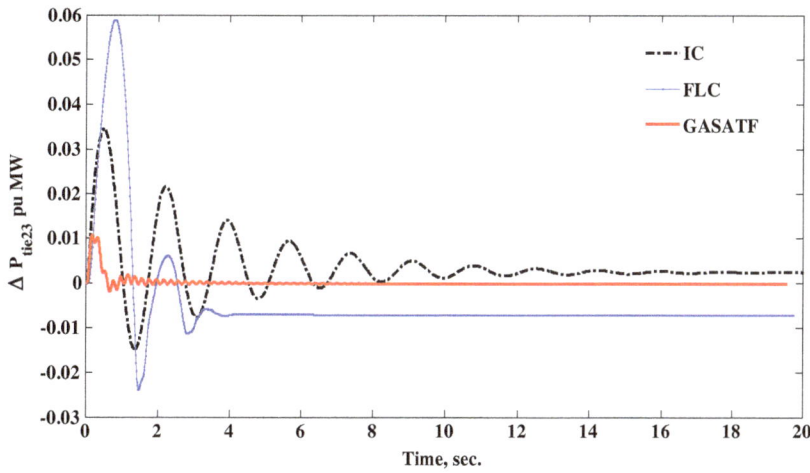

Figure 14. Dynamic response of ΔP_{tie23} for power system model.

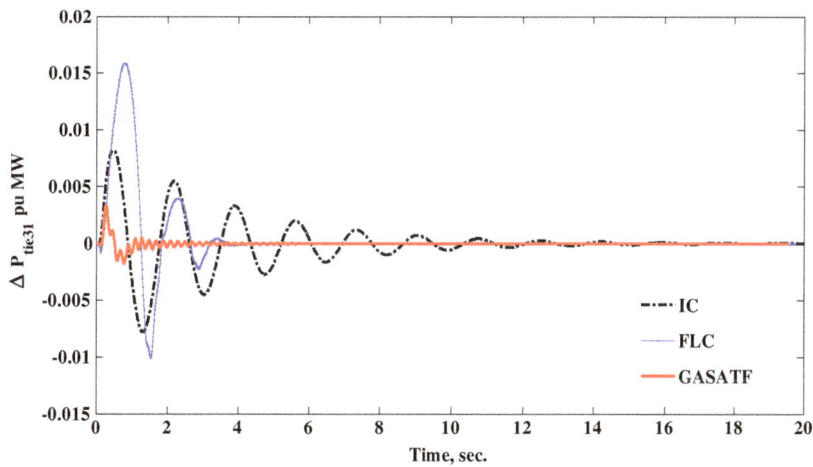

Figure 15. Dynamic response of ΔP_{tie31} for power system model.

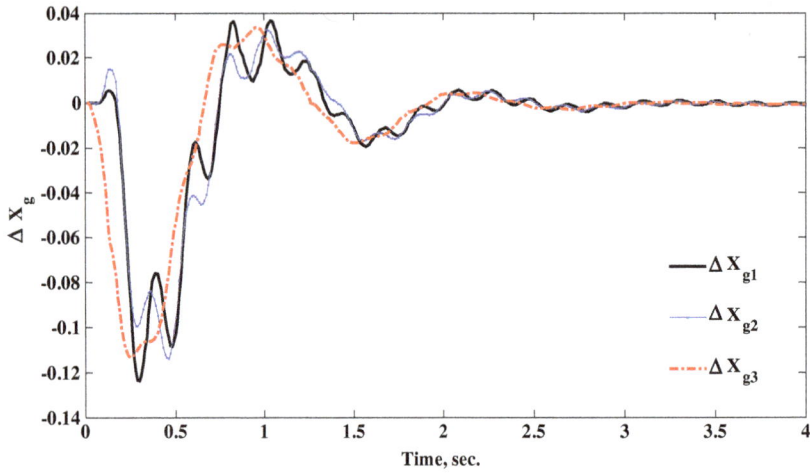

Figure 16. Dynamic response of ΔX_g with GDBs.

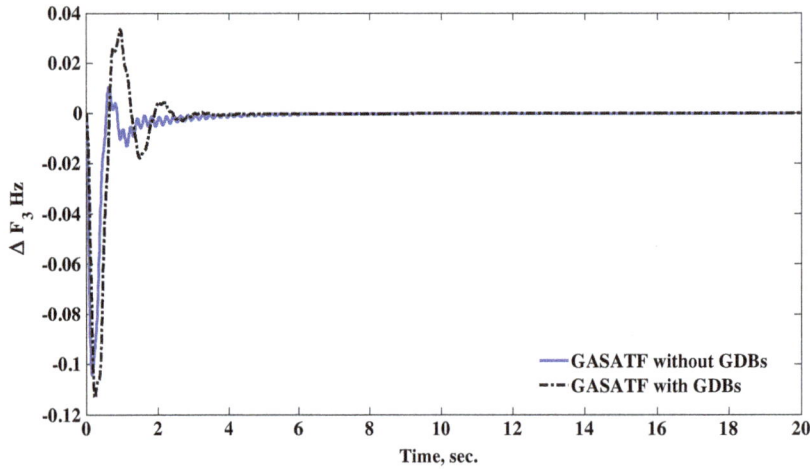

Figure 17. Dynamic response of ΔF_3 with and without GDBs.

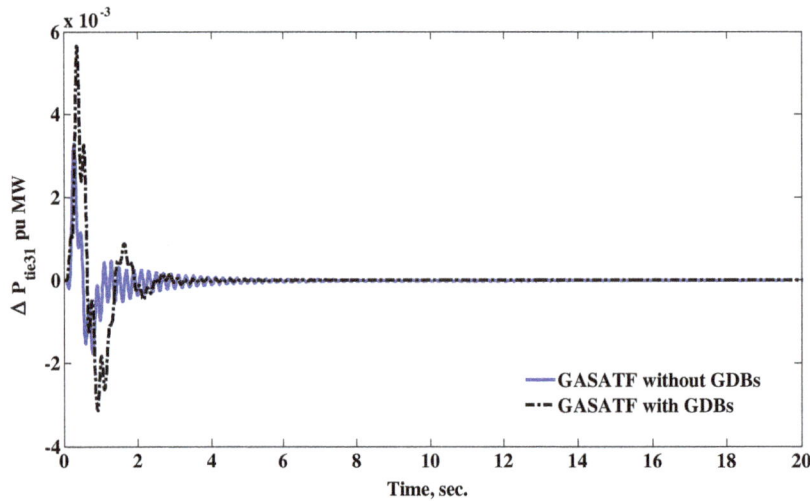

Figure 18. Dynamic response of ΔP_{tie31} with and without GDBs.

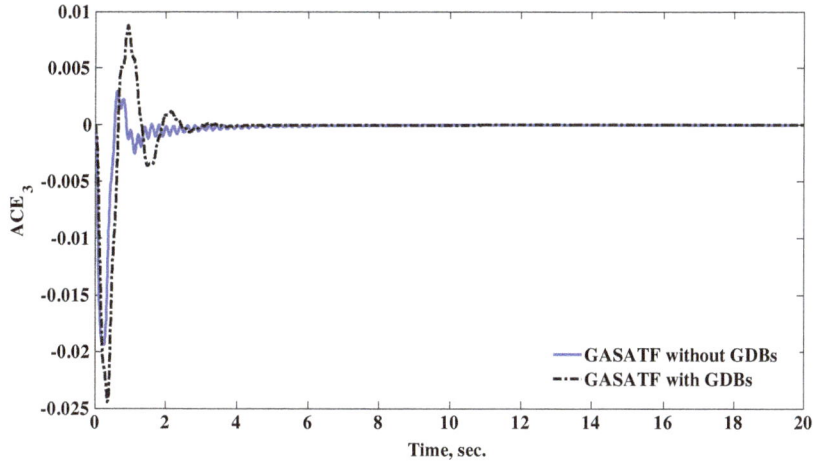

Figure 19. Dynamic response of ACE$_3$ with and without GDBs.

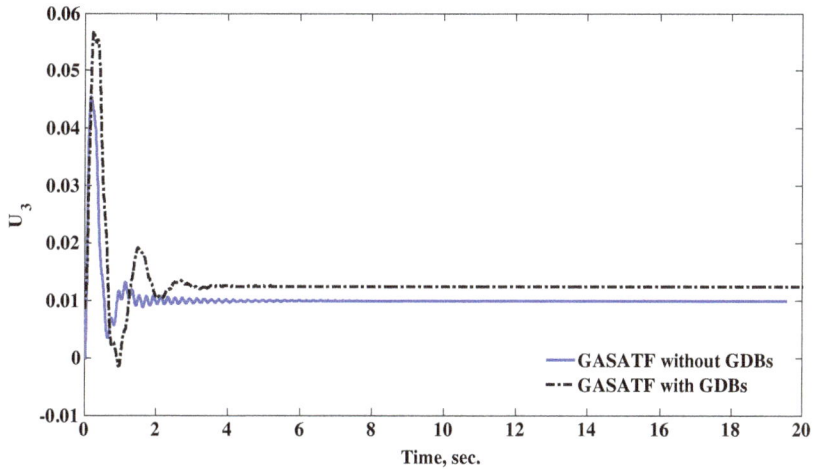

Figure 20. Dynamic response of U_3 with and without GDBs.

achieved with AGC controller based on GASATF control action settles quickly as compared to that obtained with other controllers. The responses are not only settling quickly but also the settling trend is very smooth, and associated with a few number of oscillatory modes.

6. Conclusion

In a nutshell, the present work includes a new method for calculating optimal feedback gains of AGC controller is presented using hybrid GASA-tuned fuzzy logic design technique. From the exhibited results, the dynamic performance of proposed hybrid AGC controller is found to be superior over classical and fuzzy AGC controllers. It is also observed that the GDB non-linearity produced oscillation in dynamic responses of power systems model under investigation until AGC become effective after 1–4 s. The GASATF controller gives ameliorated performance in the multi-area interconnected power systems including different sort of turbines.

The most important feature of the design of optimal AGC controller is that the proposed algorithm considers transient response characteristics as hard constraints which are strictly satisfied in the solutions. This is in contrast to the classical technique-based regulator designs where these constraints are treated as soft constraints. The proposed technique has been tested on a power system model under a disturbance condition and found good enough to give desired results. This proves to be a good alternative for optimal controller design where direct treatment of response specifications is required.

Nomenclature

ΔF_i	incremental change in frequency
$(i = 1, 2, 3)$	subscript referring to area
$\Delta X_{g1}, \Delta X_{g2}, \Delta X_{g3}$	incremental change in governor valve position
K_{gi}	speed governor gain constants
T_{gt1}, T_{gt2}	thermal speed governor time constant (s)
T_{gi}	governor dead band time constant (s)
K_{t1}	reheat thermal turbine gain constant
K_{t2}	non-reheat thermal turbine gain constant
T_{t1}, T_{t2}	turbine time constants (s)
K_{r1}, T_{r1}	reheat coefficients and reheat times (s)
T_1	speed governor time constant of hydro area (s)
T_2, T_3	time constants associated with hydro governor (s)
T_w	hydro turbine time constant (s)
K_{pi}	electric system gain constants
T_{pi}	electric system time constants (s)
B_i	frequency bias constant (p.u.MW/Hz)
R_i	speed regulation parameter (Hz/p.u.MW)
ΔP_{r1}	change in reheater output
ΔP_{di}	incremental change in load demand (p.u.MW/Hz)
ΔP_{ci} or U_i	variation in controller output
ΔP_{gi}	incremental change in power generation (MW)
$\Delta P_{tie\ i}$	change in AC tie-line power (MW)
$T_{ij\ (i\neq j)}$	synchronizing coefficient of AC tie-line
ΔF_i	incremental change in frequency
$(i = 1, 2, 3)$	subscript referring to area
$\Delta X_{g1}, \Delta X_{g2}, \Delta X_{g3}$	incremental change in governor valve position
K_{gi}	speed governor gain constants
T_{gt1}, T_{gt2}	thermal speed governor time constant (s)

Funding
The authors received no direct funding for this research.

Author details
Omveer Singh[1]
E-mail: omveers@gmail.com
ORCID ID: http://orcid.org/0000-0001-6710-7062

Ibraheem Nasiruddin[2]
E-mail: ibraheem_2k@yahoo.com
[1] Electrical Engineering Department, Maharishi
    Markandeshwar University, Ambala, Haryana, India.
[2] Electrical Engineering Department, Quassim University,
    Buraydah, Saudi Arabia.

References

Albertos, P., & Sala, A. (1998). Fuzzy logic controllers: Advantages and drawbacks. *VIII International Congress of Automatic Control, 3*, 833–844.

Anand, B., & Jeyakumar, A. E. (2009). Load frequency control with fuzzy logic controller considering non-linearities and boiler dynamics. *ICGST-ACSE Journal, 8*, 15–20.

Boroujeni, B. K. (2012). Multi area load frequency control using simulated annealing. *Journal of Basic and Applied Scientific Research, 2*, 98–104.

Chown, G. A., & Hartman, R. C. (1998). Design and experience with a fuzzy logic controller for automatic generation control (AGC). *IEEE Transactions on Power Systems, 13*, 965–970.
http://dx.doi.org/10.1109/59.709084

Devi, P., & Avtar, R. (2014). Improvement of dynamic performance of two-area thermal system using proportional integral & fuzzy logic controller by considering AC Tie line parallel with HVDC link. *ICFTEM, special issue*, 170–176.

Elgerd, O. I., & Fosha, C. E. (1970). Optimum megawatt-frequency control of multiarea electric energy systems. *IEEE Transactions on Power Apparatus and Systems, PAS-89*, 556–563.
http://dx.doi.org/10.1109/TPAS.1970.292602

Ghoshal, S. P. (2004). Application of GA/GA-SA based fuzzy automatic generation control of a multi-area thermal generating system. *Electric Power Systems Research, 70*, 115–127.
http://dx.doi.org/10.1016/j.epsr.2003.11.013

Gozde, H., & Taplamacioglu, M. C. (2011). Automatic generation control application with craziness based particle swarm optimization in a thermal power system. *International Journal of Electrical Power & Energy Systems, 33*, 8–16.
http://dx.doi.org/10.1016/j.ijepes.2010.08.010

Ibraheem, P. K., & Kothari, D. P. (2005). Recent philosophies of automatic generation control strategies in power systems. *IEEE Transactions on Power Systems, 20*, 346–357.

Pratyusha, C., & Sekhar, K. C. (2014). Adaptive fuzzy control approach for automatic generation control of two-area interconnected hydro-thermal system. *International Journal of Innovative Science, Engineering and Technology, 1*, 247–252.

Saini, J. S., & Jain, V. K. (2014). A genetic algorithm optimized fuzzy logic controller for automatic generation control for single area power system. *Journal of Inst. Eng. India Sec. B. Springer, 96*(1), 1–8.

Shayeghi, H., Shayanfar, H. A., & Jalili, A. (2009). Load frequency control strategies: A state-of-the-art survey for the researcher. *Energy Conversion and Management, 50*, 344–353.
http://dx.doi.org/10.1016/j.enconman.2008.09.014

Singh, O., & Das, D. B. (2008, December). Design of optimal state feedback controller for AGC using a hybrid stochastic search. *IEEE International Conference POWERCON*, IIT Delhi.

Singh, O., Ibraheem (2013). A survey of recent automatic generation control strategies in power systems. *International Journal of Emerging Trends in Electrical and Electronics, 7*(2), 1–14.

Singhal, P., & Bano, T. (2015). Automatic load frequency control of two-area interconnected thermal reheat power system using genetic algorithm with and without GRC. *International Journal of Engineering Research and General Science, 3*, 298–309.

Yegireddi, N. K., & Panda, S. (2013). Automatic generation control of multi-area power system using multi-objective non-dominated sorting genetic algorithm-II. *International Journal of Electrical Power & Energy Systems, 53*, 54–63.

Yousef, H. (2015). Adaptive fuzzy logic load frequency control of multi-area power system. *International Journal of Electrical Power & Energy Systems, 68*, 384–395.
http://dx.doi.org/10.1016/j.ijepes.2014.12.074

Yousef, H. A., AL-Kharusi, A. K., Mohammed, H. A. (2014). Load frequency control of a multi-area power system: An adaptive fuzzy logic approach. *IEEE Transactions on Power Systems, 29*, 1822–1830.

Appendix A

Reheat, non-reheat thermal and hydro plants

f = 50 Hz, $K_{g1} = K_{g2} = K_{g3}$ = 1.0, K_{r1} = 0.5, $K_{t1} = K_{t2}$ = 1.0, $K_{p1} = K_{p2}$ = 120, $T_{gt1} = T_{gt2}$ = 0.08 s, T_{r1} = 10 s, $T_{t1} = T_{t2}$ = 0.3 s, $T_{p1} = T_{p2}$ = 20 s, T_1 = 0.6 s, T_2 = 5 s, T_3 = 32 s, T_w = 1.0 s, K_{p3} = 20, T_{p3} = 3.76 s, $T_{g1} = T_{g2} = T_{g3}$ = 0.2 s, $R_1 = R_2 = R_3$ = 2.4 Hz/p.u.MW, $B_1 = B_2 = B_3$ = 0.425 p.u.MW/Hz, $P_{tie-max}$ = 2000 MW, $2\pi T_{ij}$ = 0.545 p.u.MW, $\delta_1 - \delta_2$ = 30°, subscript referring to area (i = 1, 2, 3) and ($i \neq j$), ΔP_{d3} = 0.01 p.u.MW.

8

Integrated fuzzy-PI controlled current source converter based D-STATCOM

M. Deben Singh[1]*, Ram Krishna Mehta[1] and Arvind Kumar Singh[1]

*Corresponding author: M. Deben Singh, Department of Electrical Engineering, North Eastern Regional Institute of Science & Technology, Nirjuli-791 109, Arunachal Pradesh, India
E-mail: mdsingh2007@gmail.com
Reviewing editor: Wei Meng, Wuhan University of Technology, China

Abstract: The distribution static synchronous compensator (D-STATCOM) is a shunt-connected current injection custom power (CP) device that can be used for improving quality and reliability of the power supplied to electric consumers under dynamic conditions of the distribution system. The CP devices such as dynamic voltage restorer and unified power quality conditioner including D-STATCOM are mostly designed and implemented using voltage source converter topology. The performance of D-STATCOM can be improved, if it is realized by using a current source converter topology which can generate a controllable current directly at its output terminals. Although the current source converter (CSC) topology offers many promising advantageous features, not much research publications with CSC-based approach in the field of CP devices has been reported over the last one decade. This paper presents a novel method of realizing a D-STATCOM using CSC topology with an integrated fuzzy-PI controller for not only mitigating voltage sag but also capable of maintaining improved DC-link voltage profile. The model has been simulated in the MATLAB/SIMULINK environment. The simulation results under steady state and dynamic load perturbation provide excellent output characteristics that support the validity of the proposed model.

ABOUT THE AUTHORS

M. Deben Singh received his BTech degree in Electrical Engineering from the North Eastern Regional Institute of Science & Technology (NERIST), Arunachal Pradesh, India, in 1997 and M Tech degree in Electronics Design & Technology from Tezpur University, Assam, in 2001. Mr. Singh is a member of IEEE, USA. His area of research interest is in Power Electronics Applications. He is presently working as an assistant professor in the Department of Electrical Engineering, NERIST.

Ram Krishna Mehta obtained his PhD degree from Jadavpur University, Kolkata in 2008. His research interests are Flight and Power system control. He is presently working as an associate professor in the Department of Electrical Engineering, NERIST.

Arvind Kumar Singh received his PhD degree in 2006 from Tezpur University, Assam, India. His research areas are power system, machines and drives. He is presently working as an associate professor in the Department of Electrical Engineering, NERIST.

PUBLIC INTEREST STATEMENT

This paper presents a novel method of realizing a D-STATCOM using current source converter topology with an integrated fuzzy PI controller for not only mitigating voltage sag but also capable of maintaining improved DC-link voltage profile which is a great challenge where an inductor is used as an energy storage element instead of a precharged capacitor as in the case of VSC. The use of CSC topology in the custom power applications which has not been the focus of many researchers for a long time due to various reasons is explored through this paper. The model has been simulated in the MATLAB/SIMULINK. The simulation results under steady state and dynamic load perturbation provide excellent output characteristics that support the validity of the proposed model. This paper will pave the way for initiating the application of CSC topology in other types of custom power devices so as to enhance power quality in the future.

Subjects: Electrical & Electronic Engineering; Engineering & Technology; Power Engineering; Systems & Controls

Keywords: current source converter; custom power device; distribution static synchronous compensator; proportional integral; fuzzy logic controller

1. Introduction

Modern electric power system networks are highly interconnected and the performances of the most of the loads connected to such systems are very much sensitive to the power quality. The application of power electronics technology in power transmission and distribution system has led to the design and implementation of new engineering systems, viz. flexible AC transmission system (FACTS) devices and custom power (CP) devices, which are capable of solving various types of power quality problems (Acha, Agelidis, Anaya-Lara, & Miller, 2006; Hingorani, 1995; Hingorani, & Gyugyi, 2013). The FACTS devices are aimed at the transmission level, whereas CP devices are meant for use at the distribution level, particularly at the point of connection of the electricity distribution system for the consumers having sensitive loads (Deben & Khumanleima, 2015; Lipi & Deben, 2013). The quality of electrical power which is supplied to the customers can be judged in terms of constant voltage magnitude, i.e. no voltage sags or swell, constant frequency, constant power factor, balanced phases and sinusoidal waveform, i.e. no harmonic content, lack of interruptions and capability to withstand faults and to recover quickly (Deben & Khumanleima, 2015). The voltage source converters (VSC) topology has been using as the basic building block of the new generation of power electronics controllers emerging from FACTS and CP initiatives due to the fact that current source converter (CSC) topology is more complex than a VSC topology in both power and control circuits. The filter capacitors which are connected at the AC side of CSC improve the quality of the output AC current waveforms but increase the overall cost of the converter. These filter capacitors resonate with the AC-side inductances. As a result, some of the harmonic components present in the output current might be amplified, thereby causing high harmonic distortion in the AC-side current (Singh, Mehta, & Singh, 2015). If power switching devices having sufficient reverse voltage withstanding capability such as Gate-Turn-Off Thyristor (GTO) are not used, a diode has to be placed in series with each of the switches in CSC. This almost doubles the conduction losses compared with the case of VSC. The DC-side energy storage element in CSC topology is an inductor, whereas that in VSC topology is a capacitor. The power loss of an inductor is expected to be larger than that of a capacitor because of the need to store energy by circulating current in inductors which are more lossy than capacitive energy storage (Ghosh & Ledwich, 2009). Thus, the efficiency of a CSC is expected to be lower than that of a VSC (Ye, Kazerani, & Quintana, 2005). The advancement made in the field of power semiconductor switching devices recently along with the emergence of new control strategies of CSC topology may overcome the demerits cited above in the near future. The CSC topology is usually more reliable and fault tolerant than a VSC due to the presence of large series inductor which limits the rate of rise of current in the event of a fault (Ghosh & Ledwich, 2009). The presence of the AC-side capacitors provides good sinusoidal voltage and current waveforms at the output terminals when CSC topology is used as the capacitors are the inherent filter for the CSC. The problem of the resonance between the capacitances and inductances on the AC-side can be overcome by careful design of the capacitor-based filter circuit and introduction of sufficient damping using proper control methods (Ye et al., 2005). Furthermore, all the switching problems faced in the early stages of CSC development can be overcome by employing tri-level switching scheme which has become a standard technique in the control of CSC (Wang, 1993). In high-power applications such as FACTS devices, Integrated Gate Commutated Thyristor which has the optimum combination of the characteristics, viz. high ratings, high reverse voltage blocking capability, low snubber requirements, lower gate-drive power requirements than GTO and higher switching speed than GTO, can be used (Steimer et al., 1999). The DC-side losses are expected to be minimized using superconductive materials in the construction of the DC-side reactor (Ye et al., 2005).

In comparison with the VSC topology, the application of CSC topology in distribution static synchronous compensator (D-STATCOM) is expected to achieve many advantages. The direct output of a CSC is a controllable AC current, whereas that of a VSC is a controllable AC voltage. When operated under sinusoidal pulse width modulation (SPWM) technique (Mohan, Undeland, & Robbins, 1989), the magnitudes of the harmonic components in both converters are directly proportional to the magnitudes of the fundamental components of their direct output quantities. Under the normal operating conditions, the current injected by D-STATCOM is a small percentage of the line current. Hence, when CSC topology is used, the current harmonics are also small. However, when VSC topology is used, for a small injected current, the output voltage of VSC is large and very close to the system voltage. This results in large voltage harmonics, leading to current harmonics that are larger than those generated by CSC, and thus more costly to filter. The other aspect of comparison is the DC-side energy storage requirement. When the D-STATCOM is realized by a CSC, the DC-side current is just larger than the peak value of the required injected current which is a small percentage of the line current. However, when a VSC is used to inject reactive power to the system, the DC-side voltage must be larger than the peak value of the system line-to-line voltage so that the reactive power can be exchanged between the D-STATCOM and the AC distribution system. Hence, the DC energy storage requirement of the CSC topology is expected to be lower than that of the VSC topology when it is used to implement a D-STATCOM system for mitigation of voltage sag (Singh et al., 2015).

The D-STATCOM has plenty of applications in low voltage power distribution systems. It can be used to prevent non-linear loads from polluting the rest of the distribution system. The rapid response of the D-STATCOM makes it possible to provide continuous and dynamic control of the power supply, including voltage and reactive power compensation, mitigation of voltage sag, swell and elimination of harmonics (Acha et al., 2006). A thorough investigation on the feasibility and performance analysis of a CSC-based D-STATCOM for mitigating voltage sag phenomenon which is considered to be one of the well-known unwanted power quality problem resulted from sudden change in load connected to a power distribution system has been fairly reported at (Singh et al., 2015); however, the paper has not discussed about the challenge of maintaining the DC-link voltage profile during the voltage sag compensation period. The magnitude of output voltage of the inverter circuits used in the FACTS and CP devices is directly proportional to the DC-link voltage. Hence, maintenance of this voltage to a fair level is an important and challenging aspect at the time of designing a CSC-based CP device because an inductor is used as energy storage element instead of a pre-charged capacitor as in the case of VSC. By incorporating fuzzy logic controller (FLC) in the control system of the proposed D-STATCOM, an attempt has also been made to maintain the improved DC-link voltage profile during the process of voltage sag mitigation. The simulation results of the proposed model reveal its effectiveness in mitigating voltage sag while maintaining improved DC-link voltage.

2. Power reliability and quality issues
The main concern of electrical energy consumers was the reliability of power supply a few years back. This reliability can be defined as the continuity of electric supply (Ghosh & Ledwich, 2009). Power quality is mainly concerned with deviations of the voltage from its ideal waveform (voltage quality) and deviations of the current from its ideal waveform (current quality). Such deviation is known as "power quality phenomenon" or "power quality disturbance" (Math & Bollen, 2013). Some examples of the power quality problem include impulsive and oscillatory transients, short duration voltage variations (sag or dip, swell and interruption), long duration voltage variation (undervoltage, overvoltage and sustained oscillation), voltage imbalance, waveform distortion (harmonics, notching and DC offset) and voltage flicker. These problems are generally caused by the nature, faults on transmission or distribution system and also by the power consumers. The power transmission lines are exposed to the forces of nature and its loadability limit is usually determined by either stability considerations or by thermal limits. Although the power quality problem is a distribution side problem, transmission lines frequently have an impact on the quality of power supplied. It is however to be noted that while most of the problems associated with transmission systems arise due to the forces of nature or due to the interconnection of power systems, individual customers are responsible for a more substantial fraction of the power quality problems in the distribution side (Ghosh & Ledwich, 2009). The FACTS and the CP devices are the two major power electronics-based initiatives to counter the power quality problems. CP focuses primarily on the reliability and power quality.

However, voltage regulation, voltage balancing and harmonic cancellation may also benefit from this technology (Deben & Khumanleima, 2015).

Amongst the various power quality problems mentioned above, this paper confines itself to the voltage sag only. The voltage sag is a power quality problem phenomenon which falls under the category of short duration voltage variation. Any variation in the supply voltage (rms) for duration not exceeding one minute is called a short duration voltage variation. Usually, such variations are caused by system faults, energization of heavy loads that require large inrush currents and intermittent loose connection in the power wiring. Voltage sag is a fundamental frequency decrease in the supply voltage for a short duration. The duration of voltage sag varies between five cycles to a minute (Ghosh & Ledwich, 2009). The interest in voltage sag is mainly due to the problems they cause on several types of equipment, viz. adjustable speed drives, process control equipment and computers, which are notorious for their sensitivity. Some pieces of equipment trip when the rms voltage drops below 90% for longer than one or two cycles. If this is the process control equipment of a paper mill, one can imagine that the damage due to voltage sags can be enormous. Of course, voltage sag is not as damaging to industry as a (long or short) interruption. But as there are far more voltage sags than interruptions, the total damage due to voltage sags is still larger. Short interruptions and most long interruptions originate in the local distribution network. However, voltage sags at equipment terminals can be due to short circuit faults 100s of kilometres away in the transmission system. Hence, voltage sag is much more of a "global" problem than an interruption. Reducing the number of interruptions typically requires improvements on one feeder and reducing the number of voltage sags requires improvements on several feeders and often even at transmission lines far away (Math & Bollen, 2013). The voltage sag occurring at the power distribution system can be mitigated using VSC-based CP devices such as dynamic voltage restorer (DVR) and D-STATCOM (Deepa & Ranjani, 2015; Singh et al., 2015).

2.1. VSC- and CSC-based D-STATCOM
When STATCOM is used in the low voltage distribution system, it is identified as D-STATCOM. Although both the STATCOM and the D-STATCOM are shunt-connected devices, there is a substantial difference in their operating characteristics. The STATCOM is required to inject a set of three balanced quasi-sinusoidal voltages that are phase displaced by 120°, but the D-STATCOM must be able to inject an unbalanced and harmonically distorted current to eliminate unbalance or distortions in the load current or the supply voltage. Hence, its control is significantly different from that of a STATCOM. In case of VSC-based D-STATCOM, it can be configured using a voltage source inverter circuit interfaced with a coupling transformer and energy storage element, viz. capacitor (Singh et al., 2015). The D-STATCOM used in CP applications uses PWM switching control as opposed to the fundamental frequency switching strategies which are preferably used in FACTS applications. PWM switching is practically used in CP applications as it is at relatively low power level (Acha et al., 2006; Deben & Khumanleima, 2015; Ghosh & Ledwich, 2009). The CSC-based D-STATCOM can be realized by modifying the basic configuration of the VSC-based one. It also consists of a CSC circuit which is interfaced with a coupling transformer and a DC-link reactor/inductor as shown in Figure 1.

3. CSC-based D-STATCOM
The proposed CSC-based D-STATCOM consists a three-leg CSC driven by SPWM, AC-side low pass filters (LPF), coupling transformer and internal control system. The simplified structure of the proposed model is depicted in Figure 2. The working principle and control of CSC for use in D-STATCOM are similar to those of the CSC employed in AC motor drives, but its design strategy is different to some extent. The CSC circuit will be subjected to voltage regulation problem on the AC side when the reactive power is varied from full inductive to full capacitive. This will affect the voltage rating of the switching devices as well as the design of the input LPF circuit (Singh et al., 2015).

As the DC-link circuit consists only inductor and its internal resistance, the electrical time constant will be high and thus affects the design process of this circuit. The design objectives of the CSC for use in D-STATCOM are not only to filter harmonics but also to achieve optimal sizing of the CSC so as

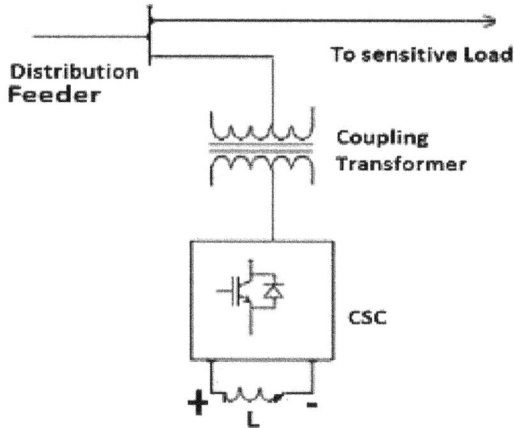

Figure 1. Schematic diagram of a CSC-based D-STATCOM.

to meet the control range requirements of the D-STATCOM in both the capacitive and inductive operation ranges (Bilgin & Ermis, 2010). The selection of switching device and modulation scheme depends on the application voltage and power ratings (Han, Moon, & Karady, 2000). At distribution-level applications, viz. D-STATCOM, IGBT switches and PWM scheme can be chosen if VSC topology is used. The converter topology for the proposed model being CSC, GTO switches having sufficient reverse voltage withstanding capability are selected. These switches are configured in bridge fashion and fed from a DC-link reactor acting as the energy storage element. The DC-link reactor approximates the DC-link current I_{dc} to a level current waveform in the steady state. The level DC-link current is converted to bidirectional current pulses, alternating at supply frequency in the AC lines of CSC by switching power semiconductor switches in accordance with a pre-specified pattern (Bilgin & Ermis, 2011; Bilgin et al., 2007). In the CSC circuit, the energy storage element is an inductor with its internal resistance. The amount of reactive power to be generated by the CSC can be computed from the relation:

$$Q = \sqrt{(3/2)}VMI_{dc}\text{Cos } \theta \tag{1}$$

where V = rms value of the fundamental component of converter input line-to-line voltage, M = modulation index, I_{dc} = mean DC-link current and θ = phase shift (Bilgin & Ermis, 2010). Equation (1) indicates that Q is independent of the DC-link inductance L_{dc}. The value of L_{dc} affects the response time of D-STATCOM against the variations of reactive power demand of the load. The time constant of the DC-link circuit τ_{dc} is L_{dc}/R_{dc}. Hence, the value of L_{dc} should be selected as small as possible for allowing rapid rise or decay of mean DC-link current against the rapid changes in Q if the phase shift angle control at fixed modulation index is used for controlling the reactive power. The output of the CSC is filtered by a three-phase LPF comprising of three capacitors connected in shunt manner. These filters ensure to provide good sinusoidal output voltage and current waveforms after separating higher order harmonic components to the coupling transformer. The leakage impedance of the coupling transformer also behaves as a part of the LPF. External series reactors (i.e. X_{tr} in Figure 2) have been used on the low voltage side of the coupling transformer for adjusting the corner frequency of the input filter to an optimum value for the fixed shunt-connected capacitor in implementing the proposed D-STATCOM. The corner frequency should be chosen as small as possible for better performance of the filter circuit. The detailed design strategies of the LPF are available at (Bilgin & Ermis, 2011).

3.1. Control system of the model
The control system of the proposed D-STATCOM consists a phase lock loop (PLL), proportional integral (PI) controller, FLC and dq0 transformation block, etc. For the sack of convenience in designing the control system, the internal control structure has been divided into two sections, viz.

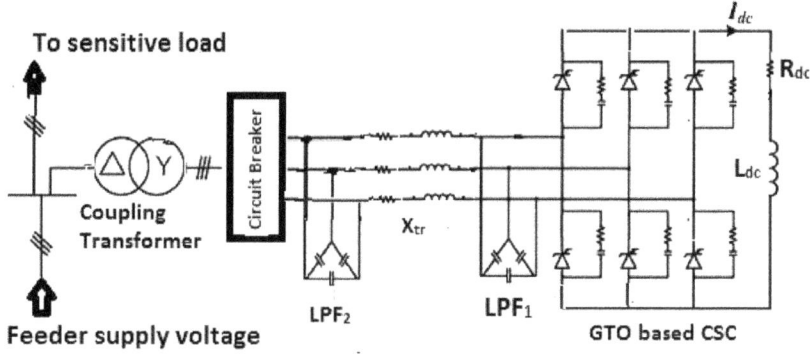

Figure 2. Structure of the proposed CSC-based D-STATCOM set-up.

conventional controller section and FLC section, which will be integrated and converted into a single controller to produce the optimum performance.

3.1.1. Conventional controller section

The task of sensing the voltage sag, calculating the required compensating voltage for the D-STATCOM and generating the reference signals for the SPWM generator to provide switching puls-es for the GTO switches used in the CSC are carried out by the internal control system. For generating proper gating signals of the GTO switches used in the CSC, SPWM control scheme is chosen for the proposed model. The SPWM switching strategy has constant switching frequency capability. This constant switching frequency reduces stress levels on the converter switches (Nagesh, Srinivas, & Mahesh, 2012). The control scheme will be able to maintain constant voltage magnitude at the point where a sensitive load is connected under system disturbances. The control system only measures the rms voltage at the point of common coupling (PCC) and no reactive power measurements are carried out. The CSC switching strategy is based on a sinusoidal PWM technique which offers simplic-ity and good response. As the distribution network operates at a relatively low power, such method offers a more flexible option than the fundamental frequency switching method which is favoured in FACTS applications. The block diagram of the control scheme designed for the proposed model is shown in Figure 3. The commonly used control schemes for the generation of reference source cur-rents in most of the VSC-based DSTATCOM include instantaneous reactive power theory, synchro-nous reference frame theory (SRFT), unity power factor based, instantaneous symmetrical components based, etc. (Bhim, Jayaprakash, & Kothari, 2008). The control scheme chosen for the proposed D-STATCOM model is based on SRFT. The load currents, I_{Load} (i_a, i_b and i_c), the PCC voltage V_{t} $_{(PCC)}$ (v_a, v_b and v_c) and the reactor DC voltage (V_{dc}) of D-STATCOM are sensed and used as feedback signals. The load currents from the a–b–c frame are first converted to the α–β–0 frame and then to d–q–0 frame using the Park's transformation relation as:

$$i_d = \frac{2}{3}[i_a \sin\theta + i_b \sin(\theta - 2\pi/3) + i_c \sin(\theta + 2\pi/3)] \qquad (2)$$

$$i_q = \frac{2}{3}[i_a \cos\theta + i_b \cos(\theta - 2\pi/3) + i_c \cos(\theta + 2\pi/3)] \qquad (3)$$

$$i_0 = \frac{1}{3}[i_a + i_b + i_c] \qquad (4)$$

where $\cos\theta$ and $\sin\theta$ are obtained using the three-phase PLL. The PLL receives signal from PCC ter-minal voltage $V_{t\,(PCC)}$ for generation of fundamental unit vectors for conversion of sensed currents to the d–q–0 reference frame. The SRF controller extracts DC quantities by a LPF and removes the har-monics from the reference signal. The distribution feeder terminal voltage $V_{t(PCC)}$ is regulated by a PI controller after comparing $V_{t(PCC)}$ with a reference terminal voltage $V_{t(ref.)}$ which produces a i_q signal and adds with i_q generated at equation (3) and acts as a reference component for the current

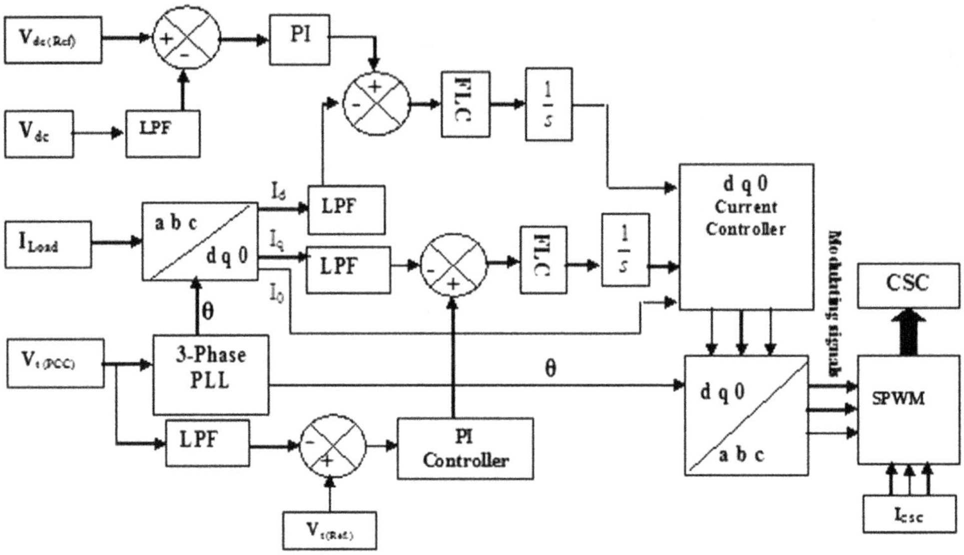

Figure 3. Block diagram of the proposed D-STATCOM control system.

controller. The error signal output after comparing DC-link reactor voltage V_{dc} with a $V_{dc(ref.)}$ is processed by another PI controller to regulate the DC-link voltage and produces i_d current component. This current is added with the i_d available at equation (1). The reference source current must be in phase with the voltage at the PCC but with no zero-sequence component and it can be obtained using reverse Park's transformation process (Bhim et al., 2008). The sensed current and reference source currents are compared and a proportional controller is used for amplifying current error in each phase in the current controller (Singh et al., 2015).

3.1.2. FLC section

The conventional PI controllers find applications in industries for a wide range of control processes and provide satisfactory performance once tuned when the process parameters are well known and there is not much variation. However, if operating conditions vary, further tuning may be necessary for good performance. Since most of the control processes are complicated and non-linear, FLC approach seems to be a good choice. The FLC has been widely used in systems with complex structure as it doesn't require mathematical model of the control system (Resul, Besir, & Fikret, 2011). The design structure of the FLC can be visualized as shown in Figure 4.

There are four main components in a FLC, viz. Fuzzification unit (the Fuzzifier), Inference engine, Rule base and Defuzzifier. The first stage in the fuzzy controller system is to transform the numeric into fuzzy sets. This operation is called fuzzification. From the point of view of fuzzy set theory, the inference engine is the heart of the fuzzy system. It is the inference engine that performs all logical

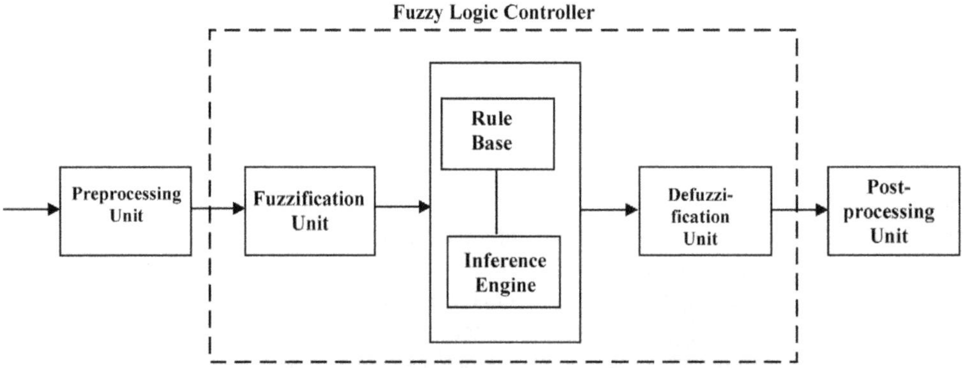

Figure 4. Various components of a fuzzy controller.

manipulations in a fuzzy system. A fuzzy system Rule base consists of fuzzy IF–THEN rules and membership functions characterizing the fuzzy sets. The result of the Inference process is an output represented by a fuzzy set, but the output of the fuzzy system should be a numeric value. The transformation of a fuzzy set into a numeric value is called defuzzification. In addition, input and output scaling factors are needed to modify the universe of discourse. Their role is to tune the fuzzy controller to obtain the desired dynamic properties of the process controller closed loop (Jan, 1998).

In the proposed integrated fuzzy PI controller, two numbers of FLC blocks are used for the error signal-d and error signal-q. Each controller has two inputs and one output. One of the input to the FLCs is the derivative of the processed output of the PI controllers and the abc to dq0 transformation blocks used in the model as depicted at Figure 5. In other words, the inputs of FLC are the errors of d- and q-axis currents and derivatives of these errors. An external integrator is used to eliminate the steady-state error in the output of FLC (Resul et al., 2011). The inputs of the FLC have been chosen as the error in DC-link voltage and the change in error in DC-link voltage and these can be represented as:

$$e(i) = V_{dc(ref.)} - V_{dc}(i) \tag{5}$$

$$de(i) = e(i) - e(i - 1) \tag{6}$$

where $e(i)$ is the error and $de(i)$ is the change in error in the ith iteration. $V_{dc(ref.)}$ is the reference DC-link voltage of the reactor and $V_{dc}(i)$ is the DC-link voltage in the ith iteration. The outputs of the FLC are chosen as the change in K_p value and the change in K_i value.

$$K_p = K_{p\,(ref.)} + \Delta K_p \tag{7}$$

$$K_i = K_{i(ref.)} + \Delta K_i \tag{8}$$

where $K_{p(ref.)}$ and K_i are ref is reference proportional and integral gain, respectively. ΔK_p and ΔK_i are changes in K_p and K_i (Harish & Mahesh, 2008). Five triangular membership functions are chosen the for input variables and the output variable, namely: NB, NS, Z, PS and PB, representing negative big, negative small, zero and positive small and positive big, respectively, in the investigation implemented using the membership function editor as depicted at Figure 6(a). A fuzzy inference system file named "dstat.fis" is developed with the triangular membership functions as shown in Figure 6(b) with the help of FIS editor available in the Matlab/Simulink.

Also, the set of 25 fuzzy rules applied while modelling the controller is depicted in Table 1. The basic rule of FLC gives the relationship between the input and output (Deepa & Ranjani, 2015). The Rule base characterizes the control goals and control policy of the domain experts by means of set linguistic control rules. Most of the FLCs are based on various methods. The widely used method in

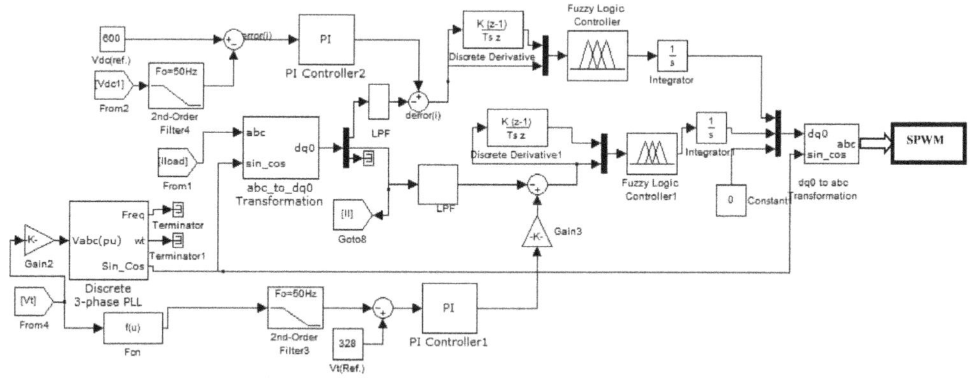

Figure 5. Simulink circuit of the integrated fuzzy PI controller.

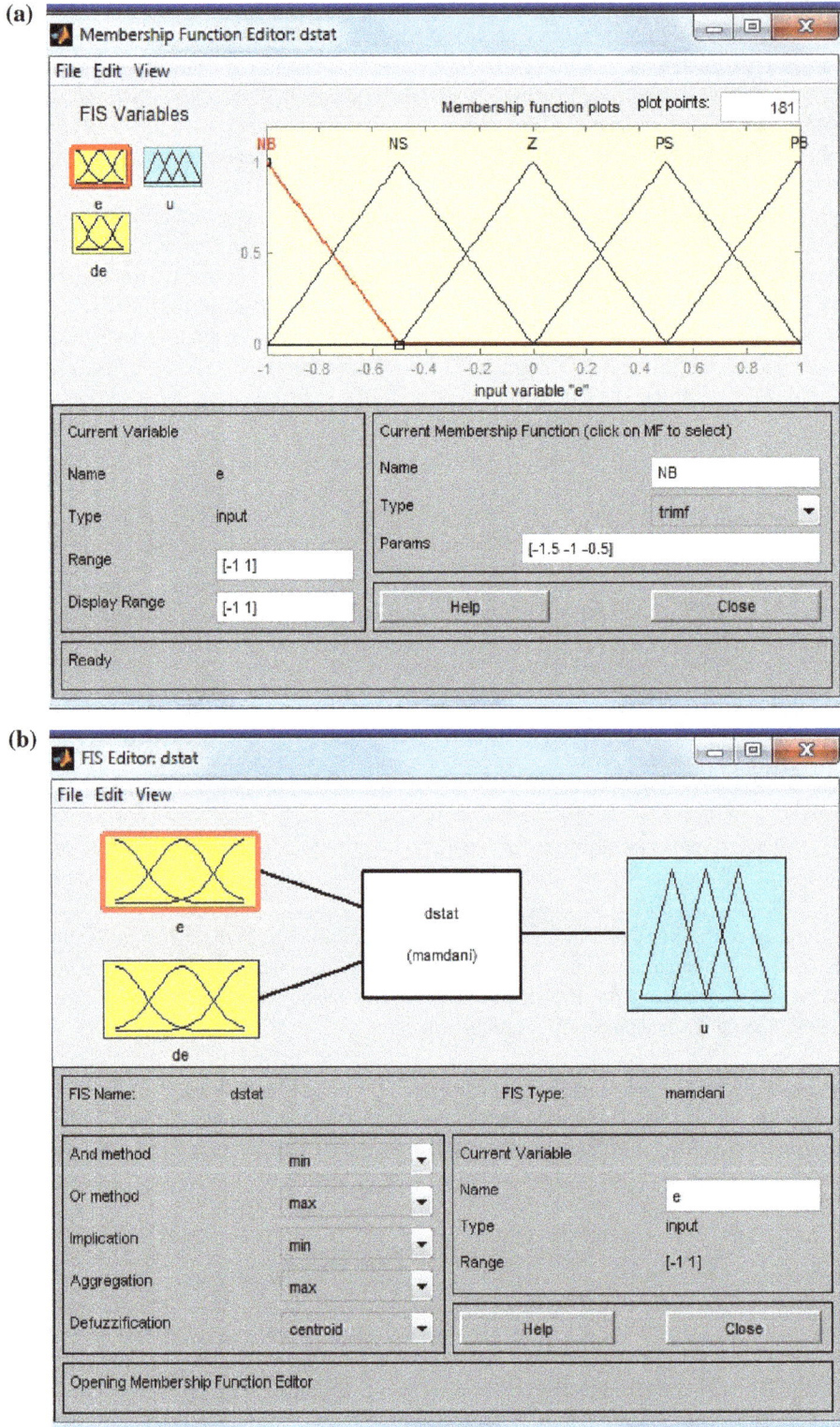

Figure 6. (a): Membership functions for the FLC of the proposed model and (b): FIS file developed for the FLC of the proposed model.

the FLC design is the Mamdani method proposed by Mamdani and his associates who adopted a min–max compositional rule of inference based on an interpretation of a control rule as a

e de	NB	NS	Z	PS	PB
NB	NB	NB	NS	NS	Z
NS	NB	NB	NS	Z	Z
Z	NS	NS	Z	PS	PS
PS	Z	PS	PS	PB	PB
PB	Z	Z	PS	PB	PB

Table 1. Rule base for fuzzy PI controller

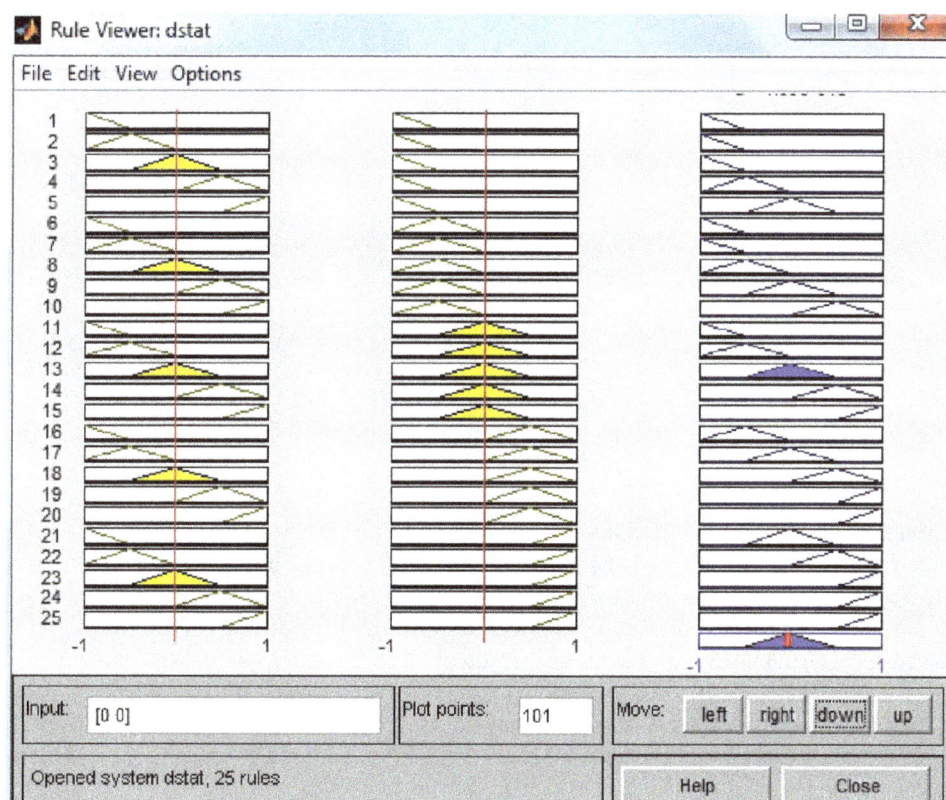

Figure 7. Rule viewer of the proposed model.

conjunction of the antecedent and consequent and this method has been used in this work. The rule viewer and the surface viewers of the FLCs used in the model are shown Figures 7 and 8, respectively. The rule viewer is used for determining the approximate reasoning of the results of FLC and the surface viewer is used to observe the pattern of decision-making based on the rules formulated. The basic fuzzy rules are framed using the rule editor available in the MATLAB environment. After compiling the basic rules, the investigation is carried out in accordance with the rule and the surface viewers. With this approach, the output coefficient of the PI controllers is able to tune to the desired levels, thereby achieving the advantages of mitigating voltage sag as well as maintaining the DC-link voltage profile at a better level. Without integrating the FLC in the control system of the CSC-based D-STATCOM, it is not possible to improve the DC-link voltage profile.

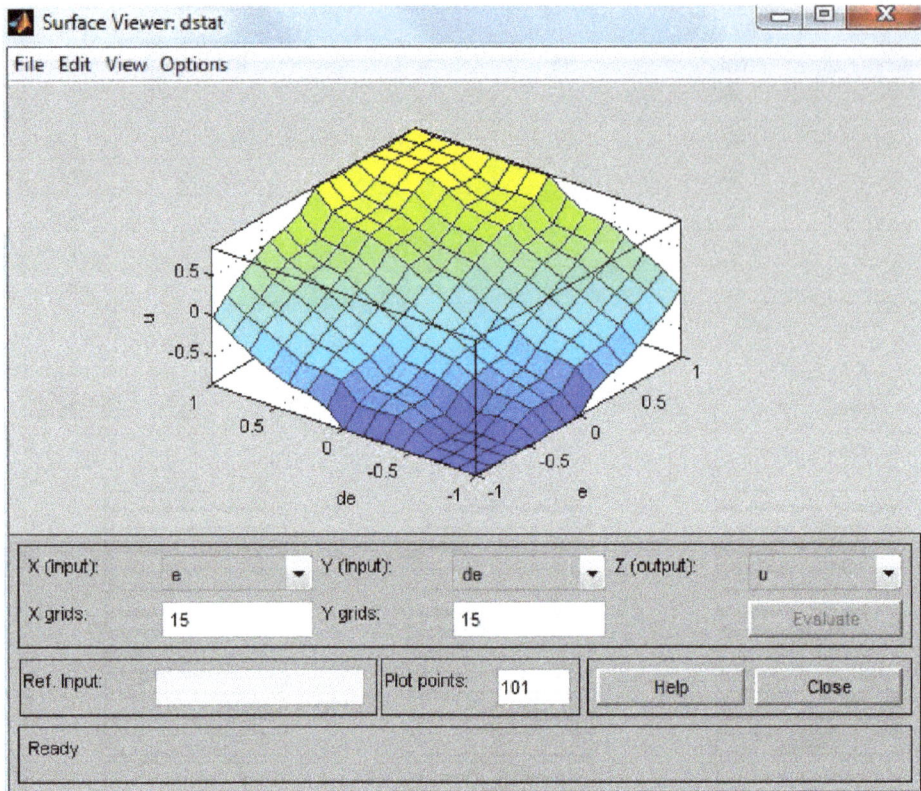

Figure 8. Surface viewer of the proposed model.

4. Simulation results and discussions

The Simulink model of the proposed integrated fuzzy PI-controlled CSC-based D-STATCOM shown in Figure 9 has been simulated under steady-state (with normal load) and dynamic (sudden change in load) conditions. The various parameter settings of the model for simulation study are listed in Table 2. The practical feeder voltage in Indian three-phase power distribution system lies in the range of 415–400 V which operates at 50 Hz supply frequency. In the proposed model, a lower of 400 V has been chosen as the feeder system voltage taking into account the introduction of CSC topology instead of VSC one in the D-STATCOM which will cause more stress on the switching devices used in the inverter circuit in practical situation.

Firstly, the model has been simulated under normal load conditions without connecting the D-STATCOM and the heavily inductive three-phase series load$_2$ by opening the three-phase circuit breakers 1 and 2, respectively. Figure 10 shows the voltage and current waveforms under normal load conditions. Figure 10(a) depicts the voltage waveform across the load and (b) shows the current waveform in the load. From this simulation result, it is observed that there is no voltage sag taking place across the load. The magnitude of voltage and current is found to be 360 V and 0.14 A, respectively.

Secondly, the model is simulated under sudden change in load conditions without operating the D-STATCOM. In this case, the D-STATCOM is disconnected by opening the circuit breaker 1and the second heavily inductive three-phase load is connected by closing the three-phase circuit breaker 2. The circuit breaker 2's closing transition time is set from 0.4 to 0.6 s. Figure 11(a) and (b) shows the load voltage and current waveforms, respectively. Under this condition, a voltage sag with a magnitude of 340 V which is 20 V less than the non-sag voltage magnitude for a duration from 0.4 to 0.6 s appears across the load as shown in Figure 11(a). Figure 12 shows the voltage and current waveforms at the PCC without operating the D-STATCOM.

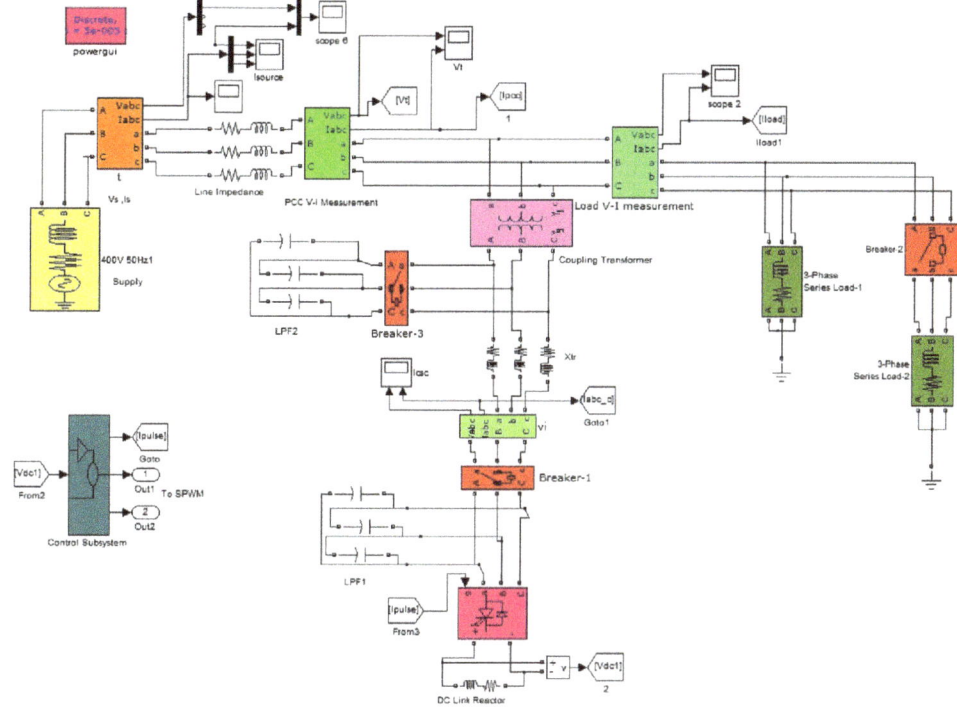

Figure 9. Simulink model of the proposed D-STATCOM.

Table 2. Parameter settings of the proposed model	
Name of parameters	**Value of parameters**
Nominal power and frequency of the coupling transformer	5,000 VA, 50 Hz
AC-side external reactor, X_{tr}	$R = 0.01\ \Omega$, $L = 1$ mH
Feeder impedance	$R_f = 0.01\ \Omega$, $L_f = 2$ mH
Feeder voltage	$V_{t(PCC)} = 400$ V(rms), 50 Hz
Feeder reference voltage	$V_{t(ref.)} = 328$ V(rms), 50 Hz
DC reference voltage	$V_{dc(ref.)} = 600$ V
DC-link reactor of CSC	$L_{dc} = 8,000$ mH, $R_{dc} = 0.01\ \Omega$
AC-side LPF_1 shunt filter capacitor of CSC	$C = 2.85$ mF
AC-side LPF_2 shunt filter capacitor of CSC	$C = 8.5$ mF
Inductive Load$_1$	$R = 0.06\ \Omega$, $L = 8,000$ mH
Inductive Load$_2$	$R = 0.06\ \Omega$, $L = 98000$ mH
PI controller gains for DC-link voltage	$K_{P(dc)} = 0.25$, $K_{i(dc)} = 0.14$
PI controller gains for AC system voltage	$K_{P(q)} = 0.4$, $K_{i(q)} = 0.5$
Switching frequency of the CSC	$f_s = 20$ kHz

Thirdly, the model has been simulated under sudden change in load conditions with operating the D-STATCOM. In order to mitigate the voltage sag taking place due to the sudden change in load, the D-STATCOM is brought to operation by closing the circuit breaker 1 with its transition time setting from 0.4 to 0.6 s. Figure 13(a) and (b) depicts the load voltage and current waveforms. It is observed that the load voltage sag during the period from 0.4 to 0.6 s has been effectively mitigated by the proposed D-STATCOM under this condition. The voltage profile of the load can be improved by connecting the LPF_2 though the operation of circuit breaker 3 of the circuit. For reducing the overall cost of the AC-side filter, LPF_2 may be eliminated or kept as optional. Under this condition, the performance of the control system of the proposed model in maintaining DC-link reactor voltage profile

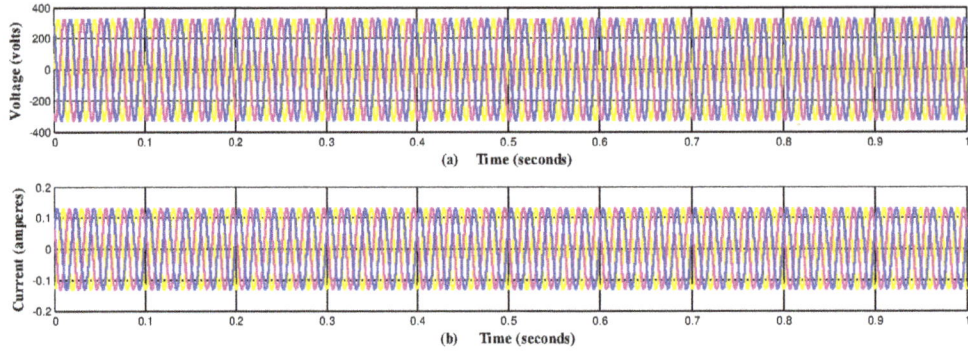

Figure 10. Voltage and current waveforms under normal load conditions.

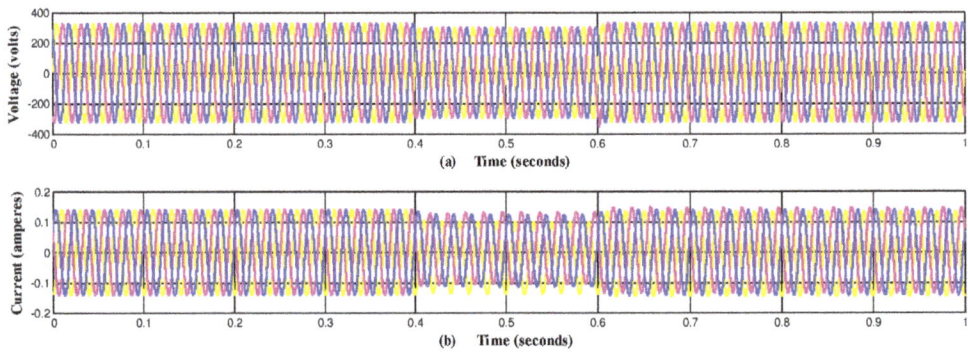

Figure 11. Voltage and current waveforms under sudden change in load conditions without operating the D-STATCOM.

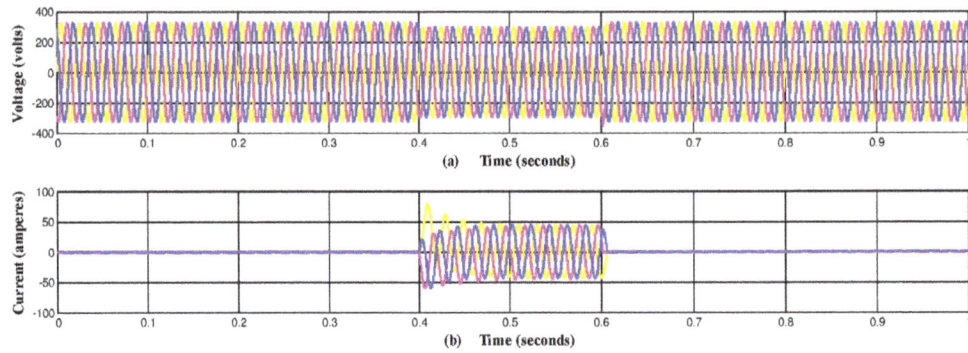

Figure 12. Voltage and current waveforms under sudden change in load conditions at PCC without operating the D-STACOM.

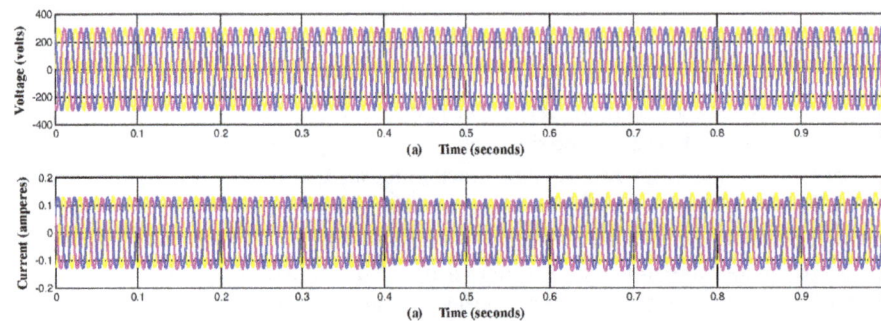

Figure 13. Voltage and current waveforms under sudden change in load conditions with the operation of D-STATCOM.

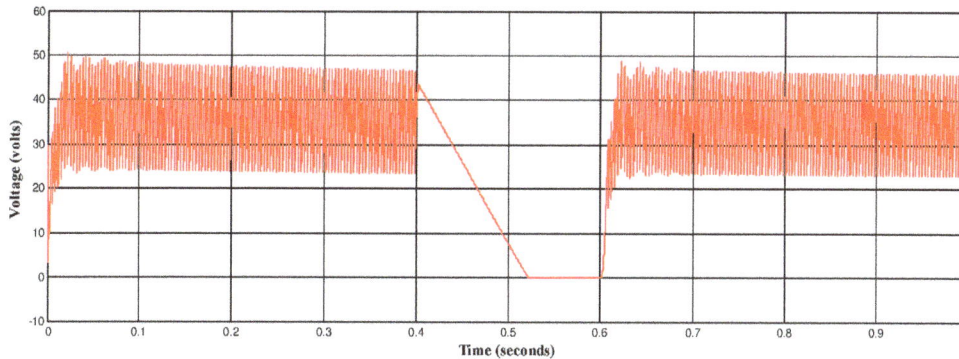

Figure 14. Voltage waveform of the DC-link reactor with integrated fuzzy PI controller.

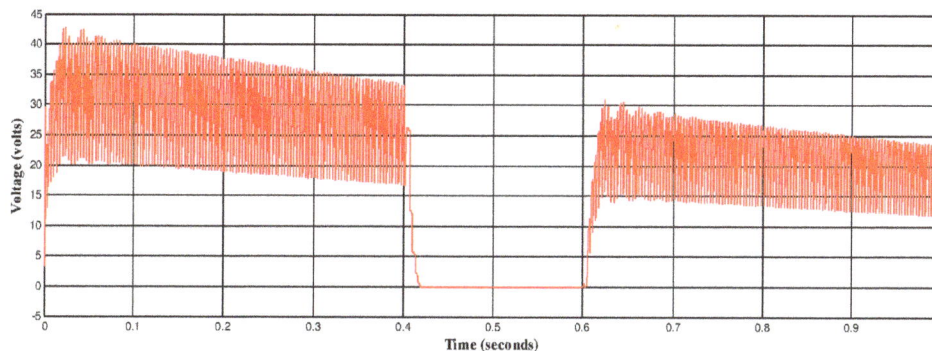

Figure 15. Voltage waveform of the DC-link reactor with PI controller alone.

during the voltage sag mitigation process has been examined at the time of simulation. Figure 14 shows the voltage waveform of the DC-link reactor with FLC integration with the conventional PI controller. With this approach, the magnitude of the voltage can be fairly maintained above 20 V (i.e. 22 V) before and after load perturbation. During the load disturbance period, from 0.4 to 0.6 s, the voltage magnitude of the voltage decreases 0 V for a duration of 0.08 s only. This reduction in voltage happens due to the fact that the inductor acts as a current source during the injection period and the voltage under steady-state condition becomes zero.

Lastly, the proposed model has also been simulated without integrating FLC, i.e. with PI controllers alone in order to observe the DC-link voltage profile under load perturbation. The voltage waveform of the DC-link reactor observed under this condition is shown in Figure 15. It is observed that the magnitude of the voltage goes on decreasing from its initial value of 20 V. During the load disturbance period, from 0.4 to 0.6 s, its value drastically reduces to 0 V with a delay of 0.02 s. From this observation, it is learnt that when CSC topology is used in other types of CP devices, such as DVR and unified power quality conditioner, integration of the FLC in the control system will definitely prove to be a good choice for maintaining an improved DC-link voltage profile.

5. Conclusions

In this paper, an integrated fuzzy PI-controlled CSC-based D-STATCOM has been modelled and simulated with the objective of mitigating voltage sag and also to maintain the improved DC-link voltage profile in a power distribution system under sudden change in load condition. The difficulty in maintaining DC-link voltage profile during voltage sag compensation period which is a challenge where inductor is used as the energy storage element instead of a pre-charged capacitor as in the case of VSC is overcome to a great extent with the integration of FLC in the control system of the proposed D-STATCOM. From this work, it is learnt that the voltage sag taking place at the distribution level under load perturbation can be successfully mitigated using a CSC-based D-STATCOM system instead of its VSC-based counterpart. This research paper will pave the way for initiating the application of CSC topology in other types of CP devices so as to enhance power quality in the future.

Funding
The authors received no direct funding for this research.

Author details
M. Deben Singh[1]
E-mail: mdsingh2007@gmail.com
ORCID ID: http://orcid.org/0000-0001-9995-977X
Ram Krishna Mehta[1]
E-mail: rkmehta.ee@gmail.com
ORCID ID: http://orcid.org/0000-0001-9995-977X
Arvind Kumar Singh[1]
E-mail: singharvindk67@gmail.com
[1] Department of Electrical Engineering, North Eastern Regional Institute of Science & Technology, Nirjuli, Arunachal Pradesh 791109, India.

References

Acha, E., Agelidis, V. G., Anaya-Lara, O., & Miller, T. J. E. (2006). *Power electronic control in electrical systems*. Newnes power engineering series. Oxford: Elsevier.

Bhim, S., Jayaprakash, P., & Kothari, D. P. (2008). A T-connected transformer and three-leg VSC based DSTATCOM for power quality improvement. *Proceedings of IEEE Transactions on Power Electronics, 23*, 2710–2718.

Bilgin, H. F., & Ermis, M. (2010). Design and implementation of a current source converter for use in industry applications of D-STATCOM. *Proceedings of IEEE Transactions on Power Electronics, 25*, 1943–2957.

Bilgin, H. F., & Ermis, M. (2011). Current source converter based STATCOM: Operating principles, design and field performance. *Elsevier Journal on Electric Power Systems Research, 81*, 478–487.

Bilgin, H. F., Ermis, M., Kose, K. N., Cetin, A., Cadirci, I., Acik, A., ... Yorukoglu, M. (2007). Reactive power compensation of coal mining excavators by using a new generation STATCOM. *Proceedings of IEEE Transactions on Industrial Applications, 43*, 97–110.

Deben, S. M., & Khumanleima, C. L. (2015). Power electronics technology for power quality improvement. *International Journal of Advanced Research in Electrical, Electronics and Instrumentation Engineering, 4*, 2073–2080. doi:10.15662/ijareeie.2015.0404031

Deepa, S., & Ranjani, M. (2015). Dynamic voltage restorer controller using gradealgorithm. *Cogent Engineering, 2*, 1017243, 1–11. doi:10.1080/23311916.2015.1017243

Ghosh, A., & Ledwich, G. (2009). *Power quality enhancement using custom power devices*. (1st Indian reprint). New Delhi: Springer International Edition.

Han, B., Moon, S., & Karady, G. (2000). Static synchronous compensator using thyristor PWM current source inverter. *Proceeding of IEEE Transactions on Power Delivery, 15*, 1285–1290.

Harish, S., & Mahesh, K. M. (2008). Fuzzy logic based supervision of DC link PI Control in a D-STATCOM. *Proceedings IEEE Conference, 2*, 453–458.

Hingorani, N. G. (1995). Introducing custom power. *IEEE Spectrum, 32*, 41–48. http://dx.doi.org/10.1109/6.387140

Hingorani, N. G., & Gyugyi, L. (2013). *Understanding FACTS: Concepts & technology of flexible AC transmission systems*. London: Wiley.

Jan, J. (1998). Design of fuzzy controllers (Technical report No. 98-E 864). Technical University of Denmark, Department of Automation, Lyngby, Denmark (pp 1–27).

Lipi, K., & Deben, S. M. (2013). *Power quality enhancement using dynamic voltage restorer system* (M. Tech thesis). Department of Electrical Engineering, NERIST, Itanagar, India.

Math, H., & Bollen, J. (2013). *Understanding power quality problems*. London: Wiley.

Mohan, N., Undeland, T. M., & Robbins, W. P. (1989). *Power electronics: Converters, applications, and design*. New York, NY: John Wiley & Sons.

Nagesh, G., Srinivas, B. K., & Mahesh, M. K. (2012). Synchronous reference frame based current controller with SPWM switching strategy for DSTATCOM applications. In *Proceedings of IEEE international conference on power electronics, drives and energy systems (PEDES-2012)*, December16–19, Bengaluru, India.

Resul, C., Besir D., & Fikret A. (2011). Fuzzy-PI current controlled D-STATCOM. *Gazi University Journal of Science, 24*, 91–99.

Singh, M. D., Mehta, R. K., & Singh, A. K. (2015). Current source converter based D-STATCOM for voltage sag mitigation. *International Journal for Simulation and Multidisciplinary Design Optimization, 6*, A5, 1–10. France. doi: 10.1051/smdo/2015005

Steimer, P. K., Gruning, H. E., Werninger, J., Carroll, E., Klaka, S., & Linder, S. (1999). IGCT—A new emerging technology for high power, low cost inverters. *IEEE Industrial Application Magazine, 5*, 12–18.

Wang, X. (1993). *Advances in pulse width modulation techniques* (PhD thesis). Deptartment of Electrical Engineering, McGill University, Canada.

Ye, Y., Kazerani, M., & Quintana, V. H. (2005). Current-source converter based STATCOM: Modeling and control. *Proceedings of IEEE Transactions on Power Delivery, 20*, 795–800.

9

RF MEMS and CSRRs-based tunable filter designed for Ku and K bands application

Ngasepam Monica Devi[1], Santanu Maity[2]*, Rajesh Saha[1] and Sanjeev Kumar Metya[2]

*Corresponding author: Santanu Maity, Electronics and Communication Engineering, National Institute of Technology, Arunachal Pradesh, Yupia 791112, India
E-mail: santanu.ece@nitap.in

Reviewing editor: Qingsong Ai, Wuhan University of Technology, China

Abstract: This paper presents the design and simulation of a reconfigurable stop-band filter on a silicon substrate based on the combination of RF microelectromechanical system and metamaterial-based technologies. The device is implemented on coplanar waveguide structure by embedding complementary split-ring resonators on the central line and an RF MEMS varactor bridge supporting the neighboring ground planes. The response characteristics of this metamaterial-based filter can be dynamically tuned, thus enhancing its usefulness. The device operates within a frequency range of 16.5–19.5 GHz, giving a tuning range of 15%, and can be tuned from Ku-frequency band to K-frequency band. It works with a comparative low pull-in voltage of 17.42 V and a faster switching time of 0.138 μs. A thorough electromechanical analysis has been done by varying various structural and material parameters. Moreover, a comparative electrical performance of silicon and glass has been shown to overcome the cons of silicon by high-resistivity glass.

Subjects: Circuits & Devices; Electrical & Electronic Engineering; Electromagnetics & Communication

Keywords: complementary split-ring resonators; coplanar waveguide; metamaterials; microelectromechanical system; tunable filters

ABOUT THE AUTHOR

Santanu Maity is working as assistant professor in the Department of Electronics and Communication Engineering at National Institute of Technology, Arunachal Pradesh, India. His research interest in RF MEMS based switch, MEMS based gas sensor, Crystalline Si solar cell, Semiconductor devices etc. He has more than four years of research experience on nano fabrication facility at Jadavpur University, IITB & IIEST.

PUBLIC INTEREST STATEMENT

Microelectromechanical system (MEMS) are types of miniature devices as well as systems fabricated on different platforms by micromachining processes. The work showed numerous parametric analyses of various parameters related to the RF MEMS varactor bridge used. Also, this work has addressed RF MEMS switch problem by achieving a lower actuation voltage of 17.42 V and a comparatively faster switching time of 0.138 μs. Also, through the material property analyses of various MEMS beams, the problem of selection of the material with respect to the requirements such as need for low pull-in voltage and high spring constant has also been solved. The tunability of the complementary split-ring resonator filter has been achieved through the implementation of MEMS varactor bridges only. In fact, this work has given a vivid idea on how MEMS technology can give good tunability through optimized device dimensions.

1. Introduction

Microelectromechanical system (MEMS) is a collection of sensors and actuators that have the ability to sense its environment and function according to the various changes in that environment with the use of a microcircuit control. Thus, MEMS has microelectronics packaging, antenna structures, micropower supply, microrelay, and microsignal processing units (Jason Yao, 2000). The popularity of MEMS is increasing day by day. Thus, numerous MEMS structures have been proposed and designed in various fields, such as microfluidics, aerospace, biomedical, chemical analysis, wireless communications, data storage, display, optics, etc. Still the feasibility of designing more complex MEMS structures is in progress.

Nowadays, many structures based on MEMS are used in RF domain. There are various types of RF micro-electromechanical system (RF MEMS) components, which can be in the form of switches, switched capacitors, tunable inductors, resonators, filters and varactors. These devices can be used instead of FETs, HEMTs and PIN diodes due to the advantages these offer, such as reduction in cost due to batch fabrication; down-scaling in dimensions, which leads to advancements in performance; reduction in size and weight to a large extent; and its integrating capability with microwave-integrated circuit through passive fabrication techniques (Jason Yao, 2000; Vu, Prigent, Ruan, & Plana, 2012). MEMS devices are fabricated using silicon micromachining technique. In silicon micromachining, the structures are made 3-dimensional by either etching silicon substrate or by forming micromechanical layers from layers or films deposited on the surface. This is a reason that MEMS has seen a fast growth in the last few decades. The use of silicon as a substrate in MEMS devices can be credited due to its material properties, cost reduction, and IC compatibility (Petersen, 1982).

The introduction of metamaterial concept in design of filters and switches is relatively new. Metamaterials are artificial electromagnetic materials which do not occur naturally. The properties of metamaterials were first predicted by Veselago (1968). Later, numerous theoretical and verifications have been carried out (Pendry, 2000; Pendry, Holden, Stewart, & Youngs, 1996; Shelby, Smith, & Schultz, 2001). As it exhibits simultaneous negative permittivity and permeability, it offers many intriguing possibilities for high-frequency circuits, such as filters and switches.

The idea of integrating MEMS with metamaterial structures started with the use of varactor diodes to achieve tunability, which is one of the important features of a metamaterial-based structures (Degiron, Mock, & Smith, 2007; Gil et al., 2004; Gorkunov & Lapine, 2004; Reynet & Acher, 2004), and to improve the performance of conventional distributed passive devices (Bonache, Gil, Garcia-Garcia, & Martin, 2006; Garcia-Garcia et al., 2005). Several works utilizing the concept of both metamaterials and RF MEMS have been carried out achieving accurate electrical models (Gil, Morata, Fernandez, Rottenberg, & De Raedt, 2011a, 2011b; Hand & Cummer, 2007; Kundu et al., 2012).

MEMS switches are basically either cantilever type or fixed–fixed beam type. A MEMS switch can be placed in either series or shunt configurations and can be a metal-to-metal contact or a capacitive contact switch. Generally, metal contacting switches are often used as serial switches, while capacitive coupling switches are used for shunt switches. The use of this configuration is primarily within the RF circuit design. DC contact switches are used for low-frequency applications (10–60 GHz), while shunt capacitive switches are used for high-frequency applications (5–100 GHz) (Rebeiz, 2003a).

Complementary split ring resonator (CSRR) has attractive features like low-cost, high-quality factor and low radiation loss. Recently, a number of filters based on CSRRs have been suggested to achieve better performance. They were composed of stepped impedance resonator and an open-circuited stub (Huang, Wen, & Huang, 2009), coupled square loop (Liu, Lin, Zeng, Yeh, & Chang, 2010), parallel micro-strip line (Luo, Qian, Ma, & Li, 2010), and circular loop by considering multi-regional objective function (Kim, Ko, Choi, & Kim, 2012). Su, Naqui, Mata-Contreras, and Martín (2015) proposed the modeling, analysis, and applications of microstrip lines loaded with pairs of electrically coupled CSRRs. They presented the equivalent lumped circuit model loaded with coupled CSRRs and a proof concept of a comparator for dielectric characterization was proposed.

In this paper, a reconfigurable band stop filter is designed by using CSRRs on the central strip and applying RF MEMS bridge on it, thus performing the function of a capacitive switch through the combination of RF MEMS capacitive tuning abilities along with CSRRs properties. The designed tunable filter can be used from Ku-frequency band to K-band. The use of MEMS configuration can be accounted due to the fact that it consumes low dc power, gives low insertion loss, high isolation, and excellent linearity. Our proposed model has very low actuation voltage and very high switching time. The problem of material selection with respect to the requirements of low pull-in voltage and high spring constant has also been solved in this paper through the material property analyses of various MEMS beam. The tunability of the CSRR filter has been achieved through the implementation of MEMS varactor bridges only. In fact, this work has given a vivid idea on how MEMS technology can give good tunability through optimized device dimensions.

2. Proposed device structure

The top and side view of the proposed device is shown in Figures 1 and 2, respectively. Initially, a coplanar waveguide (CPW) structure made of gold was designed on a high-resistivity silicon substrate. The CPW has a central signal line and two ground plane. On the central line of CPW structure, rectangular-shaped CSRR-embedded (CSRR consists of two split-type concentric structures) was designed. To achieve tunablity in frequency band, MEMS shunt switch has been designed above CPW. Two anchors of the bridge is fixed on ground plane and the metal bridge is suspended over the signal line of CPW. Table 1 shows the different parameters related to design of CPW and CSRR along with their dimensions. These dimensions of the device were calculated to achieve proper impedance matching. MEMS bridge is suspended over the signal line with an air gap of g_0 and its corresponding parameters related to length, width, and thickness are given in Table 2, respectively. The design parameters are taken from the basic CPW expression to get 50-ohm impedance. So through the impedance matching condition, the CPW parameters and CSRR structures parameters are calculated (shown in Table 1). And to get the low actuation voltage and higher switching speed, different switch parameters shown in Table 2 are calculated from basic switch expression. During actuation, the MEMS beam is 0.5 μm away from the signal line in down-state condition and 3.5 μm for up-state condition. The air-gap height has been chosen as 3 μm for obtaining a higher tuning range. There will be instability of MEMS bridges, if the gap height is less than 2 μm, as the gap becomes less than 2/3 times of original gap (Rebeiz, 2003a). A thin layer of silicon nitride (Si_3N_4) with a thickness of 0.5 μm is added on top of the central line to prevent the shorting of the CSRR with the MEMS bridge, thereby generating a capacitive effect in the down-state condition. The proposed device is very small in size in comparison to other existing devices.

3. Mechanical properties

In this section, the mechanical parameters related to the proposed device have been calculated one by one.

3.1. Spring constant (k)

The mechanical behavior can be modeled using a linear spring constant, k (N/m), since the operation of the structure is limited to small deflections. The total spring constant is given by the stiffness of the beam and the biaxial residual stress. Hence, the expression of spring constant is given as follows (Rebeiz, 2003a):

$$k = 32Ew(t/l)^3 + 8\sigma(1 - v)w(t/l) \tag{1}$$

where E is Young's modulus (E = 80 GPa for gold), σ is residual stress of the beam, and v is Poisson's coefficient (v = 0.42 for gold). In ideal case, when σ = 0 and putting the values of other parameters from Table 2, the value of k obtained is 7.8 N/m.

3.2. Pull-in voltage (V_p)

The expression for the pull-in voltage is given as (Rebeiz, 2003a)

$$V_p = \sqrt{\frac{2k}{\varepsilon_0 Ww}g^2(g_0 - g)} \tag{2}$$

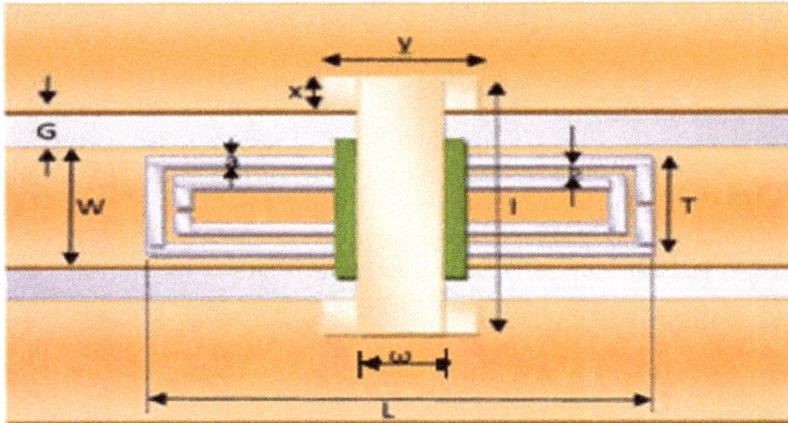

Figure 1. RF MEMS-loaded CPW along with its dimensions (top view).

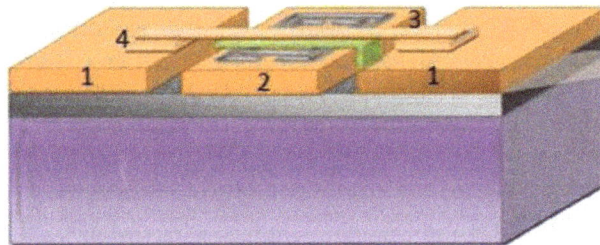

Silicon substrate (275µm)

Silicon dioxide (1µm)

Gold
1) CPW ground planes (1µm)
2) CPW central line loaded with CSRR (1µm)
3) RF MEMS bridge thickness (1µm)
4) MEMS bridge height taken initially (3.5µm)
Silicon Nitride (0.5µm)

Figure 2. Side view of the proposed device (RF MEMS loaded).

where g_0 is the initial height taken and g is the height of the beam above the signal line. The value of pull-in voltage (V_p) thus obtained from Equation 2 equals 17.42 V.

However, at exactly $(2/3)g_0$, instability occurs. This is because at $(2/3)g_0$, the increase in the electrostatic force of the beam is greater than the increase in the restoring force which may either lead to beam position becoming unstable or collapse of the beam to the down-state position. Thus, to attain stable condition, Equation 2 can be modified as (Rebeiz, 2003a)

Table 1. Dimensions of the CPW embedded with CSRR	
Parameters	**Measurements (µm)**
G/W/G of the CPW	10/200/10
Length of CSRR (L)	1,000
Width of CSRR (T)	180
Width of rings (a)	10
Spacing between rings (b)	10

Table 2. Dimensions of the MEMS bridge and the anchor	
Parameters	Measurements (µm)
Length of the MEMS bridge (l)	320
Width of the MEMS bridge (w)	100
Thickness of the MEMS bridge (t)	1
Length of the anchor (x)	50
Width of the anchor (y)	200

$$V_p = V(2/3)g_0 = \sqrt{\frac{8k}{27\varepsilon_0 Ww}g_0^3}$$ (3)

3.3. Damping coefficient (b) and quality factor (Q.F.)

The expression of damping coefficient (b) is given by (Rebeiz, 2003a)

$$b = (3/2\pi)(\mu A^2/g_0^3)$$ (4)

where μ is coefficient of viscosity of ideal gases at standard temperature and pressure (STP) and is equal to 1.85×10^{-5} Pa-s. Thus, the value of damping coefficient obtained from Equation 4 equals 1.30×10^{-4}.

The expression for the mechanical resonant frequency of beam vibration is given by (Rebeiz, 2003a)

$$f_0 = (1/2\pi)\sqrt{k/m}$$ (5)

where m is the modal mass of the beam and is given by $m = 0.35 \times (l \times w \times t) \times \rho = 2.16 \times 10^{-10}$ kg, (density of gold $\rho = 19{,}320$ kg/m³).

The equation of Q.F. for fixed–fixed beam is (Rebeiz, 2003a)

$$Q.F. = \frac{k}{w_0 b} = \frac{\sqrt{E\rho t^2}}{\mu(wl/2)^2}g_0^3$$ (6)

where w_0 is the angular frequency of beam vibration. Thus, from Equation 6, the value of Q.F. obtained equals 0.29.

3.4. Beam switching time (t_s)

The switching time for each beam to complete one up and down cycle is (Rebeiz, 2003a)

$$t_s = 3.67v_p/(v_s w_0)$$ (7)

where v_s equals $1.4v_p$ and w_0 is the angular frequency of beam vibration. Thus, the switching time obtained from Equation 7 equals 0.138 µs.

Usually, MEMS devices require high pull-in voltage of the range 30–80 V (Rebeiz, 2003b). However, this proposed device has achieved a reduced pull-in voltage of only 17.42 V. Moreover, the designed device with CPW layers and MEMS bridge made of gold gives a relatively faster switching time of 0.138 µs, which is faster than switching time of conventional electrostatic MEMS switches (Rebeiz, 2003b). Table 3 shows a comparative analysis of three different materials when used as a bridge. From this table, it is seen that Al gives the lowest pull-in voltage and spring constant, whereas Polysilicon provides the highest spring constant and pull-in voltage. Thus, there is a trade-off between the properties for the two materials. Hence, it can be inferred that the overall performance of gold is better as it gives acceptable spring constant and pull-in voltage, better quality factor and faster switching time when compared to Al and Polysilicon.

Table 3. Comparison of the beam materials with its mechanical properties			
Mechanical properties	**Au**	**Al**	**Polysilicon**
Spring constant (N/m)	7.8	6.74	16.6
Pull-in voltage (V)	17.42	17.39	27.3
Quality factor	0.29	0.109	0.16
Switching time (µs)	0.138	5.56	3.29

4. Device behavior

In this section, the electromechanical behavior of the device has been shown and analyzed. The proposed structure is designed and simulated using An soft HFSS, a simulator for 3D volumetric passive device modeling. The simulation is carried out for the loaded CPW by considering a bridge height of 3.5 µm for the up state, thus keeping an initial gap height of 3 µm. The air-gap height for the down state is chosen as $(2/3)g_0$, thus considering 2.5 µm as the bridge height. The stop-band frequency response with a reconfigurable capability gives a tuning range of 15%. The device achieves a rejection level of −24 dB at 16.5 GHz in down-state position as depicted in Figure 3. While in Figure 4, which is the up-state position, a notch level of −28.5 dB is obtained at 19.5 GHz when a dielectric thickness of 0.5 µm is given. Therefore, the resonant frequency shifts from 19.5 GHz (K-band) to 16.5 GHz (Ku band) when the switch moves from up to down state.

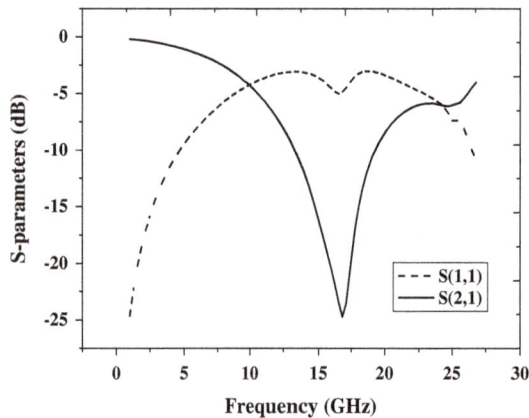

Figure 3. S-parameters performance of the proposed device in down-state position.

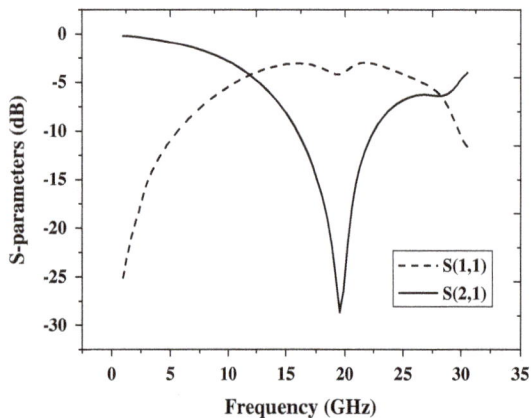

Figure 4. S-parameters performance of the proposed device in up-state position.

The effects in electrical performances with the variations in various mechanical parameters are shown below. The simulations have been carried out by giving bridge height of 0.5 μm for the down state and 3.5 μm for the up state. In fact, a dual-band response is observed in down state. However, as the device has been designed for Ku and K bands applications, the higher frequency response of down state has been considered.

4.1. Effect of bridge width

A parametric analysis has been done by varying bridge width for both up and down states for a fixed value of center conductor width (200 μm) and a dielectric thickness of 0.5 μm. It is observed that the resonant frequency varies when the bridge width is varied and the variation is more in the down state. This is due to the change in both the inductance and capacitance of the bridge. While in the up state, the change is mainly in the capacitance values resulting in more insertion loss. Figures 5 and 6 describe the parametric study of the S_{21}-parameters obtained through simulation for down and up states of the device when the bridge width is varied. Hence, at up state, tunability is obtained within K band and at down state, within Ku band.

4.2. Effect of dielectric thickness

Parametric analysis has been done by varying dielectric thickness for both up and down states keeping all other parameters fixed. It has been observed that even slight variation in Si_3N_4 thickness shifts the resonant frequency considerably. The frequency response is linear with respect to the increase

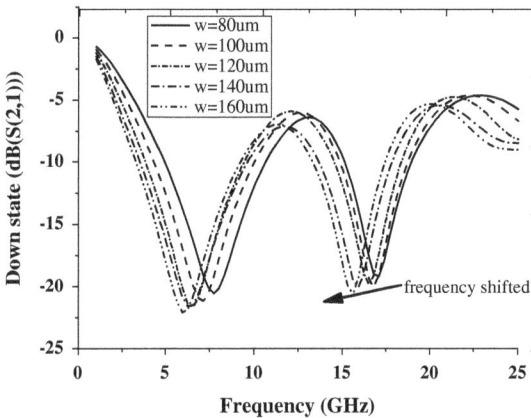

Figure 5. Parametric analysis of the bridge width (w) corresponding to the down state of the filter with t_d = 0.5 μm.

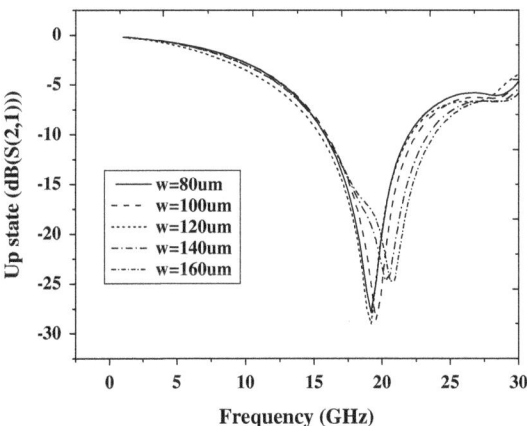

Figure 6. Parametric analysis of the bridge width (w) corresponding to the up state of the filter with t_d = 0.5 μm.

in dielectric thickness. However, to obtain a larger bandwidth and maintain a low pull-in voltage, varying the bridge width is favorable. Figures 7 and 8 show the parametric study of the S_{21}-parameters for down and up states of the device designed with the variation of dielectric thickness (t_d). Therefore, down and up state tunability has been obtained at Ku and K, band respectively. From Figure 9, it can be observed that the frequency changes almost linearly with change in dielectric thickness (t_d).

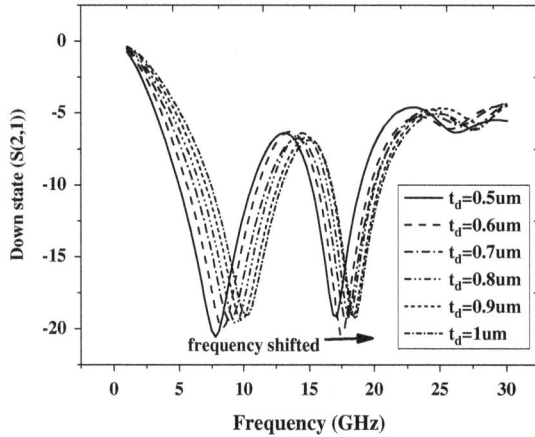

Figure 7. Parametric analysis of the dielectric thickness (t_d) corresponding to the down state of the filter with bridge width (w) = 100 μm.

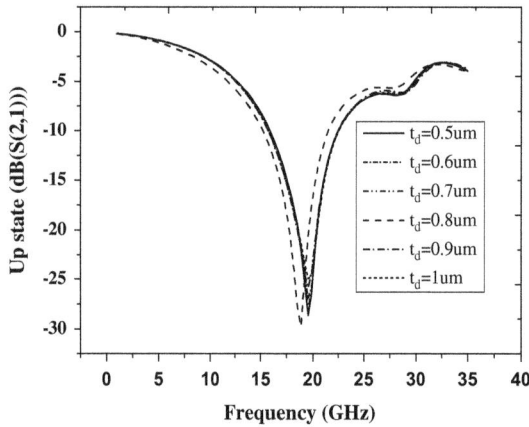

Figure 8. Parametric analysis of the dielectric thickness (t_d) corresponding to the up state of the filter with w = 100 μm.

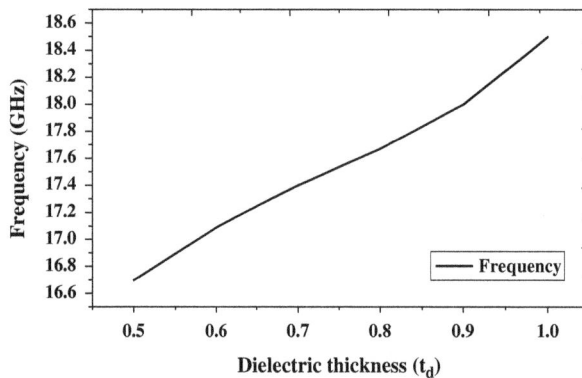

Figure 9. Simulated result showing the increase in frequency response with change in dielectric thickness.

4.3. Effect of bridge thickness

Figures 10 and 11 depict the parametric analysis with respect to bridge thickness and keeping all other parameters constant. As it is seen, varying bridge thickness does not affect the resonant frequency much. However, it is seen in Equation 2 that spring constant k is a function of t^3 and hence with increase in bridge thickness (t), the pull-in voltage increases (Sharma, Koul, & Chandra, 2007). Also, thicker beam membrane may require higher pull-in voltage which is not favorable. Hence, optimum dimension of bridge thickness should be considered for proper operation of the device. Here, at up state, the device works at greater than 20 GHz.

4.4. Effect of air-gap height

Insertion loss values for different air-gap heights have been studied for a fixed value of center conductor width of 200 μm and Si_3N_4 thickness of 0.5 μm. Figure 12 shows the variation of frequency with air-gap height.

It is observed that with increase in air-gap height, the resonant frequency shifts to the right side considerably. This can be accounted due to the fact that the capacitance decreases with the increase in air-gap height, thus shifting the resonant frequency. Therefore, by varying the air gap, the resonance

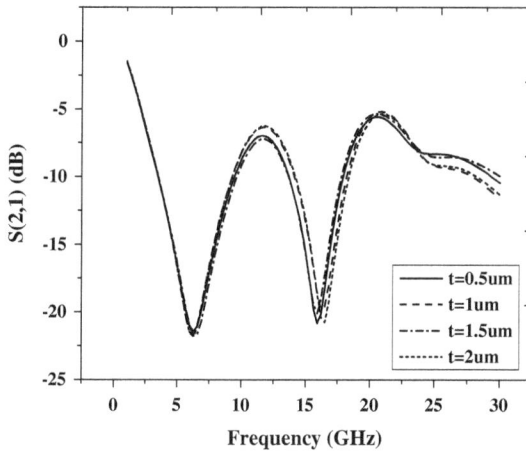

Figure 10. Parametric analysis of the bridge thickness (t) corresponding to the down state of the filter with bridge width (w) = 100 μm and dielectric thickness (t_d = 0.5 μm).

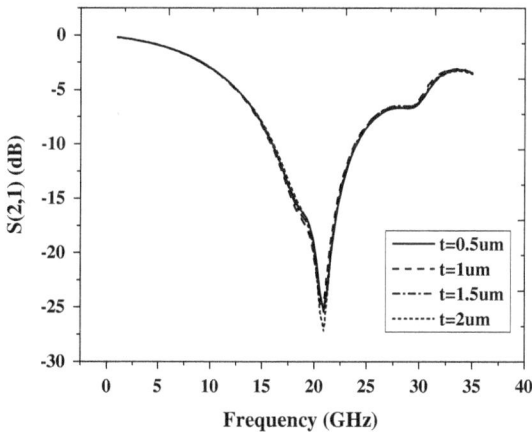

Figure 11. Parametric analysis of the bridge thickness (t) corresponding to the up state of the filter with bridge width (w) = 100 μm and dielectric thickness (t_d = 0.5 μm).

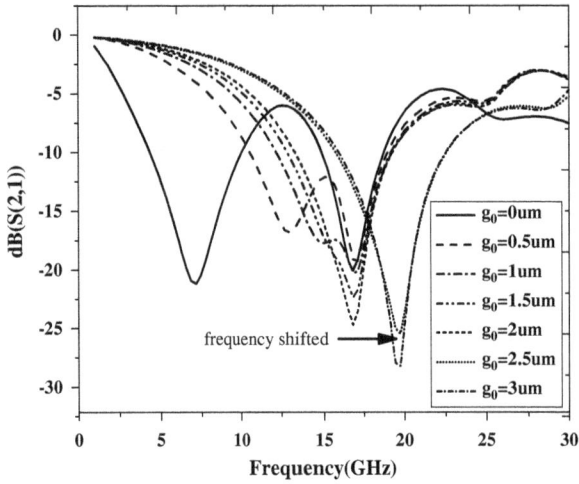

Figure 12. Simulated result showing the increase in frequency responses with increase in air-gap heights.

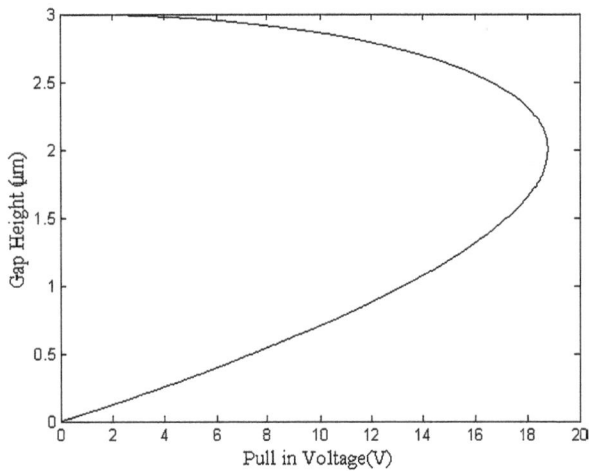

Figure 13. A graph showing change in pull-in voltage as a function of air-gap height when less than 3 μm.

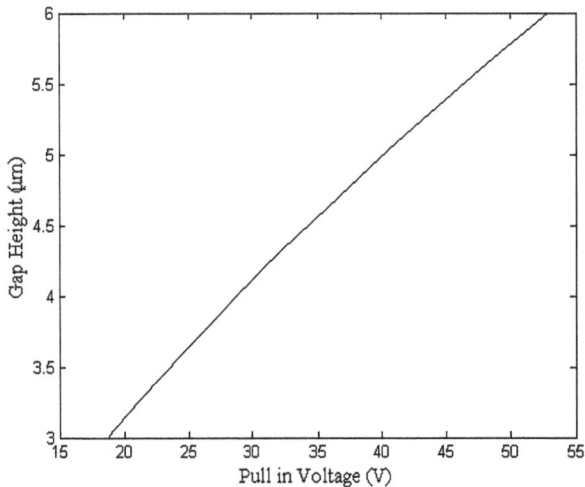

Figure 14. A graph showing change in pull-in voltage when air-gap height is more than 3 μm.

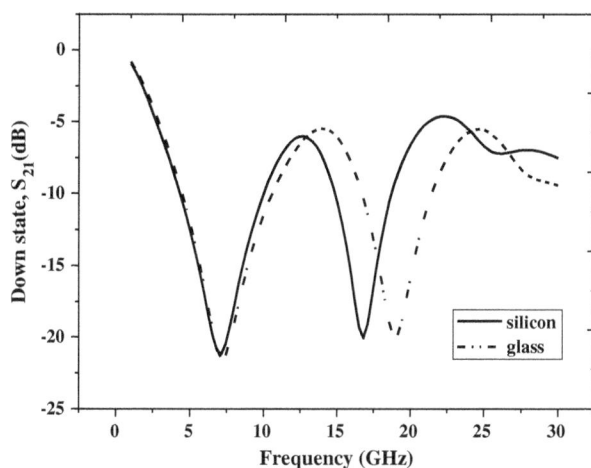

Figure 15. Simulated result showing S_{21} performances for two substrates for down-state position.

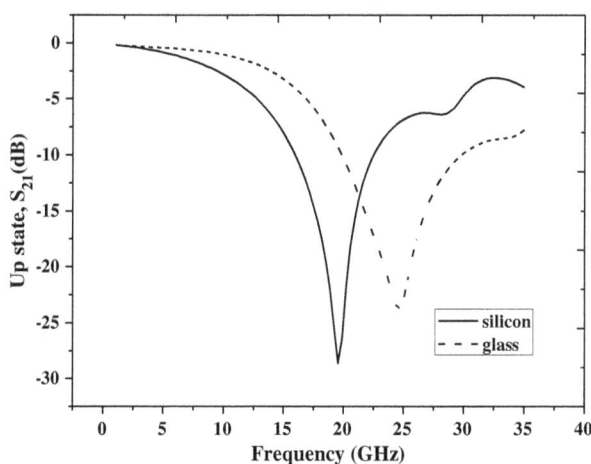

Figure 16. Simulated result showing S_{21} performances for two substrates in up-state position.

Note: Wider bandwidth and lower loss is seen for glass substrate.

frequency shifted from Ku to K band. From Figure 13, it can be observed that the pull-in voltage changes almost linearly with change in air gap height, which was also indicated in Equation 2. Figure 14 shows the linear change in pull-in voltage when air-gap height is increased from 3 µm.

5. Future developments

The total simulation was carried out using silicon as the substrate. But, due to having high value of permittivity, the excitation of surface waves takes place and hence results in decrease in bandwidth and increase in losses. However, we have also explored other materials to achieve better results. Same or higher resistivity can be achieved by using glass as substrate and the cost of this substrate is relatively less compared to pure silicon. It has been found through simulations that by using glass as substrate, one can achieve higher band width. Comparative simulation results using both the substrates have been shown in the following Figures, 15 and 16. Thus, the limitations of silicon such as losses and narrow bandwidth have been overcome by using glass. The parametric analysis of the proposed device based on CSRRs and MEMS switch on glass substrate will be done in future.

6. Conclusion

We have proposed a novel design and simulation of a high-frequency band stop filter concept based on a single MEMS capacitive structure and metamaterials. The tunable filter is designed with a tuning range of 15% over 16.5–19.5 GHz with an acceptable rejection level (around 24–28.5 dB). Electromechanical studies including the effects of variations of various parameters have been analyzed to show the good performances of the device designed related to switching time (t_s = 0.138 µs), pull-in voltage (V_p = 17.42 V), and quality factor (Q.F. = 0.24) when gold is used as the beam material. The electrical performance of glass as a substitute of silicon has also been covered. This work envisages the design and development of numerous millimeter wave devices through the integration of RF MEMS and metamaterial-based technologies.

Funding
The authors received no direct funding for this research.

Author details
Ngasepam Monica Devi[1]
E-mail: ngcanimo@gmail.com
Santanu Maity[2]
E-mail: santanu.ece@nitap.in
Rajesh Saha[1]
E-mail: rajeshsaha93@gmail.com
Sanjeev Kumar Metya[2]
E-mail: smetya@gmail.com

[1] Computer Science and Engineering, National Institute of Technology, Arunachal Pradesh, Yupia 791112, India.

[2] Electronics and Communication Engineering, National Institute of Technology, Arunachal Pradesh, Yupia 791112, India.

References
Bonache, J., Gil, I., Garcia-Garcia, J., & Martin, F. (2006). Novel microstrip bandpass filters based on complementary split-ring resonators. *IEEE Transactions on Microwave Theory and Techniques, 54,* 265–271. http://dx.doi.org/10.1109/TMTT.2005.861664

Degiron, A., Mock, J. J., & Smith, D. R. (2007). Modulating and tuning the response of metamaterials at the unit cell level. *Optics Express, 15,* 1115–1127.

Garcia-Garcia, J., Martin, F., Falcone, F., Bonache, J., Baena, J. D., Gil, I., ... Marques, R. (2005). Microwave filters with improved stopband based on sub-wavelength resonators. *IEEE Transactions on Microwave Theory and Techniques, 53,* 1997–2006. http://dx.doi.org/10.1109/TMTT.2005.848828

Gil, I., García-García, J., Bonache, J., Martín, F., Sorolla, M., & Marqués, R. (2004). Varactor-loaded split ring resonators for tunable notch filters at microwave frequencies. *Electronics Letters, 40,* 1347–1348. http://dx.doi.org/10.1049/el:20046389

Gil, I., Morata, M., Fernandez, R., Rottenberg, X., & De Raedt, W. (2011a). Characterization and modelling of switchable stop-band filters based on RF-MEMS and complementary split ring resonators. *Microelectronic Engineering, 88,* 1–5. http://dx.doi.org/10.1016/j.mee.2010.07.031

Gil, I., Morata, M., Fernandez, R., Rottenberg, X., & De Raedt, W. (2011b, March 20–23). Reconfigurable RF MEMS metamaterials filters. *Progress in Electromagnetics Research Symposium Proceeding* (pp. 1239–1242). Marrakesh.

Gorkunov, M., & Lapine, M. (2004). Tuning of nonlinear metamaterial bandgap by external magnetic field. *Physical Review B, 70,* 235109.

Hand, T., & Cummer, S. (2007). Characterization of tunable metamaterial elements using MEMS switches. *IEEE Antennas and Wireless Propagation Letters, 6,* 401–404.

Huang, J.-F., Wen, J.-Y., & Huang, M.-C. (2009). Design of a compactplanar UWB filter for wireless communication applications. In *International Conference WCSP* (pp. 1–4). Nanjing.

Jason Yao, J. (2000). RF MEMS from a device perspective, Rockwell Science Center. *Journal of Micromechanics and Microengineering, 10,* R9–R58.

Kim, K.-T., Ko, J.-H., Choi, K., Kim, H.-S. (2012). Optimum design of wideband bandpass filter with CSRR-loaded transmission line using evolution strategy. *IEEE Transactions on Magnetics, 48,* 811–814. http://dx.doi.org/10.1109/TMAG.2011.2177643

Kundu, A., Das, S., Maity, S., Gupta, B., Lahiri, S. K., & Saha, H. (2012). A tunable band-stop filter using a metamaterial structure and MEMS bridges on a silicon substrate. *Journal of Micromechanics and Microengineering, 22,* Article ID: 045004, 12 p. doi:10.1088/0960-1317/22/4/045004

Liu, J.-G.J.-C., Lin, H.-C., Zeng, B.-H., Yeh, K.-D., & Chang, D.-C. (2010). An improved equivalent circuit model for CSRR-based bandpass filter design with even and odd modes. *IEEE Microwave and Wireless Components Letters, 20,* 193–195. http://dx.doi.org/10.1109/LMWC.2010.2042548

Luo, X., Qian, H., Ma, J.-G., & Li, E.-P. (2010). Wideband bandpass filter with excellent selectivity using new CSRR-based resonator. *Electronics Letters, 46,* 1390–1391. http://dx.doi.org/10.1049/el.2010.1817

Pendry, J. B. (2000). Negative refraction makes a perfect lens. *Physical Review Letters, 85,* 3966–3969.

Pendry, J. B., Holden, A. J., Stewart, W. J., & Youngs, I. (1996). Extremely low frequency plasmons in metallic mesostructures. *Physical Review Letters, 76,* 4773–4776.

Petersen, K. E. (1982). Silicon as a mechanical material. *Proceedings of the IEEE, 70,* 420–457.

Rebeiz, G. M. (2003a). *RF MEMS—Theory, design & technology.* New York, NY: Wiley.

Rebeiz, G. M. (2003b, June 8–12). RF MEMS. In *12th International Conference on Solid State Sensors, Actuators and Microsystems.* Boston, MA. http://dx.doi.org/10.1002/0471225282

Reynet, O., & Acher, O. (2004). Voltage controlled metamaterial. *Applied Physics Letters, 84,* 1198–1200.

Sharma, P., Koul, S. K., & Chandra, S. (2007). Studies on RF MEMS shunt switch. *Indian Journal of Pure & Applied Physics, 45,* 387–394.

Shelby, R. A., Smith, D. R., & Schultz, S. (2001). Experimental verification of a negative index of refraction. *Science, 292,* 77–79. http://dx.doi.org/10.1126/science.1058847

Su, L., Naqui, J., Mata-Contreras, J., & Martín, F. (2015). Modeling and applications of metamaterial transmission lines loaded with pairs of coupled complementary split ring resonators (CSRRs). *IEEE Antennas and Wireless Propagation Letters, 7.* doi:10.1109/LAWP.2015.2435656

Veselago, V. G. (1968). The electrodynamics of substances with simultaneously negative values of permittivity and permeability. *Soviet Physics Uspekhi, 10,* 509–514.

Vu, T. M., Prigent, G., Ruan, J., & Plana, R. (2012). Design and fabrication of RF-MEMS switch for V-band reconfigurable application. *Progress in Electromagnetics Research B, 39,* 301–318. http://dx.doi.org/10.2528/PIERB12021404

10

Solving economic dispatch problem with valve-point effects using swarm-based mean–variance mapping optimization (MVMO^S)

T.H. Khoa[1]*, P.M. Vasant[1], M.S. Balbir Singh[1] and V.N. Dieu[2]

*Corresponding author: T.H. Khoa, Department of Fundamental and Applied Sciences, Universiti Teknologi PETRONAS, 31750 Tronoh, Perak, Malaysia
E-mail: trhkhoa89@gmail.com
Reviewing editor: Siew Chong Tan, University of Hong Kong, Hong Kong

Abstract: Mean–variance mapping optimization (MVMO) is a new population-based metaheuristic technique which is successfully applied for different power system optimization problems. The special feature of MVMO is the mapping function applied for the mutation based on the mean and variance of n-best population. Recently, the modified version of MVMO has been developed to become more powerful, named as swarm-based mean–variance mapping optimization (MVMO^S). This paper proposes MVMO^S as a new approach for solving the economic dispatch (ED) problem considering valve-point effects. To validate the performance of the proposed method, the MVMO^S is tested on three systems including 3, 13, and 40 thermal generating units with valve-point effects and the obtained results from MVMO^S are compared to those from other existing methods in the literature. Test results have indicated that the proposed MVMO^S is more robust and produces better solution quality than many other methods. Therefore, the MVMO^S is efficient for solving the ED with valve-point effects.

Subjects: **Computer Science; Electrical & Electronic Engineering; Power & Energy**

Keywords: **mean–variance mapping optimization; economic dispatch; valve-point effects; metaheuristic; nonconvex objective function; swarm-based mean–variance mapping optimization**

ABOUT THE AUTHOR

P.M. Vasant is a senior lecturer at Department of Fundamental and Applied Sciences, Universiti Teknologi PETRONAS in Malaysia. His research interests include Soft Computing, Hybrid Optimization, Holistic Optimization and Applications. He has co-authored research papers and articles in national journals, international journals, conference proceedings, conference paper presentation, and special issues lead guest editor, lead guest editor for book chapters' project, conference abstracts, edited books and book chapters. In the year 2009, P.M. Vasant was awarded top reviewer for the journal *Applied Soft Computing* (Elsevier).

PUBLIC INTEREST STATEMENT

In recent years, swarm intelligence has been widely applied to a variety of fields in engineering due to its outstanding characteristics for solving optimization problems with complex objective function and constraints. This paper presents a swarm intelligent approach for solving the nonconvex economic dispatch problems which is one of the important tasks in power generation systems. This problem is often related to fuel cost saving. Real-world economic dispatch problems have nonconvex objective functions with complex constraints. This leads to difficulty in finding the global optimal solution. Over the past decades, various optimization techniques have been applied to economic dispatch problems. In general, these techniques can be classified into classical calculus based methods, Artificial Intelligence techniques and hybrid methods. However, nonconvex optimization problems are still a challenge for engineers and decision-makers in the industry. Hence, there is always a need for developing new techniques for solving nonconvex problems.

1. Introduction

In power system, thermal generating units are supplied with multiple fuel sources such as coal, natural gas, and oil. The price of these fuels is highly volatile and faces depletion. Hence, the economical operations of the power systems gained importance. The economic dispatch (ED) is defined as the process of allocating the real power output of generating units to meet required load demand so as their total fuel cost is minimized while satisfying all physical and operational constraints (Dieu, Schegner, & Ongsakul, 2013).

Traditionally, the fuel cost function of each generating unit is presented as the quadratic function approximations and is solved using mathematical programming techniques such as lambda iteration method, Newton's method, gradient search, dynamic programming (Wollenberg & Wood, 1996), and quadratic programming (Fan & Zhang, 1998). However, most of these techniques are not capable of dealing with nonconvex and nonlinear ED problems. The ED problem is more practical when considering the effects of valve-point loadings. The valve-point effects (VPE) can cause the input–output curve of thermal generators to become more complicated. Therefore, the ED should be represented as nonconvex or nonsmooth optimization problem. This leads to difficulty in finding global optimum solution. More advanced optimization methods based on artificial intelligence concepts are implemented effectively to deal with ED problems such as genetic algorithm (GA) (Chiang, 2005), evolutionary programming (EP) (Sinha, Chakrabarti, & Chattopadhyay, 2003), artificial bee colony (ABC) (Hemamalini & Simon, 2010; Le Dinh, Vo Ngoc, & Vasant, 2013; Secui, 2015), ant colony optimization (ACO) (Pothiya, Ngamroo, & Kongprawechnon, 2010), evolutionary strategy optimization (ESO) (Pereira-Neto, Unsihuay, & Saavedra, 2005), and differential evolution (DE) (Noman & Iba, 2008). Recently, particle swarm optimization (PSO) is the most popular method applied for solving the ED problems, especially for nonconvex problems (Lin, Chen, Tsai, Yuan, et al., 2015; Shahinzadeh, Nasr-Azadani, & Jannesari, 2014; Vasant, Ganesan, Elamvazuthi, 2012). Several improvements of PSO method are developed for solving.

ED problem with valve-point loading effects such as modified particle swarm optimization (MPSO) (Park, Lee, Shin, & Lee, 2005), anti-predatory particle swarm optimization (APSO) (Selvakumar & Thanushkodi, 2008), self-organizing hierarchical particle swarm optimization (SOH_PSO) (Chaturvedi, Pandit, & Srivastava, 2008), simulated annealing like particle swarm optimization (SA-PSO) (Kuo, 2008), PSO with recombination and dynamic linkage discovery (PSO-RDL) (Chen, Peng, & Jian, 2007), new PSO with local random search (NPSO-LRS) (Selvakumar & Thanushkodi, 2007), improved coordinated aggregation-based particle swarm optimization (ICA-PSO) (Vlachogiannis & Lee, 2009), quantum-inspired PSO (QPSO) (Meng, Wang, Dong, & Wong, 2010), a modified hybrid PSO and gravitational search algorithm based on fuzzy logic (PSOGSA) (Duman, Yorukeren, & Altas, 2015). These improved PSO methods can obtain high-quality solutions for the problem. The PSO method is continuously improved for dealing with large-scale and complex problems in power systems. In addition, hybrid methods are also developed for solving the nonconvex ED problems by combining advantages of the single methods such as hybrid EP with sequential quadratic programming (EP-SQP) (Attaviriyanupap, Kita, Tanaka, & Hasegawa, 2002), integration particle swarm optimization with sequential quadratic programming (PSO-SQP) (Victoire & Jeyakumar, 2004), hybrid technique integrating the uniform design with the genetic algorithm (UHGA) (He, Wang, & Mao, 2008), self-tuning hybrid differential evolution (self-tuning HDE) (Wang, Chiou, & Liu, 2007), combining of chaotic differential evolution and quadratic programming (DEC-SQP) (Coelho & Mariani, 2006), hybrid GA, pattern search, and sequential quadratic programming (GA-PS-SQP) (Alsumait, Sykulski, & Al-Othman, 2010), hybrid differential evolution with biogeography-based optimization (DE-BBO) (Bhattacharya & Chattopadhyay, 2010), hybrid harmony search with arithmetic crossover operation (ACHS) (Niu, Zhang, Wang, Li, & Irwin, 2014). The hybrid methods have become among the most effective search techniques for obtaining high-quality solutions. However, the hybrid methods may be slower and more algorithmically complicated than conventional methods since they combine several operations into one technique.

The MVMO is a novel optimization algorithm which is conceived and developed by István Erlich (Erlich, Venayagamoorthy, & Worawat, 2010). This algorithm also falls into the category of the so-called "population-based stochastic optimization techniques". Recently, the extensions of MVMO has been developed by Rueda and Erlich (2013), which is named MVMO^S. The search process of MVMO^S starts with a set of particles. In addition, two parameters of MVMO including the scaling factor and variable increment parameters are extended to enhance the mapping. Hence, the ability for global search of MVMO^S is more powerful than the single particle version. In this paper, MVMO^S is proposed for solving the ED problem with valve-point effects.

The remaining organization of this paper is as follows. Section 2 presents the formulation of the ED problem with valve-point effects. The review of MVMO, extension of MVMO–MVMO^S, and implementation of the proposed MVMO^S to ED problem are exhibited in Section 3. The numerical test and results discussion are shown in Sections 4 and 5, respectively. The paper is concluded in Section 6.

2. Problem formulation

In this study, the VPE is considered as practical operation of generators. The VPE is a natural characteristic of a thermal turbine. The turbine of generating unit has many admission steam valves. The opening of these steam valves increase the throttling losses rapidly, leading to rise the incremental heat rate suddenly. The VPE is the direct result of the practical operation of thermal generating unit which produces ripples effects on the input–output curve as seen in Figure 1.

The VPE makes the fuel cost function highly nonlinear and nonsmooth containing multiple minima. The fuel cost function is described as the superposition of sinusoidal functions and quadratic functions.

The model of ED problem with VPE is formulated as follows (Dieu et al., 2013):

$$\text{Min } F = \sum_{i=1}^{N} \left(a_i + b_i P_i + c_i P_i^2 + \left| e_i \sin(f_i(P_{i,\,min} - P_i)) \right| \right) \tag{1}$$

subject to

(a) *Real power balance constraint*: The total active power output of generating units must be equal to total active power load demand plus power loss:

$$\sum_{i=1}^{N} P_i = P_D + P_L \tag{2}$$

where the power loss P_L can be approximately calculated by Kron's formula (Wollenberg & Wood, 1996):

$$P_L = \sum_{i=1}^{N} \sum_{j=1}^{N} P_i B_{ij} P_j + \sum_{i=1}^{N} B_{0i} P_i + B_{00}. \tag{3}$$

(b) *Generator capacity limits*: The active power output of generating units must be within the allowed limits:

$$P_{i,\,min} \leq P_i \leq P_{i,\,max}. \tag{4}$$

3. MVMO^S for ED problem with VPE

3.1. MVMO and MVMO^S

Mean–variance mapping optimization (MVMO) is a new optimization algorithm which falls into the category of the so-called "population-based stochastic optimization technique". The similarities between MVMO and the other known stochastic algorithms are in three evolutionary operators including selection, mutation, and crossover. The major differences between MVMO and other existing techniques are summarized in Erlich et al. (2010) as follows:

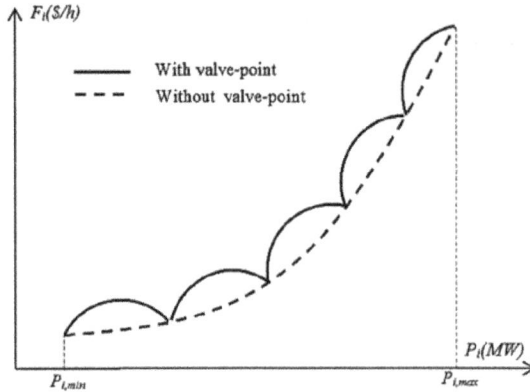

Figure 1. Fuel cost curve of units with valve-point effects.

- The key feature of MVMO is a special mapping function which applied for mutating the offspring. The mapping function is described by the mean and variance of *n* best solutions stored in the archive.
- The total space for searching of all variables is limited within the range from 0 to 1. The output of mapping function is always inside [0, 1]. However, the function evaluation is carried out always in the original scales.
- MVMO is a single-agent search algorithm because only a single offspring is generated in each iteration. Therefore, the number of fitness evaluations is identical to the number of iterations.
- A compact and dynamically updated solution archive serves as the knowledge base for guiding the search direction (i.e. adaptive memory). The normalized *n*-best are filled up in the archive progressively over iterations and sorted in a descending order of fitness.

Swarm-based mean–variance mapping optimization (MVMO[S]) is an extension of the original version MVMO. The difference between MVMO and MVMO[S] is the initial search process with particles. MVMO starts the search with single particle while MVMO[S] starts the search with a set of particles. At the beginning of the optimization process of MVMO[S], each particle performs *m* steps independently to collect a set of reliable individual solutions. Then, the particles start to communicate and to exchange information. MVMO is extended two parameters including the scaling factor f_s and variable increment Δd parameter to enhance the mapping. Therefore, the search global ability of MVMO[S] is strengthened.

3.2. Handing of constraints
Neglecting the transmission power loss, the equality constraint (2) is rewritten by:

$$\sum_{i=1}^{N} P_i = P_D. \tag{5}$$

By using the slack variable method (Kuo, 2008) to guarantee that the equality constraint (5) is always satisfied. The power output of the slack unit is calculated as follows:

$$P_s = P_D - \sum_{\substack{i=1 \\ i \neq s}}^{N} P_i. \tag{6}$$

The fitness function for the proposed MVMO[S] will include the objective function (1) and penalty terms for the slack unit if inequality constraint (4) is violated. The fitness function is as follows:

$$F_T = \left(a_i + b_i P_i + c_i P_i^2 + \left| e_i \sin(f_i(P_{i,\,min} - P_i)) \right| \right) + K.\left[\max(0, P_s - P_{s,\,max}) \right.$$
$$\left. + \max(0, P_{s,\,min} - P_s) \right]. \tag{7}$$

The penalty factor K for the slack unit is large enough and set to 1,000 for all systems.

3.3. Implemention of MVMO^s to ED
The flowchart of MVMO^s is depicted in Figure 2:

3.3.1. Initialization of algorithm
The parameters for MVMO^s have to be initialized including $iter_{max}$, n_var, n_par, $mode$, d_p, Δd_0^{ini}, Δd_0^{final} archive zize, $f^*_{s\_ini}$, $f^*_{s\_final}$, $n\_randomly$, $n\_randomly\_min$, $indep.runs(m)$, D_{min}.

Since different parameters of the proposed method have effect on the performance of MVMO^s, it is important to determine an optimal set of parameters of the proposed methods for ED problem. For each selection, one parameter is varied from the low value to higher value while the other parameters are fixed. The obtained result after one run is compared with the previous one. Multiple runs are carried out to choose the suitable set of parameters.

3.3.2. Normalization and de-normalization of variables
The search process of the MVMO^s starts with a set of particles. Initial variables is normalized to the range [0, 1] as follows:

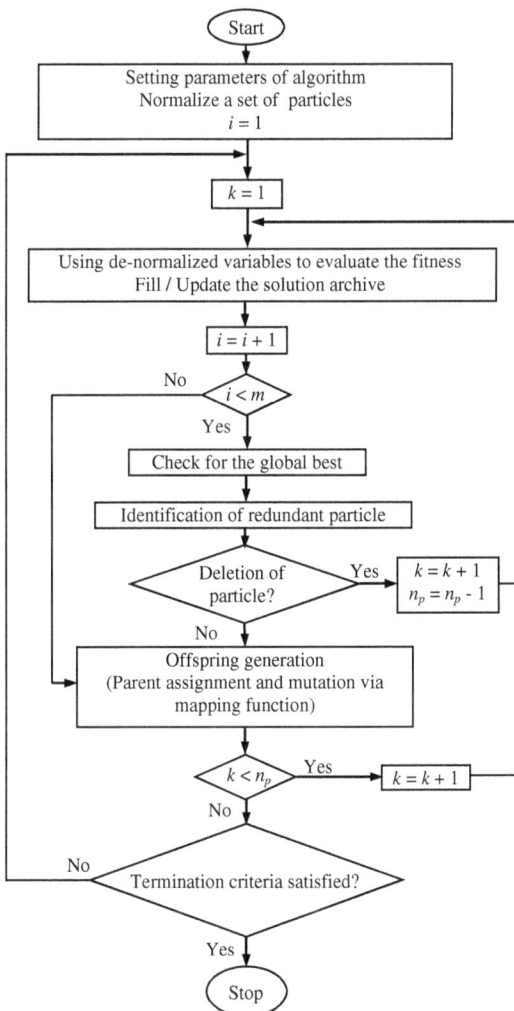

Figure 2. The flowchart of MVMO^s.

$$x\_normalized = rand(n\_par, n\_var). \tag{8}$$

However, the function evaluation is carried out always in the original scales of the problem space. The de-normalization of optimization variables is carried using (9):

$$P_i = P_{i, min} + Scaling \, x\_normalized(t, :). \tag{9}$$

where

$$Scaling = P_{i, max} - P_{i, min}.$$

After that, the power output for the slack generator is calculated using (6) to evaluate fitness function in (7), store f_{best} and x_{best} in archive.

MVMO^S utilizes swarm implementation to enhance the power of global searching of the classical MVMO by starting the search with a set of n_p particles, each having its own memory and represented by the corresponding archive and mapping function. At the beginning of the optimization process, each particle performs m steps independently to collect a set of reliable individual solutions. Then, the particles start to communicate and to exchange information.

It is worthless when particles are very close to each other since this would entail redundancy. To avoid closeness between particles, the normalized distance of each particle's local best solution $x^{lbest, i}$ to the global best x^{gbest} is calculated by Rueda and Erlich (2013). The i-th particle is discarded from the optimization process if the distance D_i is less than a certain user defined threshold D_{min}.

$$D_i = \sqrt{\frac{1}{N} \sum_{j=1}^{N} (x_j^{gbest} - x_j^{lbest, i})^2}. \tag{10}$$

where N denotes the number of optimization variables.

3.3.3. Solution archive

The best n individuals are stored in the archive table which is described as Figure 3. The archive size (n) is taken to be a minimum of two. If archive size is greater than two, the table of best individuals is filled up progressively over iterations in a descending order of the fitness. When the table is filled with n members, an update is performed only if the fitness of the new population is better than those in the table.

Mean $\bar{x}_i$ and variance v_i are computed from the archive where the n best populations are stored as follows (Erlich et al., 2010):

$$\bar{x}_i = \frac{1}{n} \sum_{j=1}^{n} x_i(j). \tag{11}$$

$$v_i = \frac{1}{n} \sum_{j=1}^{n} (x_i(j) - \bar{x}_i)^2. \tag{12}$$

#	Fitness	x_1	x_2	$\cdot$	$x_i \cdots$	x_D
1						
...						
n						
Mean $\bar{x}_i$	---					
Variance v_i	---					

Figure 3. The archive is used to store n-best population.

where j goes from 1 to n (archive size). At the beginning $\overline{x}_i$ corresponds with the initialized value of x_i and the variance is set to $v_i = 1.0$.

3.3.4. Parent assignment
The individual with the best fitness f_{best} and its corresponding optimization values, x_{best}, are stored in memory as the parent of the population for that iteration. This parent is used for creation of offspring .

3.3.5. Offspring creation
Creation of an offspring, of N dimensions involves three common evolutionary computation algorithms' operations including selection, mutation and crossover.

3.3.5.1. Selection: Among N variables of the optimization problem, d variables are selected for mutation operation. There are four strategies which are described in Erlich et al. (2010) for selecting the variables.

3.3.5.2. Mutation: For each of the d selected dimension, mutation is used to assign a new value of that variable. Given a uniform random number $x_i^* \in [0, 1]$, the transformation of x_i^* to x_i via mapping function is calculated in (13) and depicted as Figure 4. The transformation mapping function, h, is calculated by the mean $\bar{x}$ and shape variables s_{i1} and s_{i2} as in (15) (Erlich et al., 2010):

$$x_i = h_x + (1 - h_1 + h_0).x_i^* - h_0. \tag{13}$$

where h_x, h_1, and h_0 are the outputs of transformation mapping function based on different inputs given by:

$$h_x = h(x = x_i^*)h_o = h(x = 0)h_1 = h(x = 1). \tag{14}$$

$$h(\bar{x}_i, s_{i1}, s_{i2}, x) = \bar{x}_i.(1 - e^{-x.s_{i1}}) + (1 - \bar{x}_i).e^{-(1-x).s_{i2}}. \tag{15}$$

where

$$s_i = -\ln(v_i).f_s. \tag{16}$$

The scaling factor f_s in (16) is a MVMO parameter which can be used to change the shape of the function during iteration. In MVMO^s, this factor is extended for the need of exploring the search space at the beginning more globally, whereas at the end of the iterations, the focus should be on the exploitation. It is determined by (Rueda & Erlich, 2013):

$$f_s = f_s^*.(1 + rand()). \tag{17}$$

where

$$f_s^* = f_{s\_ini}^* + \left(\frac{i}{i_{final}}\right)^2 \left(f_{s\_final}^* - f_{s\_ini}^*\right). \tag{18}$$

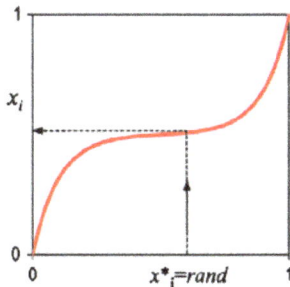

Figure 4. Variable mapping.

In (18), f_s^* denotes the smallest value of f_s and the variable i represents the iteration number. $f_{s\_ini}^*$ and $f_{s\_final}^*$ are the initial and final values of f_s^*, respectively. The recommended range of $f_{s\_ini}^*$ is from 0.9 to 1.0, and range of $f_{s\_final}^*$ is from 1.0 to 3.0. When, $f_{s\_final}^* = f_{s\_ini}^* = 1$ which means that the option for controlling the f_s factor is not used (Rueda & Erlich, 2013).

```
s_{i1} = s_{i2} = s_i
if s_i > 0 then
Δd = (1 + Δd_o + 2 . Δd_o(rand() − 0.5)
   if s_i > d_i
      d_i = d_i. Δd
   else
      d_i = d_i/Δd
   end if
   if rand() ≥ 0.5 then
      s_{i1} = s_i; s_{i2} = d_i
   else
      s_{i1} = d_i; s_{i2} = s_i
   end if
end if
```

The shape variables s_{i1} and s_{i2} in (15) are determined using the following algorithm (Rueda & Erlich, 2013):

At the start of the algorithm, the initial values of d_i (typically between 1 and 5) are set for all variables. Sometimes, the variance can oscillate over a wide range. Using the factor d_i instead of s_i which is a function of variance a smoothing effect is achieved. At every iteration, if $s_i > d_i$, d_i will be multiplied by Δd leads to increased d_i. In case $s_i < d_i$, the current d_i is divided by Δd which is always greater than 1.0 resulting in reduced value of d_i. Therefore, d_i will always oscillate around the current shape factor s_i. Furthermore, Δd is randomly varied around the value $(1 + \Delta d_0)$ with the amplitude of Δd_0 adjusted in accordance to (19), where Δd_0 can be allowed to decrease from 0.4 to 0.01 (Rueda & Erlich, 2013).

$$\Delta d_0 = \Delta d_0^{ini} + \left(\frac{i}{i_{final}}\right)^2 \left(\Delta d_0^{final} - \Delta d_0^{ini}\right). \tag{19}$$

3.3.5.3. Crossover: For the remaining un-mutated dimensions, the genes of the parent, x_{best}, are inherited. In other words, the values of these un-mutated dimensions are clones of the parent. Here, crossover is by direct cloning of certain genes. In this way, the offspring is created by combining the vector x_{best}, and vector of m mutated dimensions.

3.3.6. Termination criteria
The algorithm of the proposed MVMO^S is terminated when the maximum number of iterations $iter_{max}$ is reached.

3.4. Overall procedure
The steps of procedure of MVMO^S for the ED problem are described as follows:

Step 1: Setting the parameters for MVMO^S including $iter_{max}$, n_var, n_par, mode, d, Δd_0^{ini}, Δd_0^{final} archive zize, $f_{s\_ini}^*$, $f_{s\_final}^*$, n_randomly, n_randomly_min, indep.runs(m), D_{min}.

Set $i = 1$, i donotes the function evaluation.

Step 2: Normalize initial variables to the range [0, 1] (i.e. swarm of particles).

x_normalized = rand(n_par, n_var)

Step 3: Set $k = 1$, k denotes particle counters.

Step 4: De-normalized variables using (9), calculate power output for the slack generator using (6) to evaluate fitness function in (7), store f_{best} and x_{best} in archive.

Step 5: Increase $i = i + 1$. If $i < m$ (independent steps), go to step 6. Otherwise, go to step 7.

Step 6: Check the particles for the global best, collect a set of reliable individual solutions. The i-th particle is discarded from the optimization process if the distance D_i is less than a certain user defined threshold D_{min}. If the particle is deleted, increase $k = k + 1$, $n_p = n_p - 1$ and go to step 4. Otherwise, go to step 7.

Step 7: Create offspring generation through three evolutionary operators: selection, mutation, and crossover.

Step 8: if $k < n_p$, increase $k = k + 1$ and go to step 4. Otherwise, go to step 9.

Step 9: Check termination criteria. If stopping criteria is satisfied, stop. Otherwise, go to step 3.

4. Numerical results

This part presents results of the implementation of proposed MVMO^S and origional MVMO in solving the ED problem with valve-point effects. The obtained results by the proposed MVMO^S are compared to those from the other optimization methods for three test cases including 3-unit system, 13-unit system, and 40-unit system. For each case, the algorithm of MVMO^S is run 50 independent trials on a core i5 3.4 GHz PC with 4 GB RAM. The implementation of the proposed MVMO^S is coded in the Matlab R2013a platform.

4.1. Case 1: 3-unit system

The data of 3-unit test system with valve-point effects is taken from Sinha et al. (2003). In this case, the power load demand is 850 MW, the transmission power loss is neglected. The obtained results by MVMO and MVMO^S for this case are presented in Table 1. Figure 5 depicts the convergence characteristic of the MVMO and MVMO^S for case 1.

The parameters for MVMO^S for this system are as follows: $iter_{max} = 10,000$, n_var (generators) = 3, $n_p = 20$, *archive size* = 5, *mode* = 4, *indep.runs* (m) = 200, $n\_randomly$ = 2, $n\_randomly\_min$ = 2, $f^*_{s\_ini} = 0.9$, $f^*_{s\_final} = 3$, $d_i = 1$, $\Delta d_0^{ini} = 0.3$, $\Delta d_0^{final} = 0.01$, $D_{min} = 0$.

The min, average, and max fuel cost and CPU time obtained by the proposed MVMO^S are compared to the results of the other methods in Table 2. The best optimal solution for this case is 8,234.0717 ($/h) (Park et al., 2005). All methods obtain the minimum cost with 8,234.0717 ($/h) in Table 2. However, the proposed MVMO^S achieves the best optimal solution with a high probability (the standard deviation is 0%). For computational time, it may not be directly comparable among the methods because these methods were run and coded on different computers and programming languages. However, a CPU time comparison is used to show the efficiency of the compared methods. The computational time of MVMO^S is faster than EP, EP-PSO, PSO, CEP, FEP, MFEP, and IEEP, and close to PSO-SQP. The PSO, EP-SQP, and PSO-SQP were run on a Pentium II 500 MHz PC. There is no computer processor reported for CEP, FEP, MFEP, and IEEP.

4.2. Case 2: 13-unit system

The data of 13-unit test system with valve-point effects are referred to Sinha et al. (2003). The power load demand is 1,800 and 2,500 MW, respectively. The transmission power loss is also neglected in this case. The obtained results by the MVMO and MVMO^S corresponding to the two load demand are

Table 1. Obtained results for 3-unit system by MVMO & MVMO[s]				
		Power outputs		
		MVMO	MVMO[s]	
Unit	$P_{i, min}$ (MW)	P_i (MW)	P_i (MW)	$P_{i, max}$ (MW)
1	100	300.2669	300.2669	600
2	100	400.0000	400.0000	400
3	50	149.7331	149.7331	200
Total power (MW)		850.0000	850.0000	
Min cost ($/h)		8,234.0717	8,234.0717	
Average cost ($/h)		8,252.8227	8,234.0717	
Max cost ($/h)		8,390.8235	8,234.0717	
Standard deviation ($/h)		39.936	0.0000	
Average CPU time (s)		3.42	3.65	

Table 2. Comparisons of fuel cost for 3-unit system				
Method	Min cost ($/h)	Mean cost ($/h)	Max cost ($/h)	CPU (s)
EP (Victoire & Jeyakumar, 2004)	8,234.07	8,234.16	–	6.78
EP-PSO (Victoire & Jeyakumar, 2004)	8,234.07	8,234.09	–	5.12
PSO (Victoire & Jeyakumar, 2004)	8,234.07	8,234.72	–	4.37
PSO-SQP (Victoire & Jeyakumar, 2004)	8,234.07	8,234.07	–	3.37
CEP (Sinha et al., 2003)	8,234.07	8,235.97	8,241.83	20.46
FEP (Sinha et al., 2003)	8,234.07	8,234.24	8,241.78	4.54
MFEP (Sinha et al., 2003)	8,234.08	8,234.71	8,241.80	8.00
IFEP (Sinha et al., 2003)	8,234.07	8,234.16	8,234.54	6.78
MPSO (Park et al., 2005)	8,234.07	–	–	–
MVMO[s]	8,234.07	8,234.07	8,234.07	3.65

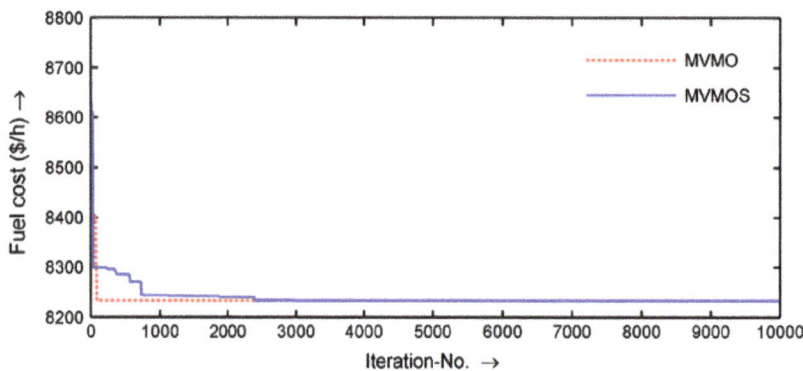

Figure 5. Convergence property of MVMO and MVMO[s] for case 1 (3 units).

shown in Table 3. Figures 6 and 7 show the convergence characteristic of the MVMO and MVMO[s] for the case of load demands 1,800 MW and the case of load demands 2,520 MW, respectively.

The parameters for MVMO[s] are as follows for all the cases of load demands 1,800 and 2,520 MW: $iter_{max}$ = 70,000, n_var (generators) = 13, n_p = 20, $archive\ size$ = 5, $mode$ = 4, $indep.runs$ (m) = 2,000, $n\_randomly$ = 5, $n\_randomly\_min$ = 4, $f_{s\_ini}^*$ = 0.95, $f_{s\_final}^*$ = 3, d_i = 1, Δd_0^{ini} = 0.4, Δd_0^{final} = 0.02, D_{min} = 0.

In Tables 4 and 5, the fuel cost and CPU time of proposed MVMO[s] are compared to those of other optimization methods for two load demands 1,800 and 2,520 MW. For the case of load demands 1,800 MW, the minimum fuel cost obtained by MVMO[s] is less than CEP, FEP, MFEP, IEEP, PSO, EP-SQP, PSO-SQP, HDE, and CGA-MU, and close to UHGA, Self-tuning HDE and GA-PS-SQP. It is noted that the mean cost obtained by MVMO[s] is less than is less than that of the others, except the UHGA. The computational time of MVMO[s] is faster than CEP, FEP, MFEP, IEEP, PSO, and EP-SQP, slower than UHGA, CGA-MU, HDE, self-tuning HDE, and GA-PS-SQP, and close to PSO-SQP. For the case of load demands 2,520 MW, the minimum total cost by MVMO[s] is less than that from the other methods in Table 5. The computational time of MVMO[s] is slower than ESO . The PSO, EP-SQP, and PSO-SQP were executed on a Pentium II 500 MHz PC. The HDE and self-tuning HDE were run on a Pentium 1.5 GHz with 768 MB of RAM. The computational time for ESO, UHGA, CGA-MU, GA-PS-SQP were from a Pentium IV PC, Pentium IV 2.99 GHz PC, Pentium III—700 PC, and Pentium III—1 GHz—256 MB of RAM, respectively. There is no computer processor reported for CEP, FEP, MFEP, and IEEP and no computational time for the other methods.

4.3. Case 3: 40-unit system

The data of the test system including 40 thermal generating units with VPE are from Sinha et al. (2003). The system load demand for this case is 10,500 MW neglecting transmission power loss. The obtained solutions by MVMO and MVMO[s] for this case are given in Table 6. The convergence characteristic of the MVMO and MVMO[s] are depicted in Figure 8 for case 3.

The parameters for MVMO[s] for this system are as follows: $iter_{max}$ = 150,000, n_var (generators) = 40, n_p = 5, $archive\ size$ = 5, $mode$ = 4, $indep.runs$ (m) = 2,000, $n\_randomly$ = 20, $n\_randomly\_min$ = 10, $f^*_{s\_ini}$ = 0.9, $f^*_{s\_final}$ = 3, d_i = 5, Δd_0^{ini} = 0.4, Δd_0^{final} = 0.02, D_{min} = 0.

Table 3. Obtained results for 13-unit system for load demands 1,800 and 2,520 MW by MVMO and MVMO[s]

Unit	$P_{i,\ min}$ (MW)	Power outputs				$P_{i,\ max}$ (MW)
		MVMO	MVMO	MVMO[s]	MVMO[s]	
		P_i (MW)	P_i (MW)	P_i (MW)	P_i (MW)	
1	0	538.5587	628.3185	628.3185	628.3452	680
2	0	149.7970	299.1741	148.2939	299.1906	360
3	0	224.5880	299.1858	224.2433	299.1924	360
4	60	159.7332	159.7325	60.0000	159.7318	180
5	60	109.8667	159.7314	109.7217	159.7299	180
6	60	109.9145	159.7268	109.8501	159.7328	180
7	60	109.8917	159.7329	109.8602	159.7314	180
8	60	109.8675	159.7317	109.8509	159.7320	180
9	60	109.5884	159.7271	109.8613	159.7077	180
10	40	60.0001	73.7967	40.0000	77.3682	120
11	40	77.7801	76.6576	40.0000	77.3731	120
12	55	55.0004	92.1942	55.0000	92.3625	120
13	55	55.0000	92.2908	55.0000	87.8023	120
Total power (MW)	1,800.0000	2,520.0000	1,800.0000	2,520.0000		
Min cost ($/h)	17,985.4638	24,170.8763	17,964.1226	24,170.0137		
Average cost ($/h)	18,126.9549	24,313.9828	18,011.0370	24,193.4933		
Max cost ($/h)	18,257.2713	24,473.9750	18,070.7615	24,226.8256		
Standard deviation ($/h)	54.7923	70.8093	26.7448	23.6363		
Average CPU time (s)	33.08	33.86	34.02	34.32		

Table 4. Comparisons of fuel cost for 13-unit system with VPE, P_D = 1,800 MW

Method	Min cost ($/h)	Mean cost ($/h)	Max cost ($/h)	CPU (s)
CEP (Sinha et al., 2003)	18,048.21	18,190.32	18,404.04	294.96
FEP (Sinha et al., 2003)	18,018.00	18,200.79	18,453.82	168.11
MFEP (Sinha et al., 2003)	18,028.09	18,192.00	18,416.89	317.12
IFEP (Sinha et al., 2003)	17,994.07	18,127.06	18,267.42	157.43
PSO (Victoire & Jeyakumar, 2004)	18,030.72	18,205.78	–	77.37
EP-SQP (Victoire & Jeyakumar, 2004)	17,991.03	18,106.93	–	121.93
PSO-SQP (Victoire & Jeyakumar, 2004)	17,969.93	18,029.99	–	33.97
UHGA (He et al., 2008)	17,964.81	17,992.92	–	15.33
QPSO (Meng et al., 2010)	17,969.01	18,075.11	–	–
HDE (Wang et al., 2007)	17,975.73	18,134.80	–	1.65
CGA-MU (Chiang, 2005)	17,975.34	–	–	21.91
ST HDE (Wang et al., 2007)	17,963.89	18,046.38	–	1.41
GA-PS-SQP (Alsumait et al., 2010)	17,964.25	18,199		11.06
MVMO[S]	17,964.12	18,011.04	18,070.76	33.86

Table 5. Comparisons of fuel cost for 13-unit system with VPE, P_D = 2,520 MW

Method	Min cost ($/h)	Mean cost ($/h)	Max cost ($/h)	CPU (s)
GA (Victoire & Jeyakumar, 2004)	24,398.23	–	–	–
SA (Victoire & Jeyakumar, 2004)	24,970.91	–	–	–
GA-SA (Victoire & Jeyakumar, 2004)	24,275.71	–	–	–
EP-SQP (Victoire & Jeyakumar, 2004)	24,266.44	–	–	–
PSO-SQP (Victoire & Jeyakumar, 2004)	24,261.05	–	–	–
UHGA (He et al., 2008)	24,172.25	–	–	–
ESO (Pereira-Neto et al., 2005)	24,177.78	–	–	1.0
SA-PSO (Kuo, 2008)	24,171.40	–	–	–
MVMO[S]	24,170.01	24,193.4933	24,226.8256	34.32

Table 7 shows the comparison of the fuel cost and CPU time of the MVMO[S] and the previously reported methods. As seen in Table 7, the best total cost obtained by the MVMO[S] is less than that from the others. The computational time of the MVMO[S] is faster than IFEP, ICA-PSO, and UHGA, and slower than the other methods. The computational time from ABC, ACO, DE, IFEP, ESO, NPSO-LRS, ICA-PSO, DEC-SQP, UHGA, self-tuning HDE, GA-PS-SQP, and DE-BBO were from Pentium IV 2.3 GHz with 512-MB of RAM PC, Pentium IV

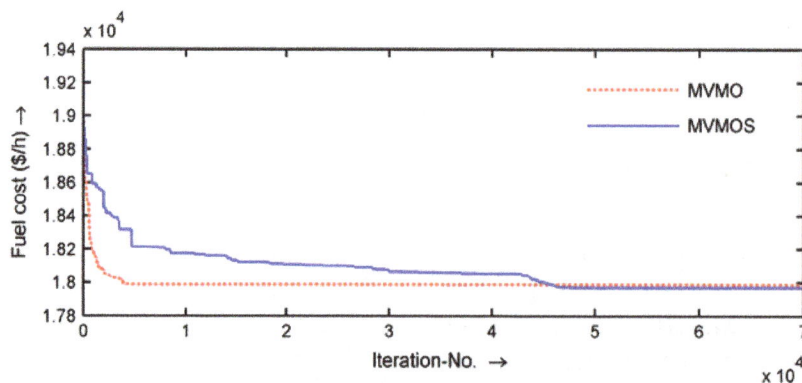

Figure 6. Convergence property of MVMO and MVMO[S] for case 2 (13 units with P_D = 1,800 MW).

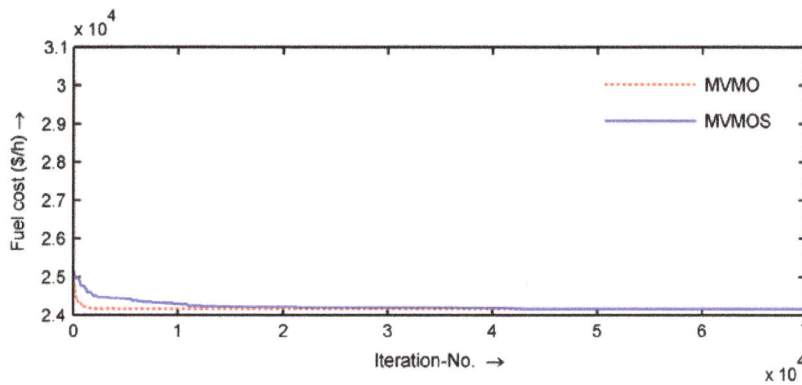

Figure 7. Convergence property of MVMO and MVMO^s for case 2 (13 units with P_D = 2,520 MW).

2.6 GHz with 1 GB of RAM PC, Intel 1.67 GHz with 1 GB of RAM PC, Pentium-II 350 MHz with 128 MB of RAM PC, Pentium IV 1.5 GHz with 128 MB of RAM PC, Pentium IV PC, Pentium IV 1.4-GHz PC, 1.1 AMD Athlon GHz with 112 MB of RAM, Pentium IV 2.99 GHz PC, Pentium 1.5 GHz with 768 MB of RAM, Pentium III 1 GHz with 256 MB of RAM, and Pentium IV 2.3-GHz PC with 512-MB RAM, respectively. There is no computational time or computer processor reported for the other methods.

5. Discussion

5.1. Advantages of MVMO^s

The advantages of MVMO^s are robustness, global solution with high probability, and easy implementation to ED problem. In this study, the MVMO and the MVMO^s are run 50 independent trials. The mean cost, max cost, average cost, and standard deviation obtained by the MVMO and MVMO^s to evaluate the

Table 6. Obtained results for 40-unit system by MVMO and MVMO^s

Unit	$P_{i,\,min}$ (MW)	Power outputs		$P_{i,\,max}$ (MW)
		MVMO	MVMO^s	
		P_i (MW)	P_i (MW)	
1	36	110.8011	110.8441	114
2	36	110.8067	110.9734	114
3	60	97.3999	97.4030	120
4	80	179.7331	179.7337	190
5	47	168.7998	88.3994	97
6	68	89.6332	168.8003	140
7	110	140.0000	140.0000	300
8	135	259.5997	259.6099	300
9	135	284.5997	284.6087	300
10	130	284.5997	284.6093	300
11	94	130.0000	130.0000	375
12	94	168.7998	168.8000	375
13	125	214.7598	214.7598	500
14	125	304.5196	394.2795	500
15	125	394.2794	304.5211	500
16	125	394.2794	394.2807	500
17	220	489.2794	489.2802	500
18	220	489.2794	489.2811	500

(Continued)

Table 6. (Continued)		Power outputs		
19	242	511.2794	511.2829	550
20	242	511.2794	511.2806	550
21	254	523.2794	523.2794	550
22	254	523.2794	523.2815	550
23	254	523.2794	523.2802	550
24	254	523.2794	523.2839	550
25	254	523.2794	523.2811	550
26	254	523.2794	523.2827	550
27	10	10.0000	10.0000	150
28	10	10.0000	10.0000	150
29	10	10.0000	10.0000	150
30	47	90.9166	91.8542	97
31	60	190.0000	190.0000	190
32	60	190.0000	190.0000	190
33	60	190.0000	190.0000	190
34	90	164.7999	164.8036	200
35	90	164.7998	164.8088	200
36	90	164.7999	164.8171	200
37	25	110.0000	110.0000	110
38	25	110.0000	110.0000	110
39	25	110.0000	110.0000	110
40	252	511.2794	511.2796	550
Total power (MW)		10,500.0000	10,500.0000	
Min cost ($/h)		121,415.4881	121,415.2346	
Average cost ($/h)		121,675.3501	121,652.7238	
Max cost ($/h)		122,006.5808	121,913.4278	
Standard deviation ($/h)		128.0542	115.3685	
Average CPU time (s)		104.57	107.98	

robustness characteristic of the proposed method for ED problems. As observed from Tables 1, 3, and 6, the power output obtained by MVMO and MVMO^S are always between the minimum and maximum generator capacity limits and the total power output of generating units equals to the power load demand. It is indicated that the equality and inequality constraints always satisfy. The proposed MVMO^S provides not only better solution but also more robust than the MVMO and the difference between the maximum and minimum costs from the proposed MVMO^S is small. Table 8 shows the ratio between the standard deviation and the minimum cost obtained by MVMO^S for all systems. The ratio between the standard deviation and the minimum cost is less than 0.149%. It clearly shows that the performance the proposed MVMO^S is robust. In addition, the comparison of the total cost obtained by MVMO^S and many other methods from Tables 2, 4, 5, and 7 shows that the MVMO^S can obtain better total fuel costs and more robust than most of other reported methods. Consequently, the MVMO^S can obtain near global solution with high probability.

5.2. Disadvantages of MVMO^S

The only disadvantage of MVMO^S is computational time. The computation time of the MVMO^S is relatively high. Similar to the original MVMO, the number of iterations in MVMO^S is equivalent to the number of offspring fitness evaluations which is usually time consuming in practical applications. The computational

Table 7. Comparison of fuel cost and CPU times for 40 unit system with VPE				
Method	Min cost ($/h)	Mean cost ($/h)	Max cost ($/h)	CPU (s)
ABC (Hemamalini & Simon, 2010)	121,441.03	121,995.82	122.123.77	32.45
ACO (Pothiya et al., 2010)	121,532.41	121,606.45	121,679.64	52.45
DE (Noman & Iba, 2008)	121,416.29	121,422.72	121,431.47	72.94
IFEP (Sinha et al., 2003)	122,624.35	123,382.00	125,740.63	1,167.35
MPSO (Park et al., 2005)	122,252.27	–	–	–
APSO (Selvakumar & Thanushkodi, 2008)	121,663.52	122,153.67	122,912.40	5.05
ESO (Pereira-Neto et al., 2005)	122,122.16	122,542.07	123,143.07	0.261
SOH_PSO (Chaturvedi et al., 2008)	121,501.14	121,853.57	122,446.3	-
SA-PSO (Kuo, 2008)	121,430.00	121,525	121,645	23.89
PSO-RDL (Chen et al., 2007)	121,468.82	–	–	–
NPSO-LRS (Selvakumar & Thanushkodi, 2007)	121,664.43	122,209.32	122,981.59	20.74
ICA-PSO (Vlachogiannis & Lee, 2009)	121,422.17	121,428.14	121,453.56	139.92
QPSO (Meng et al., 2010)	121,448.21	122,225.047	–	–
DEC-SQP (Coelho & Mariani, 2006)	121,741.98	12,295.1278	–	14.26
UHGA (He et al., 2008)	121,424.48	121,602.81	–	333.68
ST HDE (Wang et al., 2007)	121,698.51	122,304.30	–	6.07
GA-PS-SQP (Alsumait et al., 2010)	121,458.14	122.039	–	46.98
DE-BBO (Bhattacharya & Chattopadhyay, 2010)	121,420.89	121,420.90	121,420.90	1.23
BSA (Modiri-Delshad & Rahim, 2014)	121,415.61	121,474.88	121,524.96	13.12
MVMO^s	121,415.23	121,652.72	121,913.43	107.98

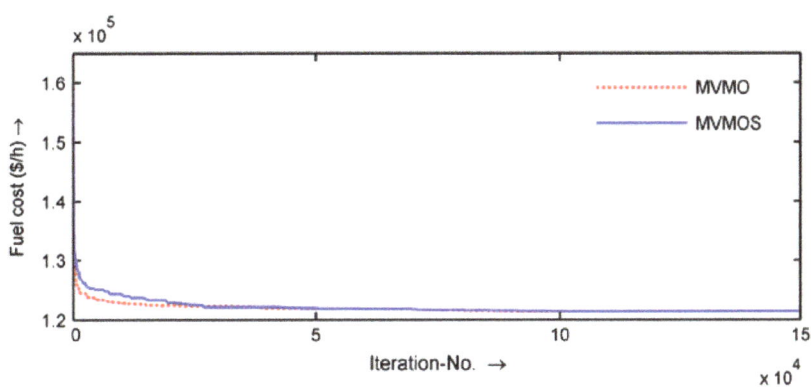

Figure 8. Convergence property of MVMO and MVMO^s for case 3 (40 units).

time of the MVMO^s is slower than the classical MVMO. This is because the MVMO^s starts the search with a set of particles while the MVMO starts the search with single particle.

6. Conclusion

This paper has presented an application of new method for solving the ED problem. The proposed MVMO^s has been successfully solved the ED problem with valve-point effects. Three test cases have been carried out to demonstrate its effectiveness and efficiency. The comparisons of numerical results have shown that the proposed MVMO^s has better performance than other optimization techniques exist in the litera-ture. It is also confirmed that the MVMO^s outperformed the classical MVMO in global search for the

System	Case 1: 3units	Case 2: 13 units		Case 3: 40 units
P_D (MW)	850	1,800	2,520	2,500
Ratio (%)	0.0	0.149	0.098	0.095

Table 8. The ratio between the standard deviation and the minimum cost obtained by MVMO^S for all systems

nonconvex problem. Therefore, the proposed MVMO^S could be favorable for solving the ED problem with valve-point effects and other nonconvex ED problems as well. In the future, the MVMO^S will be applied for solving dynamic ED, hydrothermal ED with cascaded hydro plants and emission constrained ED.

Nomenclature

N	total number of generating units, optimization variables
F	total operation cost
a_i, b_i, c_i	fuel cost coefficients of generator i
e_i, f_i	fuel cost coefficients of unit i reflecting valve-point effects
B_{ij}, B_{0i}, B_{00}	B-matrix coefficients for transmission power loss
P_D	total system load demand
P_i	power output of generator i
$P_{i, max}$	maximum power output of generator i
$P_{i, min}$	minimum power output of generator i
P_s	power output of slack unit
$P_{s, max}$	maximum power output of slack unit
$P_{s, min}$	minimum power output of slack unit
K	the penalty factor for the slack unit
P_L	total transmission loss
$iter_{max}$	maximum number of iterations
n_var	number of variable (generators)
n_par	number of particles
$mode$	variable selection strategy for offspring creation
$archive\ zize$	n-best individuals to be stored in the table
d_i	initial smoothing factor
Δd_0^{ini}	initial smoothing factor increment
Δd_0^{final}	final smoothing factor increment
$rand\ ()$	a random number in the range [0, 1]
$f_{s\_ini}^*$	initial shape scaling factor
$f_{s\_final}^*$	final shape scaling factor
D_{min}	minimum distance threshold to the global best solution
$n\_randomly$	Initial number of variables selected for mutation$i$
$ndep.runs\ m$	steps independently to collect a set of reliable individual solutions

Acknowledgment
The authors would like to sincerely thank Universiti Teknologi PETRONAS for providing the research laboratory facilities under Graduate Assistance Scheme. This research work is sponsored by the Centre of Graduate Studies with the support of the Department of Fundamental & Applied Sciences, Universiti Teknologi Petronas.

Funding
This work was supported by Graduate Assistance Scheme of Universiti Teknologi PETRONAS.

Author details
T.H. Khoa[1]
E-mail: trhkhoa89@gmail.com
ORCID ID: http://orcid.org/0000-0002-1237-0071
P.M. Vasant[1]
E-mail: pvasant@gmail.com
M.S. Balbir Singh[1]
E-mail: balbir@petronas.com.my
V.N. Dieu[2]
E-mail: vndieu@gmail.com

[1] Department of Fundamental and Applied Sciences, Universiti Teknologi PETRONAS, 31750 Tronoh, Perak, Malaysia.
[2] Department of Power Systems, HCMC University of Technology, Ho Chi Minh City, Vietnam.

References

Alsumait, J., Sykulski, J., & Al-Othman, A. (2010). A hybrid GA–PS–SQP method to solve power system valve-point economic dispatch problems. *Applied Energy, 87*, 1773–1781. http://dx.doi.org/10.1016/j.apenergy.2009.10.007

Attaviriyanupap, P., Kita, H., Tanaka, E., & Hasegawa, J. (2002). A hybrid EP and SQP for dynamic economic dispatch with nonsmooth fuel cost function. *IEEE Transactions on Power Systems, 17*, 411–416. http://dx.doi.org/10.1109/TPWRS.2002.1007911

Bhattacharya, A., & Chattopadhyay, P. K. (2010). Hybrid differential evolution with biogeography-based optimization for solution of economic load dispatch. *IEEE Transactions on Power Systems, 25*, 1955–1964. http://dx.doi.org/10.1109/TPWRS.2010.2043270

Chaturvedi, K. T., Pandit, M., & Srivastava, L. (2008). Self-organizing hierarchical particle swarm optimization for nonconvex economic dispatch. *IEEE Transactions on Power Systems, 23*, 1079–1087. http://dx.doi.org/10.1109/TPWRS.2008.926455

Chen, Y.-P., Peng, W.-C., & Jian, M.-C. (2007). Particle swarm optimization with recombination and dynamic linkage discovery. *IEEE Transactions on Systems, Man, and Cybernetics, Part B (Cybernetics), 37*, 1460–1470. http://dx.doi.org/10.1109/TSMCB.2007.904019

Chiang, C.-L. (2005). Improved genetic algorithm for power economic dispatch of units with valve-point effects and multiple fuels. *IEEE Transactions on Power Systems, 20*, 1690–1699. http://dx.doi.org/10.1109/TPWRS.2005.857924

Coelho, L. S., & Mariani, V. C. (2006). Combining of chaotic differential evolution and quadratic programming for economic dispatch optimization with valve-point effect. *IEEE Transactions on Power Systems, 21*, 989–996.

Dieu, V., Schegner, P., & Ongsakul, W. (2013). Pseudo-gradient based particle swarm optimization method for nonconvex economic dispatch. In I. Zelinka, P. Vasant, & N. Barsoum (Eds.), *Power, control and optimization* (Vol. 239, pp. 1–27). Springer International. doi:10.1007/978-3-319-00206-4_1

Duman, S., Yorukeren, N., & Altas, I. H. (2015). A novel modified hybrid PSOGSA based on fuzzy logic for non-convex economic dispatch problem with valve-point effect. *International Journal of Electrical Power & Energy Systems, 64*, 121–135.

Erlich, I., Venayagamoorthy, G. K., & Worawat, N. (2010). A mean-variance optimization algorithm. In *Evolutionary Computation (CEC), IEEE Congress*. Barcelona. Retrieved from http://ieeexplore.ieee.org/xpls/abs_all.jsp?arnumber=5586027&tag=1. doi:10.1109/CEC.2010.5586027

Fan, J.-Y., & Zhang, L. (1998). Real-time economic dispatch with line flow and emission constraints using quadratic programming. *IEEE Transactions on Power Systems, 13*, 320–325. http://dx.doi.org/10.1109/59.667345

He, D.-K., Wang, F.-L., & Mao, Z.-Z. (2008). Hybrid genetic algorithm for economic dispatch with valve-point effect. *Electric Power Systems Research, 78*, 626–633. http://dx.doi.org/10.1016/j.epsr.2007.05.008

Hemamalini, S., & Simon, S. P. (2010). Artificial bee colony algorithm for economic load dispatch problem with non-smooth cost functions. *Electric Power Components and Systems, 38*, 786–803. http://dx.doi.org/10.1080/15325000903489710

Kuo, C.-C. (2008). A novel coding scheme for practical economic dispatch by modified particle swarm approach. *IEEE Transactions on Power Systems, 23*, 1825–1835.

Le Dinh, L., Vo Ngoc, D., & Vasant, P. (2013). Artificial bee colony algorithm for solving optimal power flow problem. *The Scientific World Journal 2013*, Article ID: 159040, 9 p. doi:10.1155/2013/159040

Lin, J., Chen, C.-L., Tsai, S.-F., & Yuan, C. (2015). New intelligent particle swarm optimization algorithm for solving economic dispatch with valve-point effects. *Journal of Marine Science and Technology, 23*, 44–53.

Meng, K., Wang, H. G., Dong, Z. Y., & Wong, K. P. (2010). Quantum-inspired particle swarm optimization for valve-point economic load dispatch. *IEEE Transactions on Power Systems, 25*, 215–222. http://dx.doi.org/10.1109/TPWRS.2009.2030359

Modiri-Delshad, M., & Rahim, N. A. (2014). Solving non-convex economic dispatch problem via backtracking search algorithm. *Energy, 77*, 372–381.

Niu, Q., Zhang, H., Wang, X., Li, K., & Irwin, G. W. (2014). A hybrid harmony search with arithmetic crossover operation for economic dispatch. *International Journal of Electrical Power & Energy Systems, 62*, 237–257.

Noman, N., & Iba, H. (2008). Differential evolution for economic load dispatch problems. *Electric Power Systems Research, 78*, 1322–1331. http://dx.doi.org/10.1016/j.epsr.2007.11.007

Park, J.-B., Lee, K. S., Shin, J.-R., & Lee, K. Y. (2005). A particle swarm optimization for economic dispatch with nonsmooth cost functions. *IEEE Transactions on Power Systems, 20*, 34–42. http://dx.doi.org/10.1109/TPWRS.2004.831275

Pereira-Neto, A., Unsihuay, C., & Saavedra, O. (2005). Efficient evolutionary strategy optimisation procedure to solve the nonconvex economic dispatch problem with generator constraints. *IEE Proceedings-Generation, Transmission and Distribution, 152*, 653–660. http://dx.doi.org/10.1049/ip-gtd:20045287

Pothiya, S., Ngamroo, I., & Kongprawechnon, W. (2010). Ant colony optimisation for economic dispatch problem with non-smooth cost functions. *International Journal of Electrical Power & Energy Systems, 32*, 478–487.

Rueda, J. L., & Erlich, I. (2013). *Evaluation of the mean-variance mapping optimization for solving multimodal problems*. In *Swarm Intelligence (SIS), 2013 IEEE Symposium*. Singapore. Retrieved from http://ieeexplore.ieee.org/xpl/articleDetails.jsp?arnumber=6615153. doi:10.1109/SIS.2013.6615153

Secui, D. C. (2015). A new modified artificial bee colony algorithm for the economic dispatch problem. *Energy Conversion and Management, 89*, 43–62. http://dx.doi.org/10.1016/j.enconman.2014.09.034

Selvakumar, A. I., & Thanushkodi, K. (2007). A new particle swarm optimization solution to nonconvex economic dispatch problems. *IEEE Transactions on Power Systems, 22*, 42–51. http://dx.doi.org/10.1109/TPWRS.2006.889132

Selvakumar, A. I., & Thanushkodi, K. (2008). Anti-predatory particle swarm optimization: Solution to nonconvex economic dispatch problems. *Electric Power Systems Research, 78*, 2–10. http://dx.doi.org/10.1016/j.epsr.2006.12.001

Shahinzadeh, H., Nasr-Azadani, S. M., & Jannesari, N. (2014). Applications of particle swarm optimization algorithm to solving the economic load dispatch of units in power systems with valve-point effects. *International Journal of Electrical and Computer Engineering (IJECE), 4*, 858–867.

Sinha, N., Chakrabarti, R., & Chattopadhyay, P. (2003). Evolutionary programming techniques for economic load dispatch. *IEEE Transactions on Evolutionary Computation, 7*, 83–94. http://dx.doi.org/10.1109/TEVC.2002.806788

Vasant, P., Ganesan, T., & Elamvazuthi, I. (2012). An improved PSO approach for solving non-convex optimization problems. In *2011 9th International Conference on ICT and Knowledge Engineering (ICT & Knowledge Engineering)*. Bangkok: IEEE. Retrieved from http://ieeexplore.ieee.org/xpls/abs_all.jsp?arnumber=6152418. doi:10.1109/ICTKE.2012.6152418

Victoire, T. A. A., & Jeyakumar, A. E. (2004). Hybrid PSO–SQP for economic dispatch with valve-point effect. *Electric Power Systems Research, 71*, 51–59. http://dx.doi.org/10.1016/j.epsr.2003.12.017

Vlachogiannis, J. G., & Lee, K. Y. (2009). Economic load dispatch—A comparative study on heuristic optimization techniques with an improved coordinated aggregation-based PSO. *IEEE Transactions on Power Systems, 24*, 991–1001. http://dx.doi.org/10.1109/TPWRS.2009.2016524

Wang, S. K., Chiou, J. P., & Liu, C. W. (2007). Non-smooth/non-convex economic dispatch by a novel hybrid differential evolution algorithm. *IET Generation, Transmission & Distribution, 1*, 793–803.

Wollenberg, B., & Wood, A. (1996). *Power generation, operation and control* (pp. 264–327). New York, NY: Wiley.

11

Cognitive radio network in vehicular ad hoc network (VANET): A survey

Joanne Mun-Yee Lim[1]*, Yoong Choon Chang[2], Mohamad Yusoff Alias[3] and Jonathan Loo[4]

*Corresponding author: Joanne Mun-Yee Lim, Faculty of Engineering, SEGi University, Jalan Teknologi, Kota Damansara, Petaling Jaya 47810, Selangor, Malaysia
E-mail: jlmy555@gmail.com
Reviewing editor: Kun Chen, Wuhan University of Technology, China

Abstract: Cognitive radio network and vehicular ad hoc network (VANET) are recent emerging concepts in wireless networking. Cognitive radio network obtains knowledge of its operational geographical environment to manage sharing of spectrum between primary and secondary users, while VANET shares emergency safety messages among vehicles to ensure safety of users on the road. Cognitive radio network is employed in VANET to ensure the efficient use of spectrum, as well as to support VANET's deployment. Random increase and decrease of spectrum users, unpredictable nature of VANET, high mobility, varying interference, security, packet scheduling, and priority assignment are the challenges encountered in a typical cognitive VANET environment. This paper provides survey and critical analysis on different challenges of cognitive radio VANET, with discussion on the open issues, challenges, and performance metrics for different cognitive radio VANET applications.

Subjects: Algorithms & Complexity; Computer Science; Computer Science (General); Information & Communication Technology (ICT); Systems & Computer Architecture; Systems & Control Engineering; Technology; Transport & Vehicle Engineering

Keywords: cognitive radio network; vehicular ad hoc network (VANET); ad hoc networks; intelligent transportation system (ITS); inter-vehicle communication (IVC)

ABOUT THE AUTHOR

Joanne Mun-Yee Lim received her Bachelor of Engineering (with honors) from Monash University in 2008 and her Master of Engineering degree from University Malaya in 2011. She is also a member of the Institution of Engineers Malaysia and a professional engineer of the Board of Engineers Malaysia. Recently, she has graduated with her PhD in Engineering from Multimedia University. She joined the Department of Electrical Engineering in University Malaya as a tutor in 2010. She is currently a senior lecturer in the Department of Engineering in SEGi University. Her research interest includes vehicular ad hoc network (VANET), mobile IPv6-based networks, artificial intelligence, control systems, optimization schemes, and robotic design and applications.

PUBLIC INTEREST STATEMENT

Cognitive radio network and vehicular ad hoc network (VANET) are recent emerging concepts in wireless networking. Cognitive radio network obtains knowledge of its operational geographical environment to manage sharing of spectrum between primary and secondary users, while VANET shares emergency safety messages among vehicles to ensure safety of users on the road. Cognitive radio network is employed in VANET to ensure the efficient use of spectrum, as well as to support VANET's deployment. Random increase and decrease of spectrum users, unpredictable nature of VANET, high mobility, varying interference, security, packet scheduling, and priority assignment are the challenges encountered in a typical cognitive VANET environment. This paper provides survey and critical analysis on different challenges of cognitive radio VANET, with discussion on the open issues, challenges, and performance metrics for different cognitive radio VANET applications.

1. Introduction

Cognitive radio network emerged as the key solution to solve scarcity in spectrum. The primary idea of cognitive radio network is to allow usage of the licensed (primary) users and the unlicensed (secondary) users to share the same bandwidth without causing harmful interference. A cognitive radio device must intelligently detect the availability of spectrum and assign suitable users to occupy the spectrum bandwidth without causing interference to the authorized users (Gonzalez, Picone, & Colabrese, 2012). Cognitive radio (CR) technology is capable of making corresponding changes in certain operating parameters and adapt to its internal states by sensing its surrounding environment (Richard Yu, 2011). Cognitive radio was first proposed by Mitola and Maguire (1999) where a Radio Knowledge Representation Language with cognitive radio was proposed to actively manipulate protocol stack in order to adapt and satisfy user needs efficiently. This transforms radio nodes from being blind executors of predefined protocols into intelligent radio domain agents that actively deliver services with the realization of CR.

There are a few challenges in implementing efficient cognitive radio vehicular ad hoc network (VANET). Cognitive radio networks are motivated by studies such as Islam et al. (2008) and Datla, Wyglinski, and Minden (2009) which revealed that a large percentage of the licensed spectrum still experience lack of utilization. Issues such as common control channel, joint spectrum sensing, and cognitive implementation architecture are the open issues highlighted for cognitive radio network development work in Liang, Chen, Li, and Mähönen (2008).

On the other hand, VANET is an emerging technology which addresses issues which concern with car-to-car communication. VANET utilizes ad hoc multi-hop communication among cars with diverse mobility patterns. Vehicle-to-Infrastructure network which can also be referred to as Vehicle-to-Roadside network utilizes statically deployed Access Points or Base Stations to connect moving cars (Xu, Garrison, & Wang, 2011). The main purpose of VANET is to ensure safety on the roads. In addition to safety applications, VANET also allows users to enjoy comfort applications, such as web browsing and multimedia data downloading (Sou, Shieh, & Lee, 2011). Standards for VANET such as The Institute of Electrical and Electronics Engineers (IEEE) P1609 Wireless Access Vehicular Environment, Dedicated Short Range Communication, and IEEE 802.11p have been developed to accommodate to VANET's requirements (Wang et al., 2012).

With the growing demand for VANETs, cognitive radio network seems to be a promising solution to solve spectrum scarcity. The main challenges for cognitive radio network with VANET are to deal with high mobile nodes under dynamic channel conditions while providing fair spectrum share among nodes. In addition, varying and unpredictable nature of VANET, scheduling efficiency, security, priority assignment, and high nodes mobility are main challenges in ensuring a deployable VANET.

Framework such as Di Felice, Chowdhury, and Bononi's (2010) which allows spectrum sharing opportunities for Inter Vehicle Communication (IVC) with the use of cooperative sensing, spectrum allocation, and cooperation process is influential to promote robust primary user detection under fading conditions (Barrachina et al., 2012; Maslekar, Mouzna, Boussedjra, & Labiod, 2012).

The main contributions of this work are as follows: firstly, the overview of the advanced cognitive radio network and VANET is discussed. Secondly, the challenges and approaches toward designing cognitive radio VANET are explored. Thirdly, the open issues and research directions of cognitive radio VANET are discussed. Fourthly, the performance metrics for different cognitive radio VANET applications are analyzed.

The outline of the rest of the article is as follows. In Section 2, we discuss the overview of the current advancements of cognitive radio network and VANET. Section 3 focuses on the challenges and approaches in designing cognitive radio network in VANET. The open issues and research directions of cognitive radio VANET are discussed in Section 4. The performance metrics for different applications is discussed in Section 5. Finally, Section 6 concludes the paper.

2. Overview of advanced cognitive radio network and VANET

In order to solve spectrum starvation in VANET, cognitive radio network is a promising solution toward achieving dynamically accessed available radio spectrum in an opportunistic manner. Motivated by the above observations, this section introduces the main issues and challenges dealt by cognitive radio network and VANET and the approaches toward addressing these challenges individually as discussed in the following sections.

Figure 1 lists the challenges in cognitive radio network which include beacons, spectrum sensing, geo-location database, medium access control (MAC), and implementation of reliable test beds. On the other hand, in VANET, we encounter issues such as network selection, cross-layer design, routing, security, and implementation of reliable simulators or test beds before the actual deployment of VANET. These issues and implementation of cognitive radio VANET are discussed in detailed in the below sections.

2.1. Cognitive radio network

The main concern in cognitive radio network is to identify available free spectrum. The current spectrum managements which detect and manage the usage of spectrum are discussed in this section. The coordination of operation in cognitive radio network is mainly governed by the level of security, cross-layer design, and MAC layer. The detailed approaches and challenges toward achieving overall spectrum efficiency are discussed in the following subsections.

2.1.1. Detecting availability of a channel

In this section, ways to detect availability of a channel in a cognitive radio network are discussed. There are generally three common methods to detect availability of a channel: beacons, spectrum sensing, and geo-location database. These methods are discussed in detailed below.

2.1.1.1. Beacons. Beacons are signals which are sent to different channels to identify the channel availability. In order to transmit beacons, a communication device must have adequate radio frequency output power to disseminate the beacons to their respective locations. The main drawback of using beacons is the increase of frequency resources and additional interference sources which decrease the efficiency of the spectrum.

However, usage of beacons allows sharing of information on the spectrum availability and thus in Lei and Chin (2008), synchronization with index and payload is proposed to form an energy-efficient structure for beacon signal transmission in cognitive radio system. In Derakhshani, Le-Ngoc, and Vu (2010), beacons are used to perform study on interference caused by secondary users due to misdetection and capacity outage of the primary users in cognitive radio network.

2.1 Cognitive Radio Network
2.1.1   Detecting Availability of a Channel
2.1.1.1 Beacons *2.1.1.2 Spectrum Sensing* *2.1.1.3 Geo-Location Database*
2.1.2   Coordination of Operation in Cognitive Radio Network
2.1.2.1 Security *2.1.2.2 Cross Layer Design* *2.1.2.3 Medium Access Control (MAC)*
2.1.3 Simulation/Testbed Tools
2.2 VANET
2.2.2   Network Selection *2.2.3 Cross Layer Design* *2.2.4 Routing* *2.2.5 Security* *2.2.6 Simulators/Test beds*

Figure 1. Overview of challenges in cognitive radio network and VANET.

The different applications of beacons are explored in Vinel, Belyaev, Egiazarian, and Koucheryavy (2012), where the video-based overtaking assistant is significantly improved by exploiting information from the beacons on the position, speed, and direction of all the vehicles to ensure early detection of oncoming traffic. This ensures a warning is sent to the driver whenever needed. A simple time-based threshold policy for collective protection of primary users where the coordination of primary and secondary users is proposed to enable sharing of available opportunities is discussed in Ngoga, Yao, and Popescu (2012). In order to ensure the transmission within the given collision constraint, an analytical framework for carrier sense multiple access (CSMA)-based coexistence mechanism is integrated and the suggested model is shown to allow characterization of individual secondary users under various back off settings.

2.1.1.2. Spectrum sensing. The presence of available channels is detected in cognitive radio network with the use of spectrum sensing. If the channel is determined to be vacant and the constraints of the channel are satisfied, the secondary user occupies the respective available channel. One of the advantages of spectrum sensing in cognitive radio network is the independence of the framework quality which does not require the usage of database connection. However, the incapability of spectrum sensing to reduce false alarms due to interference reduces the efficiency of spectrum sensing.

There are two ways where spectrum sensing in cognitive radio networks could be performed: energy detection and feature detection. In Axell, Leus, Larsson, and Poor (2012) and Subhedar and Birajdar (2011), recent advances on spectrum sensing in CR such as constant false alarm rate detector, energy detection, feature detection, blind detection, multiband sensing, wideband spectrum sensing, compressive sensing, and cooperative spectrum sensing are some of the challenges and approaches discussed.

To minimize interference impact on the spectrum sensing cognitive radio networks, in Lo, Akyildiz, and Al-Dhelaan (2010), an Efficient Recovery Control Channel design is proposed to update a list of channels commonly available to neighbors. In the list of channels, each secondary user is able to efficiently establish new control channels among neighbors in response to primary user's activity changes. The proposed method is proven to be able to minimize the primary user interference.

With spectrum sensing, the presence of available channel could be detected even without the usage of database connection. A significant advantage of spectrum sensing allows spectrum sensing to be deployed in cognitive radio network. If the interference caused by devices could be resolved and the efficiency of spectrum sensing could be increased with better feature and algorithm to ease spectrum detection, spectrum sensing projects high potential to lead toward the integration of cognitive radio network.

2.1.1.3. Geo-location database. A geo-location database is necessary to determine the availability of the spectrum in their respective location to ensure operability of cognitive radio network. Once an initial access to the database is done, the available channel can be determined with the use of database enquiry. Rather than sensing spectrum or sending beacons, a database is established to respective locations to provide users an insight to the channel availability upon request. Small coverage area provides better spectrum efficiency in geo-location database framework.

In Denkovska, Latkoski, and Gavrilovska (2011), two existing approaches for television white spaces, FCC and electronic communications committee (ECC), are evaluated. It is shown that ECC is protection oriented, whereas FCC concentrates on extensive reusage of spectrum holes. Both approaches show that small coverage area of secondary networks gives significantly increased spectrum efficiency. Using geographical information, a routing protocol is also proposed in Kim and Krunz (2011). Distributed routing protocol is proposed where path selections and resource allocations are determined by receivers on a per-packet and per-hop basis. The proposed framework efficiently adapts to spectrum dynamics and node mobility.

On the other hand, in Karimi (2011), a geo-location database with detailed calculations is performed to derive location-specific maximum permitted emission levels for white space devices (WSDs) operating in Digital Terrestrial Television bands.

Geo-location database highlights the necessity of reliable information in the database to ensure smooth operation of spectrum allocation in cognitive radio network. However, the setup of a database on a wide-scale basis requires effort and cost. Geo-location database marks a leap in the cognitive radio network if the usage of database is permissible.

2.1.2. Coordination of operation in cognitive radio network

In order to ensure smooth and efficient operation in cognitive radio network, general issues such as security, cross-layer design, and MAC should be standardized and optimized to ensure efficient cognitive radio network. These issues are discussed in detailed below.

2.1.2.1. Security. Cognitive radio network is still vulnerable to attacks due to its users' mobility and open communication framework. To reduce the possibility of attacks, a centralized security can be implemented to enhance the security level in cognitive radio network. This ensures the privacy of users is well protected. A range of security breaches and attacks are discussed in Tang, Yu, Huang, and Li (2012). In order to counter Spectrum Sensing Data Falsification attacks to cognitive radio mobile ad hoc networks, a novel bio-inspired consensus-based cooperative spectrum sensing scheme is presented.

An authentication scheme using identity (ID)-based cryptography with threshold secret sharing is also proposed. In Baldini et al. (2012), a range of threats, vulnerabilities, mitigations, and protection techniques to support the feasible deployment of software-defined radio and cognitive radio, and an overview of the security threats and related protection techniques for software-defined radio and cognitive radio (CR) technologies are identified. Several security requirements such as controlled access to resources, robustness, protection of confidentiality, protection of system integrity, protection of data integrity, compliance to regulatory framework, accountability, and verification of identities are determined as the basic security standards to be achieved before cognitive radio network can be implemented. Malicious attacks reduce the reliability of spectrum sharing in cognitive radio networks.

2.1.2.2. Cross-layer design. Different layers of communication cooperate within cognitive radio network to provide accessible spectrum detection and utilization. In order to ensure the functionality of cognitive radio network, in Chen and Wyglinski (2009), a novel spectrum allocation approach which optimizes each individual wireless device and its single-hop communication links with the use of partial operating parameter and environmental information from adjacent devices within the wireless network is proposed.

The framework shows the ability to minimize the bit error rate and Out-Of-Band interference while maximizing overall throughput with the use of a multi-objective fitness function. On the other hand, to provide quality video streaming, in Tom, Li, Asefi, and Shen (2012), video streaming is formed as a cross-layer design problem where a quality of experience (QoE)-oriented video streaming framework is used. The improvement in transport control protocol (TCP) is observed in Luo, Yu, Ji, and Leung (2010) where spectrum sensing, access decision, physical layer modulation, and coding scheme and data link layer frame size are considered in a cross-layer design approach in cognitive radio networks to maximize the TCP throughput. Cross-layer in cognitive radio network in general shows an increase in performance experienced by primary and secondary users if the information from different layers of communication is collaboratively utilized.

2.1.2.3. Medium access control. MAC is the key toward enabling the deployment of technologies for dynamic spectrum allocation. To improve the accuracy of spectrum sensing and further protect the primary users, in Zhang and Hang (2010), the cooperative sequential spectrum sensing at physical layer and packet scheduling at MAC layer over wireless dynamic spectrum access networks is proposed. The Markov chain model and queuing model are used to study the proposed protocol.

To address low throughput in data transmission, a closed form expression is derived to maximize allowable power for a cognitive radio transmission based on an interference model is proposed in Salameh, Krunz, and Younis (2009). By eliminating common control channel, a MAC scheme proposed in Kondareddy (2008) explores the employment of multi-hop routing to improve the perceived throughput. The need for a common control channel is eliminated for the entire network. This eliminated the control channel saturation problem and denial of service (DoS) attacks.

Collision avoidance (CSMA/CA) is modified in Liu, Hu, Xiao, and Liu (2010), in order to make the framework feasible for cognitive radio environment. The proposed CSMA/CA is designed as such to suit the application of IEEE802.22 which suffers from uncertain spectrum and frequency width.

2.1.3. Simulation/test bed tools

In order to test the functionality of the cognitive radio network under realistic environments, several test beds are currently being implemented. In Chen, Guo, and Qiu (2011), architecture for cognitive radio network test beds with functional architecture for cognitive radio network nodes is proposed. In Qiu et al. (2010), a real-time cognitive radio test bed with considerations on the design architecture, hardware platform, and key algorithms is built. Cognitive radio network is taken into consideration for smart grid applications as well. With consideration in ECMA-392 international standard, in Franklin et al. (2010), an ultra-high frequency band cognitive radio test bed is built based on ECMA-392 international standard which defines the MAC and physical (PHY) operation for television wide space (TVWS) communication between portable devices. The test bed is verified to be operated in line with the FCC rules. In the near future, advanced test beds with complete functionalities can be used to further develop and test cognitive radio networks under realistic environment.

2.2. VANET

The recent strides made in vehicular networks allow the advancement of VANET technology to provide linkage among vehicles for safety information sharing purposes. This decreases the probability of collisions. In this section, the various VANET architectures such as IEEE802.11p standard, network selection in VANET, cross-layer design, routing, security, simulators, and test beds are discussed.

2.2.1. Network selection

To support intelligent transportation system applications, VANET for exchange of safety information between users can be implemented using networks such as wireless fidelity, WiFi IEEE802.11p, wireless access in vehicular environments (WAVE), WAVE IEEE 1609, worldwide interoperability for microwave cccess, WiMAX IEEE802.16, bluetooth, infrared, and ZigBee to ensure dynamic mobility communications. IEEE802.11p defines the advancement of IEEE802.11. An overview of the standards proposed for IEEE 802.11p is presented in Jiang and Delgrossi (2008) and IEEE Std 1609.3-2010 (2010).

The standard proposed for IEEE802.11p discusses on layers 3 and 4 of the Open System Interconnection model which includes Internet Protocol, User Datagram Protocol, and TCP elements of the Internet model. Management and data services within WAVE devices are also highlighted in the draft. The implementation of IEEE802.11p under field operational test is made possible in Hernández-Jayo et al. (2012) which supports the deployment of VANET.

Further to this standard, the implementation of IEEE802.11p has been highly researched to provide better efficiency to VANETs applications as shown in Jiang, Alfadhl, and Chai (2011) where all resources on the road side units (RSUs) are scheduled as a whole and buses are used as moving infrastructure points to reduce the burden on RSUs. Both, Variable-Length Contention Free Period for MAC and Time-To-Live values are used to guarantee data credibility on buses.

The transmission of acknowledgments for successfully received access categorized broadcast data frames allowed the transmitter to know whether the data frames are successfully received. With this information, the necessity to perform retransmission is evaluated. These are implemented in Enhanced Distributed Channel Access (Barradi, Hafid, & Gallardo, 2010). Arbitration Inter Frame Space Number parameter associated with different access categories is also implemented based on strict policies for access based on the suggested framework.

In order to increase the contention level and decrease information dissemination delay, a geo-casting packet transmission technique for IEEE802.11p is proposed to transmit safety messages in a VANET (Javed & Khan, 2011). The impact of urban characteristics, RSU deployment conditions, and communication settings on the quality of IEEE 802.11p is presented based on an extensive field test-ing campaign in Gozalvez, Sepulcre, and Bauza (2012). Results show that the streets' layout, urban environment, traffic density, presence of heavy vehicles, trees, and terrain elevation should be con-sidered when configuring urban RSUs.

A design and analysis study is conducted on providing Internet access for VANET by combining Proxy Mobile IPv6 (PMIPv6) and ETSI TC ITS Geo-Networking (GN) protocols in Sandonis, Calderon, Soto, and Bernardos (2012). Detailed performance evaluations which include the performance met-ric of packet delivery ratio, delay, handover, and overhead are analyzed. The performance metric shows that the design of PMIPv6 and ETSI TC ITS Geo-Networking (GN) protocols is feasible for em-ployment. In IEEE802.11p, the PHY and MAC layer based on existing IEEE802.11 standards are im-proved for vehicular applications. The performance of EEE802.11a-, IEEE802.11b-, and IEEE802.11 g-based VANETs is evaluated in Shen, Zou, and Liu (2009). Results showed that within certain distance, IEEE802.11a- and g-based VANET are able to achieve better stability good put, while IEEE802.11b-based VANET is better suited for long-distance communication. On the other hand, a novel adaptation of PMIPv6 for multi-hop VANET scenarios is introduced in Asefi, C'espedes, Shen, and Mark (2011) to ensure efficient network mobility management. This shows the importance of network selection in VANET toward determining the transmission efficiency.

2.2.2. Cross-layer design
The paradigm of cross-layer design in VANET has been proposed as an alternative solution toward achieving better efficiency in VANET. Cross-layer design utilizes information sharing among different layers to achieve robust protocols. In Jarupan and Ekici (2011), the architecture of cross-layer de-signs has been discussed as an alternative method to allow information to be shared across layer boundaries in order to create robust and efficient protocols. Several open research problems for, e.g., striking the balance between modularity and cross-layer design, system stability, realistic physical layer and mobility modeling, and standardization of cross-layer designs are listed.

In order to recover broken links in multicasting VANET, in Chang, Liang, Lai, and Wang (2012), a cross-layer uni-cast-type multi-hops local repair approach is proposed to increase the network throughput. Results show better successful repair rate, control message overhead, packet delivery ratio, and network good put. A cross-layer design which jointly optimizes routing and MAC functions is proposed in Jarupan (2009) to discover stable and delay efficient routes for highly dynamic com-munication networks. Various packet traffic statistics which are collected by the MAC layer from different locations in the networks are used to determine small end-to-end delay routes.

2.2.3. Routing
The process of selecting path in VANET is crucial to determine the reliability of information dissemi-nation. In Fonseca and Vaz~ao (2012), various existing position-based routing protocols and its ap-plicability to different environments are studied. The studies are characterized into vehicular network environment, topology-based protocols, and position-based protocols. One of the routing

methodologies suggested in Oliveira et al. (2012) involves information being obtained from mobility models. This information is used in practical routing algorithms to improve path duration. For high-vehicles density area, the packets are routed over the oldest links created by vehicles which are moving in the same direction. For small-vehicles density area, routings are accomplished according to the most recent links created by vehicles moving both directions. A geographic stateless routing combined with node location and digital map can also be used to provide high packet ratio with comparable latency to other geographic routing schemes (Xiang, Liu, Liu, Sun, & Wang, 2012).

A routing protocol where data gathered through passive mechanisms to calculate the reliability scores for each street edge to select the most reliable route can be considered for design of VANET routing (Bernsen & Manivannan, 2012). Routing performance in a one-way-multi-lane scenario can be enhanced by having each candidate to self determine its own priority using node degree, expected transmission count, and link lifetime (Wang & Lin, 2012). Packet delivery ratio and throughput can be increased if the individual routes through a reliable, stable, and durable routing path.

To increase delivery probability and reduced the delivery delay, a vehicle delay tolerant network routing protocol can be designed to make routing decisions based on geographical data and by combining the multiple-copy and single-copy (Soares, Rodrigues, & Farahmand, 2011). To securely and efficiently accomplish route optimization procedures in VANET, a Batch Binding Update Scheme can be used (Yeh, Yang, Chang, & Tsai, 2012). To facilitate video frame distortion and streaming start-up delay, an application-centric routing framework for real-time video transmission over urban multi-hop VANET can be employed (Asefi, Mark, & Shen, 2011). Queuing can be made based on mobility model, spatial traffic distribution, and probability of connectivity for sparse and dense VANET scenarios. Most routing algorithms used only one network path at a time. A multi-path routing can be considered to enable multi-alternative paths in VANETs.

2.2.4. Security
Since VANET involves information sharing among unknown users on the road, security becomes a main concern in ensuring safety in VANET. Thus, in Hu, Chen, and Li (2012), ring signature technique is used to perform an efficient multi-level conditional privacy preservation authentication protocol in VANET. This method offers conditional privacy preservation authentication with multi-level counter measure. However, ring signature technique increases the communication overhead directly proportional to the number of ring members. Therefore, a SMART protocol which employs data aggregation tool with message fragmentation to achieve efficient bandwidth usage is suggested in Nair, Soh, Chilamkurti, and Park (2012).

The data are first broken down and analyzed to ensure unnecessary data are eliminated. Data analyzed are stored in every node to create a well-organized data structure. In order to solve the inefficient deployment and potential RSU emulation attacks, a study has been conducted to study the impact of the inter-road side unit interference on beacon broadcasting (Ganan et al., 2012). It is shown that the broadcast performance drops when a vehicle is in range of four RSUs that are hidden from one another.

This proves the incapability of current IEEE 1609.4 medium access technique to cope with the interference caused by overlapping RSUs. This could be solved by identifying road side units emulation attack (REA) attackers and excluding them from the network. In general, security in VANET is still vulnerable to malicious attacks. To become a real technology that guarantees public safety, issues such as authentication and key management, privacy, secure positioning, and threat model are to be explored.

2.2.5. Simulators/test beds for VANET
VANET simulators are fundamentally different from mobile ad hoc network (MANET) simulators. In VANET, roadside obstacles, traffic flow models, trip models, varying vehicular speed and mobility, traffic conditions, tc. are the main considerations in designing VANET framework. Simulators for

VANET are still undergoing further development and in Francisco, Toh, Cano, Calafate, and Manzoni (2009), various simulators are compared and analyzed for research purposes.

Mobility generators such as simulation of urban mobility (SUMO), MOVE, CityMob, FreeSim, Street Random Waypoint, Netstream, VanetMobiSim, Network Simulator - 2 (ns-2), global mobile information system simulator (GloMoSim), Java in Simulation Time/Scalable Wireless Adhoc, and Sensor Nodes are found to have good software features and traffic model support. However, all networks simulators described above do not specifically address VANET scenarios and requirements. In terms of VANET simulators, Traffic and Network Simulation Environment, GrooveNet, Estinet Network Simulator, and Emulator and Realistic Simulator for VANET are widely accepted and commonly used to support VANET research.

On the other hand, test beds for VANET applications are currently being developed extensively to suit the needs of research and development. In Amoroso, Marfia, Roccetti, and Pau (2012a), preliminary results that assess the feasibility of VANET system taken from real experiments performed on Los Angeles freeways and roads in August 2011 are analyzed, whereas in Amoroso, Marfia, Roccetti, and Pau (2012b), preliminary results provided from a set of experiments on a vehicular highway accident warning system are used for verification purposes. A vehicular test bed called University of California at Los Angeles Campus Vehicular Testbed (UCLA C-VeT) campus test bed developed to validate the models and protocols before being deployed in a large scale proves to be useful for framework verification (Gerla, Weng, Giordano, & Pau, 2012).

The concept uses two case studies carried out in the UCLA open vehicular test bed which validates the model by experimenting through emulation or simulation and performed parallel experiments to compare different network configurations under the same mobility and external interferences. It is suggested that the spectrum sensing performance with cognitive radio is to be implemented in the near future. A design framework which considers roadside infrastructure as well as emergency vehicle warning system which utilizes the IVC to signal other vehicles about the route information is suggested in Buchenscheit, Schaub, Kargl, and Weber (2009). The prototype has been tested under a real environment with emergency vehicles and traffic light. VANET simulators and test beds provide a good simulation and testing environment for VANET deployment.

3. Designing cognitive radio in VANET: challenges and approaches

VANET deployment faces bandwidth allocations challenges due to the random number of users in the application. As such, the implementation of cognitive radio network technology in VANET has been explored to further enhance the spectrum distribution model in VANET. Recent works on cognitive radio network in VANET as shown in Figure 2 are discussed in this section.

3.1. Cognitive radio with VANET—A general idea (IEEE standards)

VANET is a special class of MANET, with nodes in VANET generally representing highly mobile vehicles. Cognitive radio network on the other hand is a method which addresses the spectrum scarcity in the network. While mobile nodes move in a random manner in VANET, spectrums are being utilized in a high density environment. A general idea of how to incorporate cognitive radio network with VANET is discussed in this section. A large amount of spectral congestion due to high vehicle density might affect the performance of the network. Therefore, in Kirsch and O'Connor (2011), a

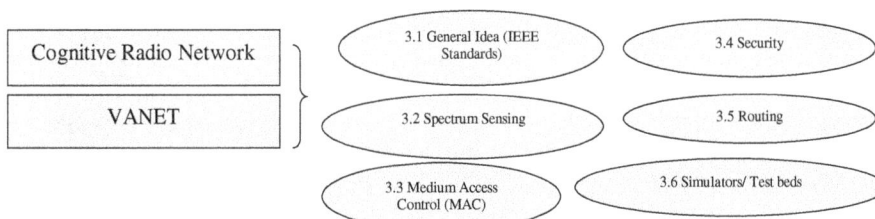

Figure 2. Current advancements of cognitive radio with VANET.

cognitive radio system is proposed to spatially and temporarily add additional radio channels to VANET when there is a high vehicle density. This framework allows high priority safety messages and secondary VANET applications to be transmitted successfully without much delay and with increased performance. A distributed channel coordination scheme that exploits the data transmission rate and the range of various frequencies is proposed for vehicle-to-vehicle communication (Tsukamoto, Matsuoka, Altintas, Tsuru, & Oie, 2009). A channel utilization model which utilizes each channel changes temporally and spatially for both primary and secondary usage is also developed. Even under temporal and spatial changes, the proposed scheme is able to utilize the unused frequency reliably.

Cognitive network principles are used to increase spectrum allocated to the control channel of WAVE protocols (Fawaz, Ghandour, Olleik, & Artai, 2010). The proposed system employs data sent by the cars to road side units to forward the aggregated data to a processing unit which created the data contention locations and generated spectrum scheduled to be dispatched to the passing cars. Advanced wireless architectures and its applications on vehicle networks are addressed in cognitive radio by seeking better spectrum reuse via Peer-to-Peer (P2P), ad hoc, and multi-hop solutions (Gerla & Kleinrock, 2011).

Negative impacts of cognitive radio technology in vehicular networks include the consequences of users not detecting a primary (licensed) user and thus causing interference to the user and the modulation techniques based on system signal to interference and noise ratios are discussed in Nyanhete, Mzyece, and Djouani (2011). To improve the performance of cognitive radio technology in vehicular networks, in Kakkasageri and Manvi (2011), a model which uses cognitive agent concept for realizing intelligent information dissemination is proposed.

In order to minimize channel allocation time and management overhead, the limited bandwidth allocated to a region is divided into prefixed overlapping spatial clusters, whereas the channel in each cluster is divided into time slots in Tomar and Verma (2011). These time slots are allocated to vehicles according to the priority of request and the availability of the channel. The contention delay experienced by cars can be monitored on a control channel (Ghandour, Fawaz, & Artail, 2011). If the contention delay exceeds a delay threshold, the RSU increases the spectrum allocation to the control channel using cognitive network, whereas if the contention delay is measured below delay threshold, the measured values are used as reference input for the controller. Dedicated protocols and frequency resources show the potential of cognitive radio network in VANET.

3.2. Spectrum sensing in cognitive radio VANET
Cognitive radio network spectrum sensing in VANET is capable of effectively detecting other transmissions, and identifying their current transmission data, and location for efficient spectrum allocation to take place. In Xiao Yu Wang and Pin-Han Ho (2010), a novel framework of coordinated spectrum sensing in cognitive radio-enhanced vehicular ad hoc networks (CR-VANETs) is introduced. The proposed sensing coordination framework uses spectrum sensing architectures, stand-alone, and cooperative sensing to achieve better sensing accuracy, efficiency, and fairness. The spatial correlation can be utilized using message passing among neighboring vehicles in order to perform collaborative spectrum sensing (Li & Irick, 2010). This is achieved using belief propagation (BP) which is applied to perform distributed observation and exploit the redundancies in both space and time.

In VANET, differing densities of primary users are common. In Pu Wang, Jun Fang, Ning Han, and Hongbin Li (2010), the channels between the primary user, cognitive user, and the variance of the noise seen at the cognitive user are used to develop a generalized likelihood ratio test to detect the presence/absence of the primary user, with the assumption that no prior knowledge of the primary user's signaling scheme. Since speed of detection is important and detection based on small number of samples is beneficial for vehicular applications, the proposed method is advantageous for VANET applications. Wavelet transform and wavelet packets can also be used in cognitive radio applications with vehicular communication for spectrum sensing and adaptive multicarrier transmission in order to solve channel impairments, especially in high relative velocity communication peers (Maurizio & Vlad, 2011).

A cognitive vehicle-to-vehicle framework which leverages the cooperation among vehicles in order for the vehicles to be aware of the spectrum conditions on future positions along its path is proposed in Di Felice et al. (2010). A collaborative sensing and decision algorithm can be designed to enable vehicles to share spectrum information and to know the spectrum availability in advance along their motion paths to increase sensing accuracy (Felice, Chowdhur, & Bononi, 2011). To improve spatial reuse and avoid interference in vehicular networks, a cognitive channel hopping (CCH) protocol is proposed to select channels based on channel quality measurements in order to significantly improve the network performance (Choi, Im, Lee, & Gerla, XXXX).

A framework with three components: opportunistic access to shared use channels, reservation of exclusive use channel, and cluster size control, is developed for a channel access management framework for cluster-based communication among vehicular nodes in Niyato, Hossain, and Wang (2011). This framework is developed to maximize the utility of the vehicular nodes in the cluster, to minimize the cost of reserving exclusive use channel, and to meet the QoS requirements of data transmission and the constraint on probability of collision with licensed users. To address the opportunistic channel access and joint exclusive use channel reservation plus cluster size control, two constrained Markov decision process (CMDP) formulations are used.

A database-assisted sensing using parameters such as vehicular density, base station radio coverage, and the trade-offs between local and database-assisted sensing analysis is proposed in Doost-Mohammady and Chowdhury (2012) to minimize the cost of operation and limiting the resulting errors in spectrum detection. A spectrum sensing using energy detection over Gamma-shadowed Nakagami-mcomposite fading channel can be used to overcome small- and large-scaled fading (Rasheed & Rajatheva, 2011).

Throughput can be used as the performance metrics for spectrum sensing. Throughput maximization problem in cognitive VANET is investigated in Pan, Li, and Fang (2012). Using the features of cooperative communications and the availability of licensed spectrum, the link is classified into cooperative links and general links. Depending on the available bands, extended link-band pairs are developed to form a three-dimensional cooperative conflict graph. The performance of link scheduling with appropriately selected transmission mode is better than purely relying on one transmission mode. In general, a central database can be used with spectrum sensing to further enhance the performance of cognitive radio networks in VANET.

3.3. Medium access control (MAC) protocols for cognitive radio VANET
Medium access control addresses channel control mechanisms in cognitive radio VANET. A multichannel MAC design which supports concurrent transmissions by allocating the channel with every beacon interval may not be feasible in a fast-fading VANET environment. A MAC design based on opportunistic spectrum access that selects channel at every individual transmission cannot provide fair share of spectrum among devices.

Therefore, in Chung, Yoo, and Kim (2009), a cognitive MAC for VANET which employs both, long-term and short-term spectrum access, provides fair share and exploits multi-user diversity while achieving high throughput. A MAC for WAVE is proposed in Shah, Habibi, and Ahmad (2012) to improve the channel utilization and reliability of safety messages by adopting EDCA and cognitive radio concept. The quality of channel can be accessed prior to transmission by employing dynamic channel allocation and negotiation algorithm using an efficient multichannel QoS cognitive MAC (MQOG) (Ajaltouni, 2012). To achieve significant increase in channel reliability, throughput, and delay while simultaneously addressing quality of service, an efficient MAC protocol for cognitive radio VANET should be implemented.

3.4. Security in Cognitive Radio VANET

VANET provides access for random users to communicate and share safety information on the road. Cognitive radio network provides efficient spectrum sharing among vehicle users. Therefore, security in cognitive radio network VANET is the key toward making cognitive radio network in VANET a reality. In Muraleedharan and Osadciw (2009), information disseminated using distributed sensor technology is implemented as part of the cognitive security protocol for VANET. This protocol ensures reliability and optimality of the protocol by employing a distributed sensor technology while prioritizing prevention of data aging, efficient quality of service (QoS), and robustness against denial of service (DoS) attacks.

VANET in general is vulnerable to DoS attacks. Jamming attacks interfere with legitimate wireless communications and degrade the overall quality of service (QoS) of network. A jammer transmits only when valid radio activity is signaled from its radio hardware to detect a particular class of jamming attack (Hamieh, Ben-Othman, & Mokda, 2009). The presence of jamming has been successfully detected using the proposed model. Security in cognitive radio VANET is still scarce and thus, the potential of further advancement in order to have a standardized security framework to ensure the reliability of the deployment of cognitive radio network in VANET should be developed.

3.5. Routing in cognitive radio VANET

In cognitive radio VANET, by employing different allocations of spectrum in the network, safety messages are routed across the network to reach other designated vehicles. The problem of routing in cognitive radio network which revolves around the identification and maintenance of the optimal path from source to destination using the available common channel is addressed in Barve and Kulkarni (2012).

The characteristics features limiting factors of the existing routing protocols are also investigated in Kim, Oh, Gerla, and Lee (2011), where the geographical location and sensed channel information are used to perform cognitive routing, yielding in a drastic improvement in throughput. The proposed protocol used unlicensed band and operates in an ad hoc, multi-hop mode which is different from the conventional cognitive radio strategies. In Liu, Ren, Xue, and Chen (, 2012), an expected path duration-maximized routing (EPDM-R) algorithm is developed in cognitive radio VANET (CR-VANET) where the expected link duration for each link is calculated.

Consequently, the maximum bottleneck algorithm is used to solve for the route that achieves the longest expected path duration. The proposed algorithm is shown to have larger average path duration than the Dijkstra-based schemes. A cross-layer channel assignment and routing algorithm which addresses the interference avoidance issues in cognitive radio network and maximizes the throughput by making use of the adjacent hop interference information are proposed in Zhan, Ren, Zhang, and Li (2012). In the above-mentioned techniques, routing in cognitive radio VANET commonly deals with achieving less delay and high throughput in communication. Further research on routing topologies to suit the deployment of cognitive radio VANET holds a great potential for further research investigation.

3.6. Simulators/test beds for cognitive radio VANET

To perform realistic experiments with the use of a few real vehicular resources in order to test new strategies and performances of VANET protocols, a test bed is proposed in Marfia et al. (2011). VANET protocols are tested as a function of different frequencies and interface switching delays in order to deal with scarcity of spectrum in dynamic environments. A set of preliminary results based on a highway accident warning system and a cognitive network using Microsoft Software Radio (SORA) technology is also presented. Test beds for VANET application hold a promising field for further research. Meanwhile, for cognitive radio VANET research purposes, simulators for VANET as discussed in section 2.2.5 can be used to perform simulations on cognitive radio VANET to address the open issues and research directions of cognitive radio VANET.

4. Designing cognitive radio network in VANET: open issues and research directions

In Steenkiste, Sicker, Minden, and Raychaudhuri (2009), several open issues and features of cognitive radio networks such as spectrum sensing, policies specifying how the radio can operate, and physical limitations of radio operation, databases configuration, radio self-configuration, adaptive algorithms, and security issues are addressed as part of the future directions of cognitive radio network.

Several characteristics and features of cognitive radio VANET are also discussed in Di Felice, Doost-Mohammady, Chowdhury, and Bononi (2012). These characteristics and features include integration with spectrum databases, impact of mobility, roles of cooperation, and presence of a common control channel. Challenges with vehicular communication based on IEEE 802.11 such as vehicular speed, distance, handover and mobility management, and unique multi-hop inter-vehicular communication are specified in Saeed, Naemat, Aris, Mat Khamis, and Awang (2010). Further research work, open issues, and research directions are discussed in this section where insights on the possible research directions are identified. Figure 3 shows the summary of the open issues and directions of cognitive radio VANET. The open issues and research directions are discussed in detailed in the following sections.

4.1. Coordination between licensed networks users and unlicensed users

One of the main issues in establishing cognitive radio network among VANET vehicles is to guarantee a smooth coordination among licensed network users. Secondary users in cognitive radio network in VANET should not interrupt the utilization of licensed networks users (primary users). To ensure efficient operation among cognitive users, a licensee can employ cognitive radio network internally within its own network. This method increases the efficiency of radio networks usage. On the other hand, the secondary users can utilize multiple licensed services by proper scheduling. Mutual interference can be avoided with efficient coordination.

4.2. Duration of spectrum opportunities

Random switching of spectrum affects the performance of a communication. Therefore, a vehicle user should be able to choose the spectrum which guarantees the best connectivity within the longest time frame. A database can be established in each cluster to maintain the current information of each spectrum. With database application, before the allocation of spectrum to vehicles can be established, the database provides insight into the best, nearest, and longest duration of spectrum opportunities for secondary users to evaluate and select the best spectrum to establish connection. This guarantees spectrum utilization as well as the duration of spectrum availability and efficient communication connection.

4.3. Random movements of vehicles

Vehicles in VANET move in a random manner. Therefore, the possibility of a vehicle to stay put in the same location is minimal. The duration of spectrum usage is generally short and thus, spectrum switching occurs frequently. This random movement of VANET vehicles increases the challenges of implementing cognitive radio network with VANET. VANET users must be able to detect the best available spectrum and establish a quick connection before the vehicle moves out from the cluster. The vehicle should also be able to detect and make connections with the next spectrum before the vehicle moves out from the cluster. The handover process that takes place should not imply a huge

Open Issues and Research Directions	
4.1	Coordination between Licensed Networks Users and Unlicensed Users
4.2	Duration of Spectrum Opportunities
4.3	Random Movements of Vehicles
4.4	Multiple Points of Observation
4.5	Interference
4.6	Delay
4.7	Security

Figure 3. Open issues and research directions.

effect on the communication links. The handover scheme in VANET holds a promising research possibility to ensure the viability of the implementation of cognitive radio network in VANET.

4.4. Multiple points of observation

Cognitive radio network (Kaur, 2014) offers multiple point spectrum observation for VANET users. Due to the ever-changing environment, the information obtained from each spectrum cannot be guaranteed as clusters are located near each other and the possibility of interference is relatively high. These uncertain outdoor factors cause false alarms in VANET where conditions such as available spectrums are not detected and non-idle spectrums are selected due to false data retrieved. Cognitive radio network in VANET is severely affected if the wrong dissemination of information is circulated. To avoid false alarms in multiple points of spectrum observation, a framework that ensures the reliability of data dissemination on each spectrum should be designed to ensure the efficiency and feasibility of cognitive radio network in VANET. A proper selection mechanism based on parameters which reflect the conditions of the current cognitive radio network conditions as well as the current VANET users locations should be integrated.

4.5. Interference

With primary and secondary users, interference between cognitive radio networks is the main issues to be faced. One of the solutions proposed in Kim and Gerla (2012) suggests a cognitive multi-channel, multi-radio multi-cast protocol known as CoCast. CoCast is developed to overcome interference issues between Wi-Fi users as well as inter-vehicle users. Two additional features such as parallel frame transmission over orthogonal frequency division multiplexing (OFDM) sub-channels and network coding are used to exploit spectral diversity and sub-channel frames, respectively. Results show that the external interference effectively reduces in a multicast scenario, especially in a channel environment where channels overlap in frequency. With vehicles in VANET moving in a random manner and multiple points of spectrum observation (Lim, Chang, Alias, & Loo, 2015; Lim, Chang, & Yusoff, 2016), interference is an avoidable issue and thus, a proper framework as discussed in Zhao, Zhu, Chen, Zhu, and Li (2013) and Jia, Lu, and Wang (2014) should be designed to ensure secondary users can transmit under minimal interference environment without disturbing the physical process of adjustment in primary users.

4.6. Delay

VANET is initially developed to ensure safety messages can be sent across vehicles in VANET to minimize collisions and promote safety on the road. These safety messages must be transmitted without further delay. Any overdue safety messages losses its functionalities. Under cognitive radio network, spectrum sharing and spectrum allocation should be deployed under specified delay constraints. These delay constraints become stringent, especially in VANET applications. Therefore, a suitable framework should be employed to ensure delay in cognitive radio VANET is minimal. Parameters such as geo-location of VANET vehicles can be employed to ensure if the locations of VANET vehicles are identified under cognitive radio network framework and thus, proper spectrum allocation can be made to minimize transmission delay.

4.7. Security

In cognitive radio VANET, vehicles connect to an available spectrum and communicate from one vehicle to another vehicle randomly. This framework leads toward the vulnerability of the primary and secondary users to succumb to different malicious attacks (Mohammed & Al-Daraiseh, 2014). Attacks such as DoS and jamming should be identified before transmission begins to ensure safety of users are not violated. The detection and prevention of attacks in cognitive radio VANET should be properly designed to ensure the feasibility of cognitive radio VANET deployment in a real-time framework.

5. Performance metrics

In this section, some of the performance metrics used to evaluate cognitive radio VANET are discussed. In Section 5.1, the common performance metrics are discussed. In section 5.2, with the use of Table 1, the performance metrics are listed according to their applications.

5.1. List of performance metrics

(a) *Throughput* - Throughput is used to measure the average packet transmitted per second.

(b) *End-to-End Delay* - End-to-end delay is used to measure the time it takes to get a packet transmitted from the source to the destination node.

(c) *Transmission Overhead* - Transmission overhead measures the excess time/excess data as compared to the normal packet transmission.

(d) *Number of Messages per Node* - Number of messages per node measures the number of messages a node received/sent.

(e) *Minimum Number of Hops* - Minimum number of hops measures the minimum number of routers/devices a packet goes through before reaching its destination node.

(f) *Beacon Interval* - Beacon interval measures the total time required for a beacon to get transmitted.

(g) *Beacons per second* - Beacons per second measure the number of beacons sent/broadcasted over a period of time in seconds.

(h) *Density Estimation* - Density estimation measures/predicts/estimates the node density of a certain area defined by user.

(i) *Forwards per Route* - Forwards per route measure the number of forwarded packet per path/route.

(j) *Warning Notification Time* - Warning notification time measures the time it takes to notify or warn a user.

(k) *Path Duration* - Path duration measures the amount of time it takes to send/route a packet.

(l) *Packet Success Rate* - Packet success rate measures the percentage of a successfully transmitted packet for every attempt the packet is transmitted.

(m) *Spectrum/Bandwidth Utilization* - Spectrum or bandwidth utilization measures the percentage of spectrum/bandwidth used by the user on average.

(n) *Computational Time* - Computational time measures the complexity of the scheme used.

Table 1. Performance metrics	
Applications	**Performance metrics**
Cognitive Radio	
Beacons	a, b, c, d, e, f, g, h, i
Spectrum Sensing	a, b, c, h, m, n, o, p, r
Geo-Location Database	h, k, n, p
Cross-Layer Design	a, b, c, l, n
MAC	a, b, c, d, e, f, g, h, I, j, k, l, m, n, o, p, q, r
Test Beds	a, b, c, d, e, f, g, h, I, j, k, l, m, n, o, p, q, r
VANET	
Network Selection	a, b, c, d, e, f, g, h, I, j, k, l, m, n, o, p
Cross-Layer Design	a, b, c, l, n
Routing	a, b, c, d, e, h, I, k, l, n, p, q
Security	a, b, c, j, l, n, p, q, r
Simulators/Test beds	a, b, c, d, e, f, g, h, I, j, k, l, m, n, o, p, q, r

(o) *Transmitted/Received Power* - Transmitted power measures the amount of power used to send a packet from source to destination node, whereas received power measures the amount of power when the packet reaches its destination node.

(p) *Probability of Detection* - Probability of detection measures the probability of a user getting detected.

(q) *Recover Time* - Recovery time measures the time it takes to recover a system.

(r) *False Alarm Probability* - False alarm probability measures the probability of getting false warnings or alarms.

5.2. Performance metrics and its applications

In this section, the performance metrics are listed according to its applications. In Table 1, the applications are divided into cognitive radio and VANET. The sub-applications are listed below. Table 1 is listed according to Figure 1 as discussed in the previous sections.

6. Conclusion

As described in this paper, despite the fact that considerable amount research work has been carried out in cognitive radio VANET, many practical problems still exist, which are yet to be solved. Several highlighted problems such as coordination between licensed and unlicensed users, duration of spectrum opportunities, multiple points of observation, interference, delay, and security are the main concerns in cognitive radio VANET. In addition, the current cognitive radio VANET uses mainly static networking parameters, even though cognitive radio network VANET is not static, and constantly exposed to ever-changing nature, fluctuating interference, high mobility, improper scheduling, and security breach. The lack of adaptivity and optimization of the current cognitive radio VANET makes cognitive radio VANET less efficient in terms of communication. Currently, cognitive radio VANET still has many loose ends which prevent the proper deployment of the framework. In order to ensure the applicability of cognitive radio VANET, proper framework which addresses the issues discussed in this paper should be developed. We are currently developing an integrated model which targets the deployment of a proper database. The database proposed uses parameters such as the locations of VANET vehicles, power model, and signal-to-noise ratio to address the issue of spectrum allocation in cognitive radio VANET. The proposed integrated model focuses on a make-before-break concept to ensure delay does not limit the transmission quality during a spectrum handover.

Funding
This research was funded in part by Motorola Solutions.

Author details
Joanne Mun-Yee Lim[1]
E-mail: jlmy555@gmail.com
Yoong Choon Chang[2]
E-mail: ycchang@utar.edu.my
Mohamad Yusoff Alias[3]
E-mail: yusoff@mmu.edu.my
Jonathan Loo[4]
E-mail: J.Loo@mdx.ac.uk
[1] Faculty of Engineering, SEGi University, Jalan Teknologi, Kota Damansara, Petaling Jaya 47810, Selangor, Malaysia.
[2] Lee Kong Chian Faculty of Engineering & Science, Universiti Tunku Abdul Rahman Sungai Long Campus, Jalan Sungai Long, Bandar Sungai Long, Kajang 43000, Selangor, Malaysia.
[3] Faculty of Engineering, Multimedia University, Persiaran Multimedia, Cyberjaya 63100, Selangor, Malaysia.
[4] School of Science and Technology, Middlesex University, London NW44BT, UK.

References
Ajaltouni, H. E. (2012). *An efficient QoS MAC for IEEE802.11p over cognitive multichannel vehicular networks* (thesis). University of Ottawa.
Amoroso, A., Marfia, G., Roccetti, M., & Pau, G. (2012a). To live and drive in L. A.: Measurements from a real intervehicular accident alert test (pp. 328–332). Conference on wireless communications and networking, Paris.
Amoroso, A., Marfia, G., Roccetti, M., & Pau, G. (2012b). *Creative testbeds for VANET research: A new methodology* (pp. 477–481). Conference on consumer communications and networking, Las Vegas.
Asefi, M., C'espedes, S., Shen, X., & Mark, J. W. (2011). *A seamless quality-driven multi-hop data delivery scheme for video streaming in urban VANET scenarios* (pp. 1–5). Conference on communications, Kyoto.
Asefi, M., Mark, J. W., & Shen, X. (2011). *An application-centric inter-vehicle routing protocol for video streaming over multi-hop urban VANETs* (pp. 1–5). Conference on communications, Kyoto.
Axell, E., Leus, G., Larsson, E. G., & Poor, H. (2012). Spectrum sensing for cognitive radio : State-of-the-art and recent advances, IEEE Signal Processing Magazine, *29*, 101–116. http://dx.doi.org/10.1109/MSP.2012.2183771
Baldini, G., Sturman, T., Biswas, A. R., Leschhorn, R., G'odor, G., & Street, M. (2012). Security aspects in software defined radio and cognitive radio networks: A survey and a way ahead. *IEEE Communications, Surveys & Tutorials, 14,* 355–379.

Barrachina, J., Garrido, P., Fogue, M., Martinez, F. J., Cano, J. C., Calafate, C. T., & Manzoni, P. (2012). VEACON: A vehicular accident ontology designed to improve safety on the roads. *Journal of Network and Computer Applications, 35,* 1891–1900.
http://dx.doi.org/10.1016/j.jnca.2012.07.013

Barradi, M., Hafid, A. S., Gallardo, J. R. (2010). *Establishing strict priorities in IEEE 802.11p WAVE vehicular networks* (pp. 1–6) Conference on IEEE global telecommunications, Miami.

Barve, S. S., & Kulkarni, P. (2012). A performance based routing classification in cognitive radio networks. *International Journal of Computer Applications, 44,* 11–21.
http://dx.doi.org/10.5120/6370-8762

Bernsen, J., & Manivannan, D. (2012). RIVER: A reliable inter-vehicular routing protocol for vehicular ad hoc networks. *Computer Networks, 56,* 3795–3807.
http://dx.doi.org/10.1016/j.comnet.2012.08.017

Buchenscheit, A., Schaub, F., Kargl, F., & Weber, M. (2009). *A VANET-based emergency vehicle warning system* (pp. 1–8). Conference on vehicular networking, Tokyo.

Chang, B. J., Liang, Y. H., Lai, J. T., & Wang, J. W. (2012). *Cross-layer-based real-time local repair for multicasting in distributed vehicular ad hoc networks* (pp. 1–6). Conference on world telecommunications congress, Beijing.

Chen, S., & Wyglinski, A. M. (2009). Efficient spectrum utilization via cross-layer optimization in distributed cognitive radio networks. *Computer Communications, 32,* 1931–1943.
http://dx.doi.org/10.1016/j.comcom.2009.05.016

Chen, Z., Guo, N., & Qiu, R. C. (2011). *Building a cognitive radio network testbed* (pp. 91–96). In Proceedings of IEEE southeastcon, Nashville.

Choi, B. S. C., Im, H., Lee, K. C., & Gerla, M. (XXXX). CCH: Cognitive channel hopping in vehicular ad hoc networks, Conference on vehicular technology, *201,* 1–5.

Chung, S., Yoo, J., Kim, C. (2009). *A cognitive MAC for VANET based on the WAVE systems* (pp. 41–46). Conference on advanced communication technology, Phoenix Park.

Datla, D., Wyglinski, A. M., & Minden, G. J. (2009). A spectrum surveying framework for dynamic spectrum access networks. *IEEE Transactions on Vehicular Technology, 58,* 4158–4168.
http://dx.doi.org/10.1109/TVT.2009.2021601

Denkovska, M., Latkoski, P., & Gavrilovska, L. (2011). *Geolocation database approach for secondary spectrum usage of TVWS* (pp. 369–372). Conference on 19th telecommunications forum, Las vegas.

Derakhshani, M., Le-Ngoc, T., Vu, M. (2010). *Interference and outage analysis in a cognitive radio network with beacon* (pp. 261–264). Conference on 25th biennial symposium on communications, Stockholm.

Di Felice, M., Chowdhury, K. R., & Bononi, L. (2010). *Analyzing the potential of cooperative cognitive radio technology on inter-vehicle communication* (pp. 1–6). Conference on wireless days, Venice.

Di Felice, M., Doost-Mohammady, R., Chowdhury, K. R., & Bononi, L. (2012). Smart radios for smart vehicles: Cognitive vehicular networks. *IEEE Vehicular Technology Magazine, 7,* 26–33.
http://dx.doi.org/10.1109/MVT.2012.2190177

Doost-Mohammady, R., Chowdhury, K. R. (2012). *Design of spectrum database assisted cognitive radio vehicular networks.* Conference on cognitive radio oriented wireless networks, Sydney.

Fawaz, K., Ghandour, A., Olleik, M., & Artai, H. (2010). *Improving reliability of safety applications in vehicle ad hoc networks through the implementation of a cognitive network.* (798–805). Conference on telecommunications, Singapore.

Felice, M. D., Chowdhur, K. R., & Bononi, L. (2011). *Cooperative spectrum management in cognitive vehicular ad hoc networks* (pp. 47–54). Conference on vehicular networking, St. Petersburg.

Fonseca, A., & Vaz~ao, T. (2012). Applicability of position-based routing for VANET in highways and urban environment, *Journal of Network and Computer Applications, 36,* 961–973.

Francisco, J. M., Toh, C. K., Cano, J. C., Calafate, C. T., & Manzoni, P. (2009). A survey and comparative study of simulators for vehicular ad hoc networks (VANETs). *Wireless Communications and Mobile Computing, 11,* 813–828.

Franklin, A. A., Pak, J., Jung, H., Kim, S., You, S., Um, J., ... Kim, C.(2010). *Cognitive radio test-bed based on ECMA-392 international standard* (pp. 1026–1030). Conference on 7th international symposium on wireless communications systems, York.

Ganan, C., Loo, J., Ghosh, A., Esparza, O., Rene1, S., Munoz, J. L. (2012). *Analysis of inter-RSU beaconing interference in VANETs.* Conference on international workshop on multiple access communications, Ireland.

Gerla, M., & Kleinrock, L. (2011). Vehicular networks and the future of the mobile internet. *Computer Networks, 55,* 457–469.
http://dx.doi.org/10.1016/j.comnet.2010.10.015

Gerla, M., Weng, J. T., Giordano, E., & Pau, G. (2012). Vehicular testbeds - model validation before large scale deployment. *Journal of Communications, 7,* 451–457.

Ghandour, A. J., Fawaz, K., & Artail, H. (2011). *Data delivery guarantees in congested vehicular ad hoc networks using cognitive networks* (pp. 871–876). Conference on wireless communications and mobile computing, Doha.

Gonzalez, A. L., Picone, M., & Colabrese, S. (2012). *Cognitive radio and applications.* Rome: Sapienza University Di Roma.

Gozalvez, J., Sepulcre, M., & Bauza, R. (2012). IEEE 802.11p vehicle to infrastructure communications in urban environments. *IEEE Communications Magazine, 50,* 176–183. http://dx.doi.org/10.1109/MCOM.2012.6194400

Hamieh, A., Ben-Othman, J., & Mokda, L. (2009). *Detection of radio interference attacks in VANET* (pp. 5077–5081). Proceedings of the 28th IEEE conference on global telecommunications, Honolulu.

Hernández-Jayo, U., Sainz, N., Iglesias, I., Elejoste, M. P., Jiménez, D., & Zumalde, I. (2012). *IEEE802.11p field operational test implementation and driver assistance services for enhanced driver safety* (pp. 305–310). Conference on consumers communications and networking, Las vegas.

Hu, X., Chen, Z., & Li, F. (2012). Efficient and multi-level privacy-preserving communication protocol for VANET. *Computers and Electrical Engineering, 38,* 573–581.

IEEE Std 1609.3-2010. (2010). *IEEE standard for wireless access in vehicular environments (WAVE) - networking services* (pp. 1–144). Revision of IEEE standard 1609.3-2007.

Islam, M. H., Koh, C. L., Oh, S. W., Qing, X., Lai, Y. Y., Wang, C., ... Toh, W. (2008). *Spectrum survey in Singapore: Occupancy measurements and analyses* (pp. 1–7). 3rd International Conference on cognitive radio oriented wireless networks and communications, Luxemborg.

Jarupan, B. (2009). *Cross-layer design for location and delay aware communication in vehicular networks.* The Ohio State university thesis, Columbus.

Jarupan, B., & Ekici, E. (2011). A survey of cross-layer design for VANETs. *Ad Hoc Networks, 9,* 966–983.
http://dx.doi.org/10.1016/j.adhoc.2010.11.007

Javed, M. A., & Khan, J. Y. (2011). *A geocasting technique in an IEEE802.11p based vehicular ad hoc network for road traffic management* (pp. 1–6). Conference on Australasian telecommunication networks and applications, Melbourne.

Jia, D., Lu, K., & Wang, J. (2014). A disturbance-adaptive design for vanet-enabled vehicle platoon. *IEEE Transactions on Vehicular Technology, 63*, 527–539. http://dx.doi.org/10.1109/TVT.2013.2280721

Jiang, D., & Delgrossi, L. (2008). *IEEE 802.11p: Towards an international standard for wireless access in vehicular environments* (pp. 2036–2040). Conference on IEEE vehicular technology, Singapore.

Jiang, T., Alfadhl, Y., & Chai, K. K. (2011). *Efficient dynamic scheduling scheme between vehicles and roadside units based on IEEE 802.11p/WAVE communication standard* (pp. 120–125).Conference on ITS telecommunications, St. Petersburg.

Kakkasageri, M. S., & Manvi, S. S. (2011). Intelligent infromation dissemination in vehicular ad hoc networks. *International Journal of Ad hoc, Sensor and Ubiquitous Computing, 2*, 112–123. http://dx.doi.org/10.5121/ijasuc

Karimi, H. R. (2011). *Geolocation databases for white space devices in the UHF TV bands: Specification of maximum permitted emission levels* (pp. 443–454). Conference on new frontiers in dynamic spectrum access networks, Aachen.

Kaur, S. (2014). Intelligent in wireless networks with cognitive radio networks. *IETE Technical Review, 30*, 6–11.

Kim, W., & Gerla, M. (2012). Cognitive multicast with partially overlapped channels in vehicular ad hoc networks. *Ad hoc networks,11*, 2016–2025.

Kim, J., Krunz, M. (2011). *Spectrum-aware beaconless geographical routing protocol for mobile cognitive radio networks* (pp. 1–5). Conference on IEEE global telecommunications, Istanbul.

Kim, W., Oh, S. Y., Gerla, M., & Lee, K. C. (2011). *CoRoute: A new cognitive anypath vehicular routing protocol* (pp. 766–711). Conference on wireless communications and mobile computing, Istanbul.

Kirsch, N. J., O'Connor, B. M. (2011). *Improving the performance of vehicular networks in high traffic density conditions with cognitive radios* (pp. 552–556). Conference on IEEE intelligent vehicle symposium, Baden.

Kondareddy, Y. R. (2008). *MAC and routing protocols for multi-hop cognitive radio networks*. Auburn university thesis, Auburn.

Lei, Z., & Chin, F. (2008). A reliable and power efficient beacon structure for cognitive radio systems. *IEEE Transactions on Broadcasting, 54*, 182–187.

Li, H., & Irick, D. K. (2010). *Collaborative spectrum sensing in cognitive radio vehicular ad hoc networks: Belief propagation on highway* (pp. 1–5). Conference On vehicular technology, Taipei.

Liang, Y. C., Chen, K. C., Li, G. Y., & Mähönen, P. (2008). Cognitive radio networking and communications: an overview. *IEEE Transactions on Vehicular Technology, 60*, 3386–3407.

Lim, J. M. Y., Chang, Y. C., Alias, M. Y., & Loo, J. (2015, June). Joint optimization and threshold structure dynamic programming with enhanced priority scheme for adaptive VANET MAC. *Wireless Networks, 1–17*.

Lim, J. M. Y, Chang, Y. C., & Yusoff, M. (2016, March 27–28). *A Novel Optimization Scheme for Human-Computer Cognitive VANET Interaction with Information System Integration*. 5th International Conference on Human Computing, Education and Information Management System, Sydney.

Liu, Y., Hu, S., Xiao, Y., & Liu, X. (2010). *CSMA/CA-based MAC protocol in cognitive radio network* (pp. 163–167). Conference on 3rd IET wireless, mobile and multimedia networks, China.

Liu, J., Ren, P., Xue, S., & Chen, H. (2012). *Expected path duration maximized routing algorithm in Cr-Vanets* (pp. 723–727). Conference on communications in China: wireless networking and applications, China.

Lo, B. F., Akyildiz, I. F., & Al-Dhelaan, A. M. (2010). Efficient recovery control channel design in cognitive radio ad hoc networks. *IEEE Transactions on Vehicular Technology, 59*, 4513–4526. http://dx.doi.org/10.1109/TVT.2010.2073725

Luo, C, Yu, F. R., Ji, H., & Leung, V. C. M. (2010). Cross-layer design for TCP performance improvement in cognitive radio networks. *Vehicular Technology, IEEE Transactions, 59*, 2485–2495.

Marfia, G., Roccetti, M., Amoroso, A., Gerla, M., Pau, G., & Lim, J. H. (2011). *Cognitive cars: Constructing a cognitive playground for VANET research testbeds*. Proceedings of the 4th International Conference on Cognitive Radio and Advanced Spectrum Management, Spain.

Maslekar, N., Mouzna, J., Boussedjra, M., & Labiod, H. (2012). CATS: An adaptive traffic signal system based on car-to-car communication, *Journal of Network and Computer Applications, 36*, 1308–1315.

Maurizio, M., & Vlad, P. (2011). Cognitive radio communications for vehicular technology – wavelet applications. *Vehicular Technologies: Increasing connectivity, 13*, 223–238.

Mitola, J., & Maguire, G. Q. (1999). Cognitive radio: making software radios more personal. *IEEE Personal Communications, 6*, 13–18. http://dx.doi.org/10.1109/98.788210

Mohammed, M. A., & Al-Daraiseh, A. A. (2014). Toward secure vehicular ad hoc network: a survey. *IETE Technical Review, 29*, 80–89.

Muraleedharan, R., & Osadciw, L. A. (2009). *Cognitive security protocol for sensor based VANET using swarm intelligence* (pp. 288–290). Conference on signals, systems and computers, Pacific Grove.

Nair, R., Soh, B., Chilamkurti, N., & Park, J. J. J. H. (2012). Structure-free message aggregation and routing in traffic information system (SMART). *Journal of Network and Computer Applications, 36*, 974–980.

Ngoga, S. R., Yao, Y., & Popescu, A. (2012). *Beacon-enabled cognitive access for dynamic spectrum access* (pp. 49–56). Conference on next generation internet, Karlskrona.

Niyato, D., Hossain, E., & Wang, P. (2011). Optimal channel access management with qos support for cognitive vehicular networks. *IEEE Transactions on Mobile Computing, 10*, 573–591. http://dx.doi.org/10.1109/TMC.2010.191

Nyanhete, E. R., Mzyece, M., & Djouani, K. (2011). *Operation and performance of cognitive radios in vehicular environments*. Conference on SATNAC, Tshwane.

Oliveira, R., Luís, M., Furtado, A., Bernardo, L., Dinis, R., & Pinto, P. (2012). Improving path duration in high mobility vehicular ad hoc networks. *Ad hoc Networks, 11*, 89–103.

Pan, M., Li, P., & Fang, Y. (2012). Cooperative communication aware link scheduling for cognitive vehicular networks. *IEEE Journal on Selected Areas in Communications, 30*, 760–768. http://dx.doi.org/10.1109/JSAC.2012.120510

Pu Wang, P., Jun Fang, J., Ning Han, N., & Hongbin Li, H. (2010). Multiantenna-assisted spectrum sensing for cognitive radio. *IEEE Transactions on Vehicular Technology, 59*, 1791–1800. http://dx.doi.org/10.1109/TVT.2009.2037912

Qiu, R. C., Chen, Z., Guo, N., Son, Y., Zhang, P., Li, H., & Lai, L. (2010). *Towards a real-time cognitive radio network testbed: architecture, hardware platform, and application to smart grid* (pp. 1–6). Conference on IEEE workshop on networking technologies for software defined radio networks, Boston.

Rasheed, H., & Rajatheva, N. (2011). Spectrum sensing for cognitive vehicular networks over composite fading. *International Journal of Vehicular Technology, 1–9*. http://dx.doi.org/10.1155/2011/630467

Richard Yu, F. (2011). *Cognitive radio mobile ad hoc networks.* Berlin: Springer.

Saeed, R. A., Naemat, A. B. H., Aris, A. B., Mat Khamis, I., & Awang, M. K. B. (2010). Evaluation of the IEEE 802.11p-based TDMA MAC method for road side-to-vehicle communications. *International Journal of Network and Mobile Technologies, 1*, 81–87.

Salameh, H. B., Krunz, M., & Younis, O. (2009). MAC protocol for opportunistic cognitive radio networks with soft guarantees. *IEEE Transactions on Mobile Computing, 8*, 1339–1352.
http://dx.doi.org/10.1109/TMC.2009.19

Sandonis, V., Calderon, M., Soto, I., & Bernardos, C. J. (2012). Design and performance evaluation of a PMIPv6 solution for geonetworking-based VANETs, *Ad hoc Networks, 11*, 2069–2082

Shah, N., Habibi, D., & Ahmad, I. (2012). *Multichannel cognitive medium access control protocol for VANET.* (pp. 222–233). Conference in vehicular technology, New York, NY.

Shen, Y., Zou, Q., & Liu, Z. (2009). *Performance evaluation of IEEE802.11-based adhoc wireless networks in vehicular environments* (pp. 1–4). Conference on wireless communications, networking and mobile computing, Beijing.

Soares, V. N. G. J., Rodrigues, J. J. P. C., & Farahmand, F. (2011). GeoSpray: A geographic routing protocol for vehicular delay-tolerant networks. *Information Fusion, 15*, 102–113.

Sou, S. I., Shieh, W. C., & Lee, Y. (2011). *A video frame exchange protocol with selfishness detection mechanism under sparse infrastructure-based deployment in VANET* (pp. 498–504). Conference on wireless and mobile computing, networking and communications, Wuhan.

Steenkiste, P., Sicker, D., Minden, G., & Raychaudhuri, D. (2009). *Future directions in cognitive radio network research* (pp. 1–37). NSF workshop report.

Subhedar, M., & Birajdar, G. (2011). Spectrum Sensing Techniques in Cognitive Radio Networks: A Survey. *International Journal of Next-Generation Networks, 3*, 37–51.
http://dx.doi.org/10.5121/ijngn

Tang, H., Yu, F. R., Huang, M., & Li, Z. (2012). Distributed consensus-based security mechanisms in cognitive radio mobile ad hoc networks. *IET Communications, 6*, 974–983.
http://dx.doi.org/10.1049/iet-com.2010.0553

Tom, H. L., Li, S., Asefi, M., & Shen, X. (2012). Quality of experience oriented, video streaming in challenged wireless networks: Analysis, protocol design and case study. *IEEE Communications Society, 7*, 9–12.

Tomar, R. S., & Verma, S. (2011). RSU assisted channel allocation in VANETs, *International Journal of Contemporary Research in Engineering and technology, 1*, 25–36.

Tsukamoto, K., Matsuoka, S., Altintas, O., Tsuru, M., & Oie, Y. (2009). Distributed channel coordination in cognitive wireless vehicle-to-vehicle communications (pp. 59–64). Conference on the institute of electronics, information and communication engineers (IEICE), Tokyo.

Vinel, A., Belyaev, E., Egiazarian, Karen., & Koucheryavy, Y. (2012). An overtaking assistance system based on joint beaconing and real-time video transmission. *IEEE Transactions on Vehicular Technology, 61*, 2319–2329.
http://dx.doi.org/10.1109/TVT.2012.2192301

Wang, S. S., & Lin, Y. S. (2012). PassCAR: A passive clustering aided routing protocol for vehicular ad hoc networks, *Computer Communications, 36*, 170–179.

Wang, X. Y., & Ho, P. H. (2010). A novel sensing coordination framework for CR-VANETs. *IEEE Transactions on Vehicular Technology, 59*, 1936–1948.
http://dx.doi.org/10.1109/TVT.2009.2037641

Wang, R., Rezende, C., Ramos, H. S., Pazzi, R. W., Boukerche, A., & Loureiro, A. A. F. (2012). *LIAITHON: A location-aware multipath video streaming scheme for urban vehicular networks* (pp. 436–441). conference on ieee symposium on computers and communications, Cappadocia.

Xiang, Y., Liu, Z., Liu, R., Sun, W., & Wang, W. (2012). GeoSVR: A map-based stateless VANET routing. *Ad hoc Networks, 11*, 2125–2135.

Xu, K., Garrison, B. T., Wang, K. C. (2011). *Throughput modeling for multi-rate IEEE 802.11 vehicle-to-infrastructure networks with asymmetric traffic* (pp. 299–306). In Proceedings of the 14th ACM international conference on modeling, analysis and simulation of wireless and mobile systems, New York, NY.

Yeh, L. Y., Yang, C. C., Chang, J. G., & Tsai, Y. L. (2012). A secure and efficient batch binding update scheme for route optimization of nested network mobility (NEMO) in VANETs, *Journal of Network and Computer Applications, 36*, 284–292.

Zhan, M., Ren, P., Zhang, C., & Li, F. (2012). *Throughput-optimized cross-layer routing for cognitive radio networks.* Conference on communications in China: wireless networking and applications, 821–826.

Zhang, X., & Hang, S. (2010). CREAM-MAC: Cognitive radio-enabled multi-channel MAC protocol over dynamic spectrum access network. *IEEE Journal of Selected Topic on Signal Processing, 5*, 110–123.

Zhao, Q., Zhu, Y., Chen, C., Zhu, H., & Li, B. (2013). When 3G meets VANET: 3G-assisted data delivery in VANETs. *IEEE Sensors Journal, 10*, 3573–3584.

12

Comparative evaluation of photovoltaic MPP trackers: A simulated approach

Barnam Jyoti Saharia[1], Munish Manas[1]* and Bani Kanta Talukdar[2]

*Corresponding author: Munish Manas, Department of Electronics Communication Engineering, Tezpur University, Assam 784028, India
E-mail: munish@tezu.ernet.in
Reviewing editor: Kun Chen, Wuhan University of Technology, China

Abstract: This paper makes a comparative assessment of three popular maximum power point tracking (MPPT) algorithms used in photovoltaic power generation. A 120 W_p PV module is taken as reference for the study that is connected to a suitable resistive load by a boost converter. Two profiles of variation of solar insolation at fixed temperature and varying temperature at fixed solar insolation are taken to test the tracking efficiency of three MPPT algorithms based on the perturb and observe (P&O), Fuzzy logic, and Neural Network techniques. MATLAB/SIMULINK simulation software is used for assessment, and the results indicate that the fuzzy logic-based tracker presents better tracking effectiveness to variations in both solar insolation and temperature profiles when compared to P&O technique and Neural Network-based technique.

Subjects: Electrical & Electronic Engineering; Engineering & Technology; Power Engineering

Keywords: PV; boost converter; perturb and observe; fuzzy logic; neural networks; tracking factor

1. Introduction

Power generation through photovoltaic (PV) generation system is one of the most sought after forms of renewable energy sources, popularity of which has been in the rise due to its non-polluting, renewable and inexhaustible nature (Eltawil & Zhao, 2013). Immense demand for finding feasible and environmental-friendly renewable energy sources to meet the future energy requirements as

ABOUT THE AUTHORS

Barnam Jyoti Saharia completed his MTech in power electronics and his research area is concerned with converter topology for MPPT algorithms. Currently he works as an assistant professor in ECE Department of Tezpur Central University in India.

Munish Manas is the recipient of meritorious UGC BSR fellowship for his research in microgrid design and optimization. Currently he works as an assistant professor in ECE Department of Tezpur Central University in India.

Bani Talukdar is a professor in Assam Engineering College and does his research in power system deregulation and system design.

PUBLIC INTEREST STATEMENT

I am sending this manuscript for consideration for publication in your esteemed journal to assist the scientific community in the area of Comparative Evaluation of Photovoltaic MPP Trackers: A Simulated Approach. This paper makes a comparative assessment of three popular maximum power point tracking (MPPT) algorithms used in photovoltaic power generation. A 120 Wp PV module is taken as reference for the study that is connected to a suitable resistive load by a boost converter. Two profiles of variation of solar insolation at fixed temperature and varying temperature at fixed solar insolation are taken to test the tracking efficiency of three MPPT algorithms based on the Perturb and Observe, Fuzzy logic, and Neural Network techniques. MATLAB/SIMULINK simulation software is used for assessment, and the results indicate that the fuzzy logic-based tracker presents better tracking effectiveness to variations in both solar insolation and temperature profiles when compared to Perturb and Observe technique and Neural Network-based technique.

fossil fuel reserves deplete. Solar energy is a viable substitute to fossil fuels among other available renewable energy sources such as wind, hydroelectric, and geothermal power (Sivasubramaniam, Faramus, Tilley, & Alkaisi, 2014). Solar PV generators have been used in small-scale, stand-alone systems at low voltage levels as well as the high power installations, connected in grid mode and operating at medium or high voltage levels. The drawback of PV generators is its low conversion efficiency (Sivasubramaniam et al., 2014). Efficiencies of typical crystalline PV cells are in the range of 12–18%, although experimental cells have been constructed that are capable of efficiency over 30% (Zaheeruddin & Manas, 2015a). The PV generators exhibit nonlinear current–voltage (I–V) and power–voltage (P–V) characteristics (Liu, Liu, & Gao, 2013), a phenomenon that is more serious in partially shaded condition due to more than one maximum power point (MPP). For optimum use of PV generation, it is important to operate the system at the maximum available power production state for an available solar irradiation, temperature, and load. Thus implementation of maximum power point tracking (MPPT) control techniques, in order to maximize the available output from a PV generator becomes an essential constituent of any PV system.

The MPPT involves operating a DC–DC controlled converter such that it leads to operates a PV system to its optimum power production state for given set of external atmospheric conditions and state of loading(Ahmed & Miyatake, 2008). Different algorithms employing tracking techniques for maximum power from PV generators are found in literature. Jain & Agarwal, 2007; and Kim, 2007 uses sliding mode current controlled for power point tracking, by varying load voltage in order to obtain either the current derivative or voltage derivative equal to zero. Chih-Lyang, Li-Jui, & Yuan-Sheng, 2007 and Venelinov, Leonardo, Vincenzo, Francesco, & Okyay, 2007 make use of artificial intelligence techniques (fuzzy logic, neural network, and genetic algorithm) to determine the change in voltage and current of load, and in turn vary duty cycle for PWM control of the converter for effective tracking at any external atmospheric condition and loading. While most research works are focused on the development on the tracking algorithms, Bhattacharjee and Saharia (2014) make a comparative study on the individual converter topologies namely buck, boost and buck-boost converters used as interface in the PV power generators. They conclude that boost converter, when used as the power converter interface for MPPT, operates with better tracking effectiveness when operated at higher values of insolation. Tsang & Chan, 2015 make use of current sweeping-based approach for obtaining global maximum under partial shaded condition, which aids in controlling the large oscillations near the panel operating MPP and overcome methods otherwise problematical to implement using artificial intelligent techniques. Li, 2015 proposes a MPPT method with variable weather parameters considered specially from the input resistance perspective, in order to improve the tracking speed and adaptability to varying weather condition for PV systems. Hassan Fathabadi, 2015 proposes a Lambert W-function-based technique to evaluate the MPP of PV panels based on the current–voltage (I–V), power–voltage (P–V), and power–current (P–I) relationship of the PV module.

Artificial intelligence-based MPPT applications for solar energy have also been implemented for maximum power extraction from PV systems. (Iqbal, Ahmed, Abu-Rub, & Sinan, 2010; Kottas, Boutalis, & Karlis, 2006; Manas, 2015; de Medeiros Torres, Antunes, & dos Reis, 1998; Mellit & Kalogirou, 2008; Mellit, Kalogirou, Hontoria, & Shaari, 2008; Muralidhar & Susovon, 2015; Saleh, Chaaben, & Ammar, 2008; Sher et al., 2015; Zaheeruddin & Manas, 2015b;) Among them, one of the most frequently preferred MPPT techniques is the perturb and observe (P&O) algorithm. Its advantages mainly include the low-cost hardware, the easy implementation and the good performance without solar irradiance and temperature varying quickly with time. Although highly popular due to its simplicity of implementation through electronic circuits, Hohm and Ropp (2003) (Hohm & Ropp, 2003) noted that only if properly optimized the P&O algorithm can overcome the shortcomings associated with it including its slow tracking speed and oscillation around the MPP.

Most of the methodology of implementation of tracking algorithm involves complex circuitry and increase in the number of measured parameters. In order to negate the issue of complexity in implementation circuitry, Salah and Ouali (2011)(Manas, 2015) proposed fuzzy logic and neural network controllers. The controllers used the climatic parameters of insolation and temperature as inputs and estimated the optimum duty cycle corresponding to the system maximum power (P_{max}). PV systems exhibit non-linear characteristics, which is appropriate for fuzzy logic-based control. Accurate rule base formulation between membership function of fuzzy logic controller leads to dynamic and quick to respond output value of duty cycle that ensures MPP tracking. While neural networks present the advantage of not requiring knowledge of the internal systems parameters, less computation time, and absence of complex algorithms for determining the optimum duty ratio. However, the effectiveness of this technique depends on the training of the system. The better the trained model, more accurate it is in predicting the operating duty cycle for MPP operation.

The MPPT algorithms works to automatically find the voltage at maximum power (V_{mpp}), the current at maximum power point (I_{mpp}) at which the PV array connected to a converter should operate to obtain the maximum power output under a given conditions of atmospheric insolation and temperature. The variation in the V_{mpp} and I_{mpp} with the change in insolation and temperature is a complex problem. This relation becomes more pertinent when there is partial shading, as it is possible to have multiple local minima, often at times resulting in error in the tracking of the MPP.

The three most popular and widely used MPPT tracking algorithms have been assessed in this study. The review of literature shows that although significant work has already been covered in development of P&O technique, fuzzy logic-based- and neural network-based controllers for MPPT applications, a comprehensive work on comparative assessment of the three techniques on a single PV system is not found in the literature. This article attempts to make a comparison based on the tracking effectiveness of each of the above-mentioned techniques to validate which tracking algorithms can give optimum performance under changing atmospheric condition.

The paper is presented in the following chronological order. Section 2 presents description on modeling of the PV module in SIMULINK and the theoretical foundation of the working of DC–DC boost converter. The methodologies related to the working of P&O technique, fuzzy logic-based controller design and neural network controllers are covered in Section 3. Section 4 presents two scenarios where the controllers are tested with respect to their tracking effectiveness in simulation environment. Section 5 presents a short discussion on the tracking performance of the three reference algorithms and Section 6 draws conclusion to the work.

2. PV System modeling and Operation of Boost Converter

In this section, the mathematical modeling of the PV panel considered as the reference in this simulation study, namely the KYOCERA KC 120-1(Kyocera KC120-1 multi-crystalline photovoltaic module datasheet, 2015) is presented. The theoretical background on the operation of DC–DC boost converter is also presented here.

2.1. PV Module Modeling

A solar cell is basically a p–n junction fabricated in a thin wafer of semiconductor. The electromagnetic radiation of solar energy can be directly converted into electricity through the PV effect. When exposed to sunlight, photon with energy greater than the band gap of the semiconductor creates the electron-hole pairs proportional to the incident radiation which is responsible for the generation of photocurrent.

Figure 1 shows the equivalent circuit of a PV cell. The current source I_{ph} represents the photocurrent. R_{sh} and R_{s} are the intrinsic shunt and series resistances of the cell, respectively. Usually the value of R_{sh} is very large and hence they may be neglected to simplify the analysis.

Each PV cell, when grouped together in a combination of parallel and series cells constitute a PV module and PV arrays. The mathematical Equations (1)–(4) are used for the modeling of the reference PV module KYOCERA KC 120-1(Kyocera KC120-1 multi-crystalline photovoltaic module datasheet, 2015).

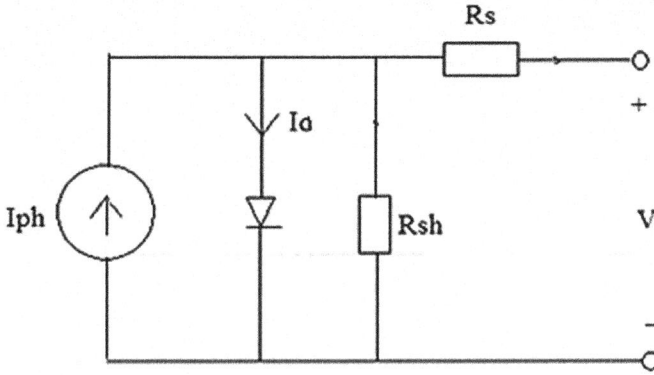

Figure 1. PV cell as a diode circuit.

Module Photocurrent (I_{ph}) is expressed by:

$$Iph = [Iscr + Ki(T - 298)]\lambda/1000 \tag{1}$$

Module reverse saturation current (I_{rs}) is given by:

$$Irs = Iscr /[\exp(qVoc/NsKAT) - 1] \tag{2}$$

The module saturation current (I_o) varies with the cell temperature, which is expressed as:

$$Io = Irs[T /Tr]^3 exp(qEgo/AK)[\frac{1}{Tr} - \frac{1}{T}] \tag{3}$$

The current output of the PV module (I_{pv}) is represented by:

$$Ipv = NpIph - NpIo[exp\{ \frac{q(Vpv + IpvRs)}{NsAKT} \} - 1] \tag{4}$$

Where I_{scr} is the PV module short-circuit current (A) at 1 kW/m^2 and 25 °C, K_i is the short-circuit cur-rent temperature coefficient at I_{scr} (0.0017A/°C), T is the module operating temperature in Kelvin (K), λ is the PV module illumination (kW/m^2), I_{rs} is the reverse saturation current of the module (A), q is Electron charge (1.6 × 10^{-19} C), V_{oc} is the open circuit voltage of the PV panel (V), N_s is the number of cells connected in series in the PV module, k is Boltzmann's constant having the value of 1.3805 × 10^{-23} J/K, A is an ideality factor having value of 1.2, I_o is the PV module saturation current (A), T_r is the reference temperature in Kelvin (298 K), E_{go} is the band gap for silicon having value of 1.1 eV, I_{pv} is output current of a PV module (A), V_{pv} is the output voltage pf the PV module (V), N_p is the number of cells connected in parallel for the PV module. In the mathematical model the cells in se-ries and the cells in parallel have values of N_s = 36 and N_p = 1.

Table 1 lists the electrical specifications of the Kyocera KC120-1 PV module (Kyocera KC120-1 multi-crystalline photovoltaic module datasheet, 2015) specified at standard testing conditions (i.e. at a irradiation of 1000 W/m^2, 25° C temperature and AM 1.5) which has been considered as the reference module in this paper for investigation. Figure 2 depicts the current voltage (I–V) and power voltage (P–V) characteristics of the PV module at STC of the simulated model indicating that the model is able to predict accurately the PV module characteristics.

Table 1. Electrical Characteristics of Kyocera KC120-1 PV model (Kyocera KC120-1 multi-crystalline photovoltaic module datasheet, 2015).

Parameter	Rating
Maximum Power	120 W
Voltage at maximum power	16.9 V
Current at maximum power	7.10 A
Short circuit current	7.45 A
Open circuit voltage	21.5 V
Total number of cells in series	36
Total no of cells in parallel	1
Band Energy	1.12 eV
Ideality factor	1.2

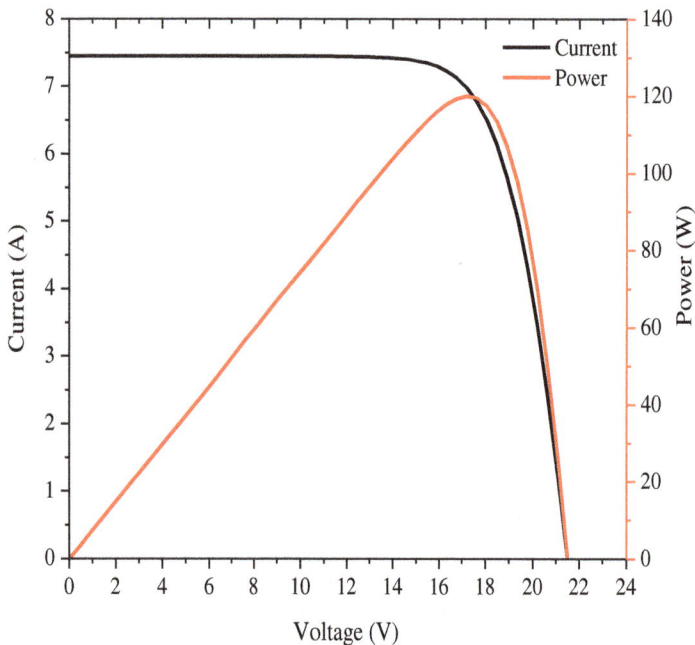

Figure 2. I–V and P–V curve of the PV module at 1 kW/m² and 25 °C.

2.2. Boost Converter Operation

The purpose of any DC–DC switched mode converter is to convert an unregulated dc input to a regulated or controlled dc output at a desired voltage level. In a PV system, a DC–DC converter is used to act as MPPT interface between the source and the load. The converter works to adjust its duty cycle to match the requirements of the system.

The voltage transformation in a DC–DC converter (switched mode) is done by controlling the operation of the switches. This is achieved by controlling the operation period, i.e. the on time (t_{on}) and off time (t_{off}) of the switches by PWM (pulse width modulation) technique as shown in Figure 3. The switching period $T_s = t_{on} + t_{off}$ is held constant while the ratio of the on time to the switching time (i.e. duty ratio) is varied. Using the switch mode control in the circuit, the output voltage V_o will be a constant pulse as shown in the figures below. Because of the inductive and capacitive circuit elements in the converter topologies the output voltage should be constant (given as the dashed line) V_o.

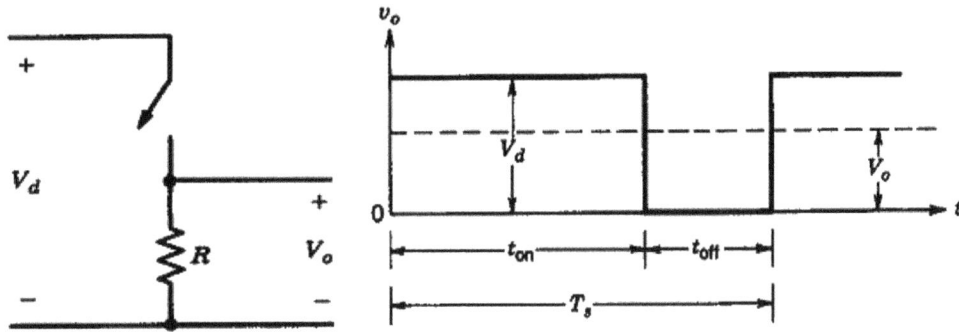

Figure 3. Switch mode operation of converters.

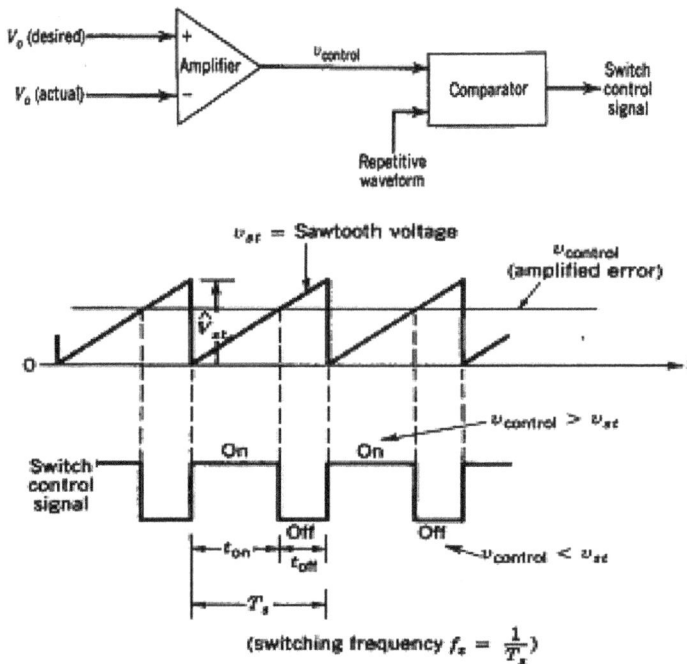

Figure 4. PWM block diagram and comparator signal.

The switch control signal is generated by comparing a control value (which mostly is a signal generated as an error signal) to a repetitive waveform V_{st}. The control value may be the difference between the actual and the desired output voltage V_o as seen by the Figure 3. The effects of comparison are when V_{ctr} ($V_{control}$) > V_{st}, switch is on and vice versa.

Hence, we can now define duty ratio (cycle) as

$$D = \frac{t_on}{T_s} = \frac{V_c tr}{V_s t} \tag{5}$$

The frequency (1/T_s) can also be varied in a PWM switching mode. This method however might make it hard to filter the ripple components in the converter waveforms.

In a DC–DC converter with an optimal design, it is assumed that the switching ripples are very small compared to the average values often less than 1% of the quantities. This is often referred to as the small or linear ripple approximation. The boost converter produces an output voltage which is higher than the input voltage.

This topology also has different circuit schemes depending on the state of the switch. When the switch is on the output stage it is isolated from the input caused by the reverse biased diode. The input will supply the inductor with constant voltage and the inductor current will increase. When the switch is off the output will be supplied by both the input and the inductor, and the current through the inductor will decrease because of this energy transfer.

The voltage and current graph of the inductor through one time period has been shown in Figure 4 above along with the individual effective circuit for t_{on} and t_{off} in Figure 5. The shapes are equal to those of the buck converter, but the voltage of the inductor is different due to the placement of the switch and the diode(Saharia, 2014). The analysis of steady-state performance for the buck converter, we have also for the boost converter,

$$(i_L)_{on} = (i_L)_{off} \text{ and } \int_0^{T_s} v_L \, dt = 0, \text{ which gives us}$$

$$\frac{V_in}{V_o} = (1 - D) = \frac{I_o}{I_in}. \tag{6}$$

4. MPPT Algorithms

Figure 6 shows the basic layout of a PV generation system connected by a DC–DC boost converter to a load. Solar insolation and temperature serve as the inputs to the PV panel. The MPP tracking algorithm is fed with the necessary parameters that it requires voltage, current, insolation, or

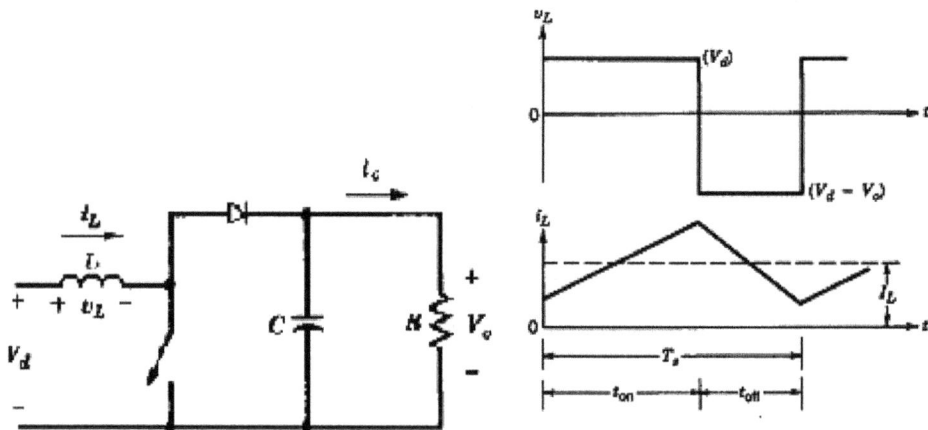

Figure 5. Boost converter operation.

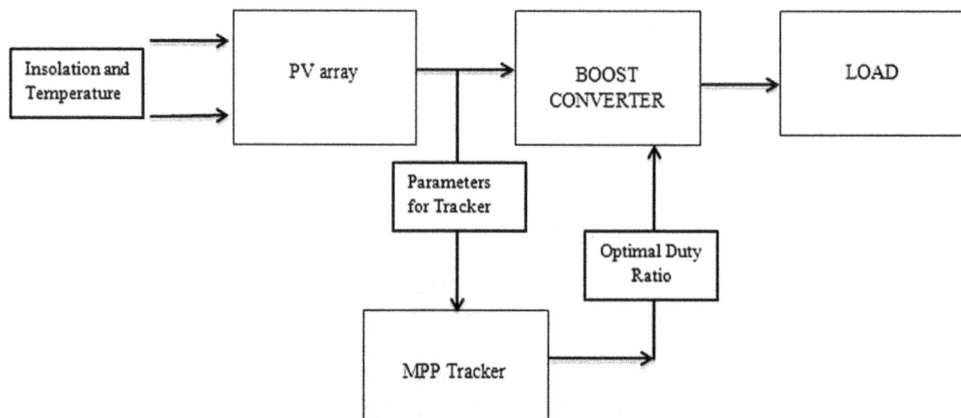

Figure 6. PV generation system with MPPT control.

temperature, and it sends a control signal in the form of duty ratio to operate the system at maximum power point. Discussions on the three MPPT algorithms follow in the ensuing subsections:

4.1. Perturb and Observe MPPT Algorithm

The most basic form of P&O algorithm compares the power previously delivered with the one after disturbance. If the comparison results such that the new power increases, the voltage is increased and vice versa. The P&O algorithm involves varying the voltage of the panel periodically with small incremental steps to reduce the oscillations around the MPP or a desired step. This algorithm is widely found to have been implemented with commercial system due to reduced circuitry and low number of depended parameters. There are four variations in voltage and the possible change in the resulting power as a reflection of the voltage change, which are tabulated in Table 2 (Esram & Chapman, 2007). The outcome of the perturbation is positive in the next step when the power change is positive, and negative in the reverse case.

Figure 7 shows the Simulink block diagram of the MPPT based on the P&O technique. The subsystem consists of Voltage and current as inputs to the tracking algorithm and the duty ratio as the output. Figure 5 shows the MATLAB/SIMULINK system model used for analysis of the MPP tracking algorithm using the boost converters. The output current and voltage of the PV panel act as input signals to the MPPT controlling subsystem. The contoller tracks the MPP using P&O technique generating a reference duty ratio command. This signal is compared with a repeating sequence of sawtooth waveform to generate the PWM signal which is fed to the MOSFET switch of the boost DC–DC converter.

Table 2. Summary of P&O technique (Esram & Chapman, 2007)		
Perturbation	**Change in power**	**Next perturbation**
Positive	Positive	Positive
Positive	Negative	Negative
Negative	Positive	Negative
Negative	Negative	Positive

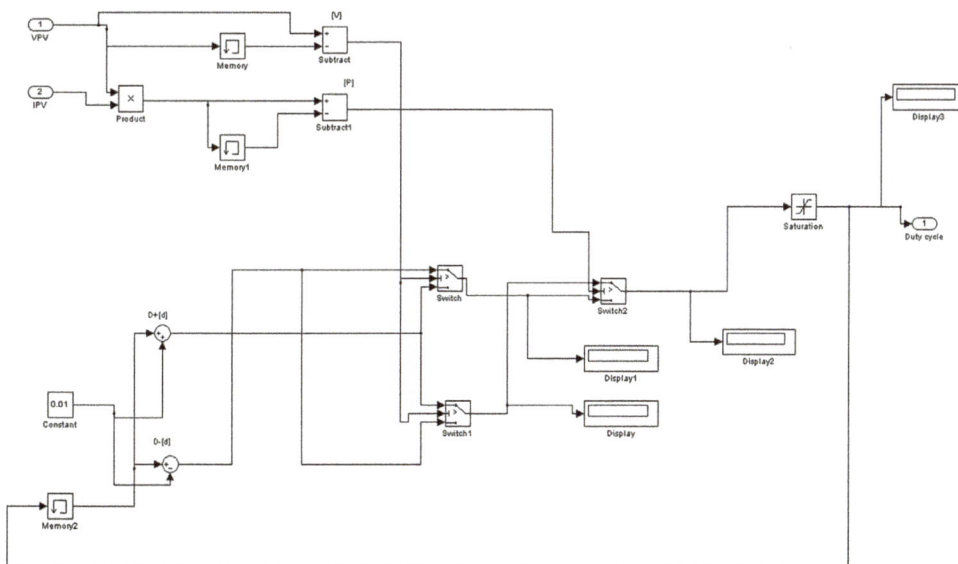

Figure 7. P&O technique developed in SIMULINK using switch logic.

4.2. Fuzzy Logic control

The fuzzy logic and neural network-based controllers for MPPT involves the use of climatic data of insolation and temperature. Hence there needs to be an adequate knowledge of the behavior of the PV system connected to a dc load by a DC–DC converter working based on either fuzzy logic or neural network to generate the optimal duty ratio to extract maximum power from the system, for any given external atmospheric conditions.

Fuzzy logic-based design of MPPT controllers involves three components in its design. First the knowledge base of the designer, the fuzzification step, the inference diagram and the defuzzification process (Manas, 2015; Saleh et al., 2008). According to the input parameters of solar insolation and temperature, the output duty ratio corresponding to the MPP operation of a PV system, the designer must have knowledge of the relation between the input and output parameters. There must be sectional division of the solar insolation and temperature, as well as the output duty ratio.

Table 3. Control Rules for Fuzzy logic-based MPPT				
Insolation\temperature	**Duty ratio**			
	Small	**Medium**	**Large**	**Very Large**
Small	Small	Small	Small	Medium
Medium	Medium	Medium	Medium	Large
Large	Medium	Large	Very Large	Very Large
Very Large	medium	Very Large	Very Large	Very Large

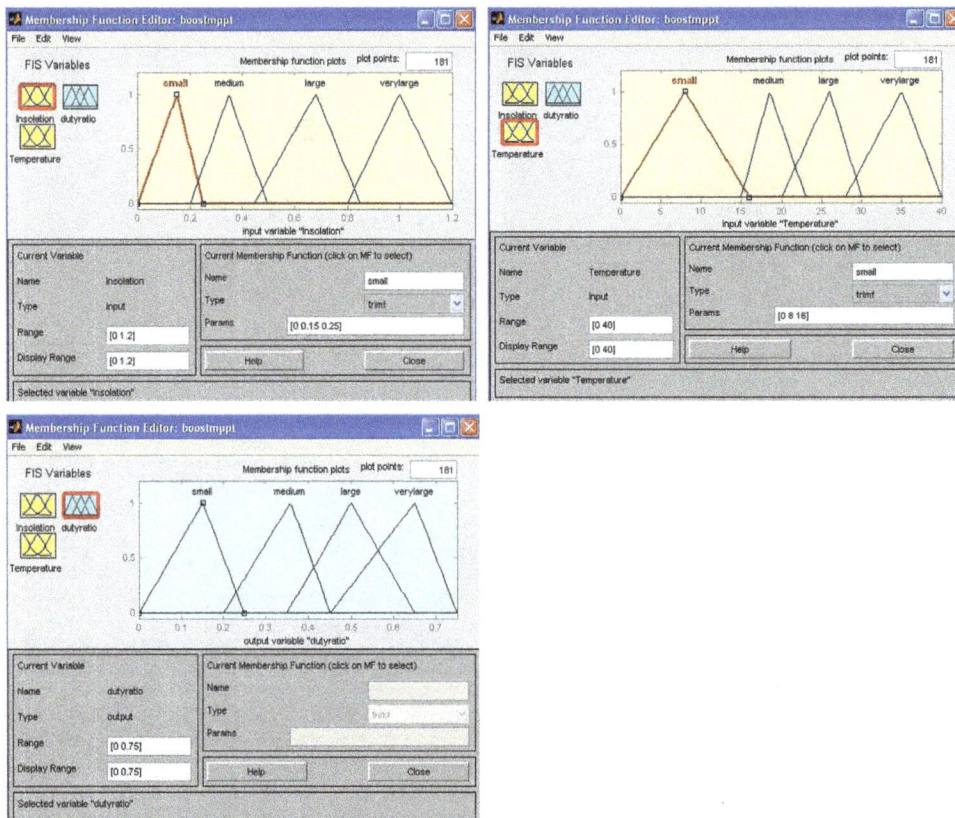

Figure 8. The membership functions of the fuzzy controller for insolation, temperature, and duty ratio.

Solar radiation is divided into four fuzzy subset sections namely small, medium, large, and very large, covering insolation values between 0 and 1.2 kW/m². The temperature values also have four sections of small, medium, large, and very large, membership functions ranging from 0 to 40 °C, similarly the duty ratio also has similar values of fuzzy subsets small, medium, large, and very large, membership function ranging from 0 to 0.75. The control relationship is tabulated in Table 3 as where the insolation and temperature serve as inputs and the duty ratio serves as the output for MPPT.

As shown in Figure 8, For the fuzzification step of the algorithm, the membership functions for solar insolation, temperature, and duty ratio are considered to be of symmetric triangular type. Based on the membership functions, a Mamdani based rule base is constructed having 4 × 4 = 16 rules. The rules aggregations are given by computing the minimum norm conjuction implication of fuzzy subset of the optimum duty ratio. As the rules are averaged, the defuzzification consists of use of the centrod method to get the duty ratio.

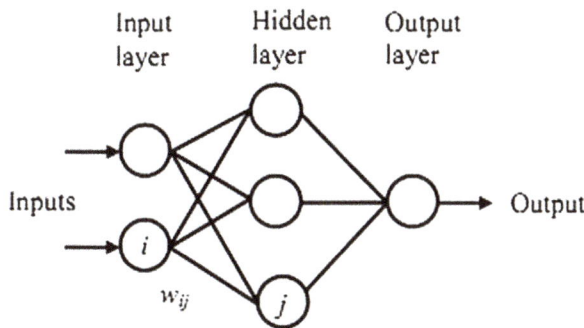

Figure 9. Schematics of a neural network indicating the input, hidden, and output layer (Eltawil & Zhao, 2013).

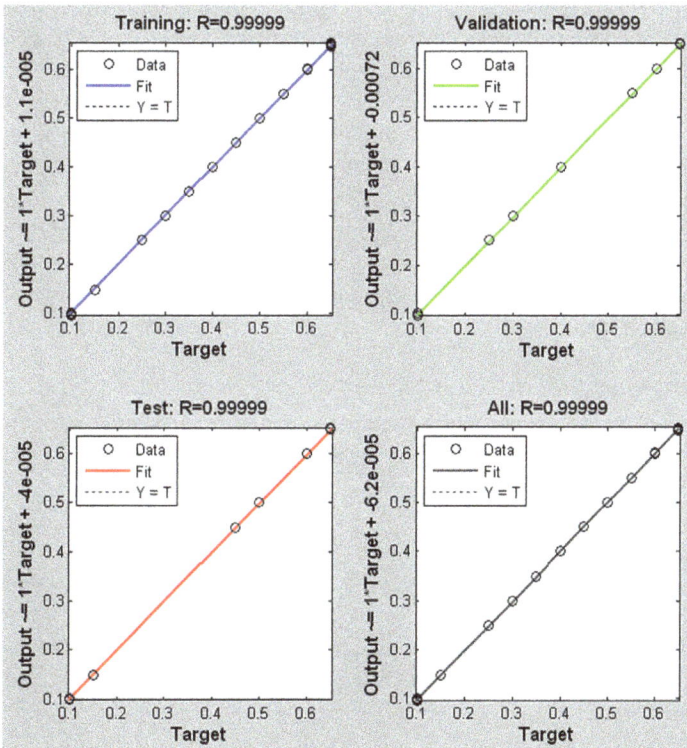

Figure 10. Fitting data of the Neural Network model.

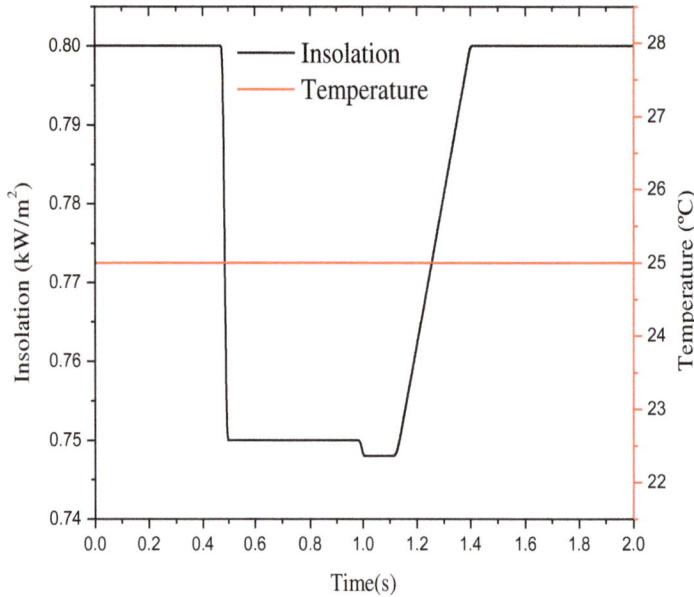

Figure 11. Insolation changes during simulation period at a fixed temperature of 25 °C.

4.3. Neural Networks

Neural network-based controllers generally have three layers: input, the hidden and the output layers as shown in Figure 9. The number of nodes (*i* – input node and *j* – hidden layer node) in each layer vary and are usually user dependent. Input variables are usually atmospheric parameters like solar insolation and temperature, or PV array paremeters like Voc and Isc or a combination of these paremeters. The output is generally one or several reference signal like duty ratio signal used to drive the DC–DC power converter interface to make it operate close to the MPP. The accuracy of the tacking algorithm based on neural network depends on the training algorithm used and the training of the network with data. As most of the PV panels have different characteristics, a neural network has to be specifically trained for the PV array with which it will be implemented. The PV array characteristics may change with time, and as such the neural network may need periodic tuning in the training for a much robust and effective working (Esram & Chapman, 2007). In this study, neural network has been designed such that it takes solar insolation and temperature as the data inputs and produces duty ratio corresponding to the input conditons as the output. The values of insolation have been varied in steps of 50 W/m^2 from zero to 1200 kW/m^2 and temperature in steps of 5 °C from zero to 40 °C. A total of 203 sets of input data were selected and the optimum duty ratio for each of the sets was calculated as shown in Figure 10. The neural network was then trained using these datasets by backpropagation technique (Haykin, 1999).

5. Simulation Procedure and Results

To test the efficiency of tracking of the three algorithms mentioned above, they were modeled in simulation environment. The tracking factor that symbolizes the effectiveness of the tracking algorithm is given by the relation:

$$\eta = \frac{P_inst}{P_mpp} \tag{7}$$

Where η is the tracking factor, P_{inst} is the instantaneous power at the operating point of the PV module, and P_{mpp} is the instantaneous MPP of the PV module under given condition of insolation and temperature.

The MATLAB/SIMULINK software platform was used for modeling and implementation of the algorithms. There were two scenarios considered, one in which the temperature was kept constant and

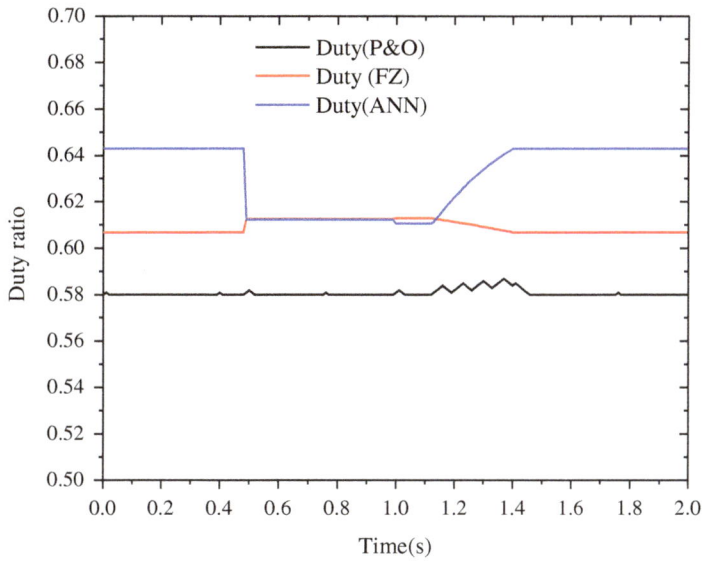

Figure 12. Duty ratio for varying insolation and fixed temperature of 25 °C.

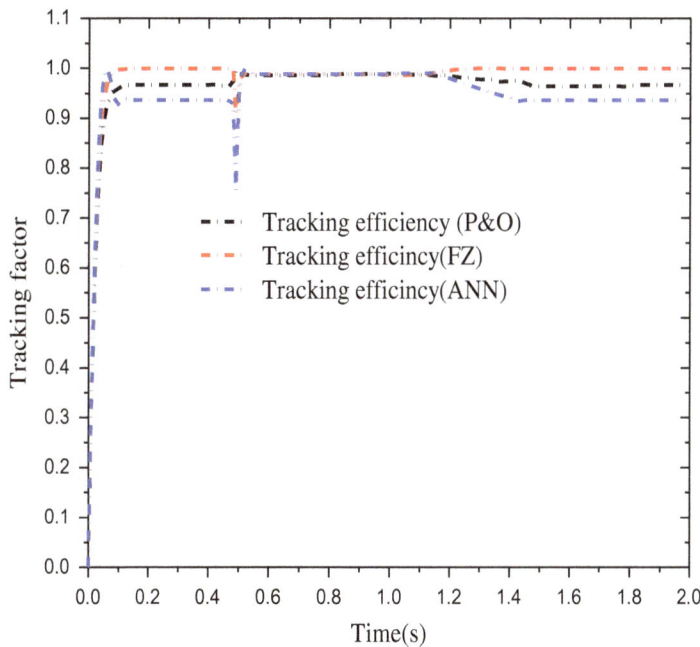

Figure 13. Tracking factor for varying insolation and fixed temperature of 25 °C.

the insolation on the PV panel was changed. Next to ascertain the response of the algorithms to change in temperature, the insolation was fixed and a varying temperature was given as the input. The two cases and the behavior of the three algorithms tracking ability for a set of external conditions are detailed in the following sections:

Table 4. Tracking efficiency of the three algorithms for insolation change	
MPPT Technique	**Tracking factor**
Perturb and observe	0.9628
Fuzzy logic	0.98252
Neural Network	0.94737

5.1. Case I: Tracking Response to Step and Linear Change in Insolation at Fixed Temperature

In the first scenario, to evaluate the three tracking algorithms, we exposed the PV model to a solar radiation of 800 W/m² at a fixed temperature of 25 °C to start with. The temperature values remain fixed for the simulation interval. The isolation takes a step change from 800 to 750 W/m² at 0.49 s. It remains fixed at this value of radiation till 1.05 s, when it slightly decreases and at 1.125 s starts to rise linearly and reaches a value of 800 W/m² at 1.4 s and holds this value up to the end of simulation time of 2 s.

The variation in duty ratio as a result of the individual tracking algorithms to attain MPP operation is shown in figure. It is observed that the highest fluctuation in the duty ratio takes place for the neural network-controlled tracker indicating that it is the most sensitive of the three algorithms. The fuzzy logic-controlled duty ratio and P&O duty ratio are much more stable and their variation due to the fluctuation in the insolation change indicates a stable system in the dynamic sense.

From the point of view of tracking efficiency of the three algorithms, we observe that the fuzzy controller is most effective in tracking the MPP for the change in solar insolation, followed closely by the P&O algorithm evaluated as per Eq. (12). The neural network algorithm is seen to be having the least efficient tracking. The convergence speeds of the algorithms also vary. The fuzzy logic controller is seen to achieve tracking efficiency fastest as shown in Figure 12 and Figure 13. This is closely followed by P&O technique and the neural network-based MPPT algorithm. The average tracking efficiency for the three algorithms is tabulated in Table 4.

5.2. Case II: Tracking Response to Step and Linear Change in Temperature at Fixed Insolation

In the second case, to evaluate the three tracking algorithms, we exposed the PV model to a solar radiation of 800 W/m² and a fixed temperature of 24.6 °C to start with. The temperature values change accordingly as shown in Figure 12 while insolation is kept constant at 800 W/m² The temperature takes a step change from 24.6 to 23.6 °C at 0.4 s. It rises a little to 23.8 °C at 0.705 s. remaining fixed at this value of radiation till 1.08 s, when it slightly decreases to 23.6 and starts to rise linearly and reaches a value of 24.2 °C at 1.46 s and then decreases to reach the value of 22.8 °C at 1.85 s. The temperature is held to this value up to the end of simulation time of 2s as shown in Figure 14.

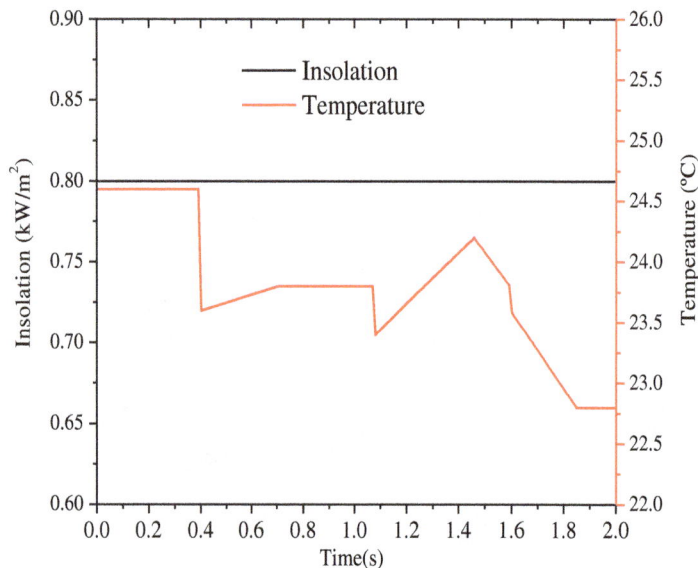

Figure 14. Temperature changes during simulation period at a fixed insolation of 1 kW/m².

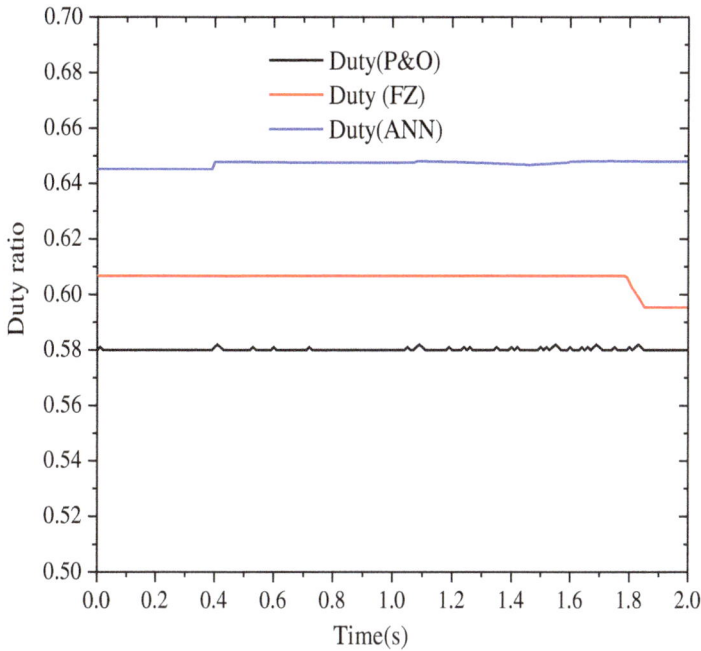

Figure 15. Duty ratio variation for varying temperature at fixed insolation.

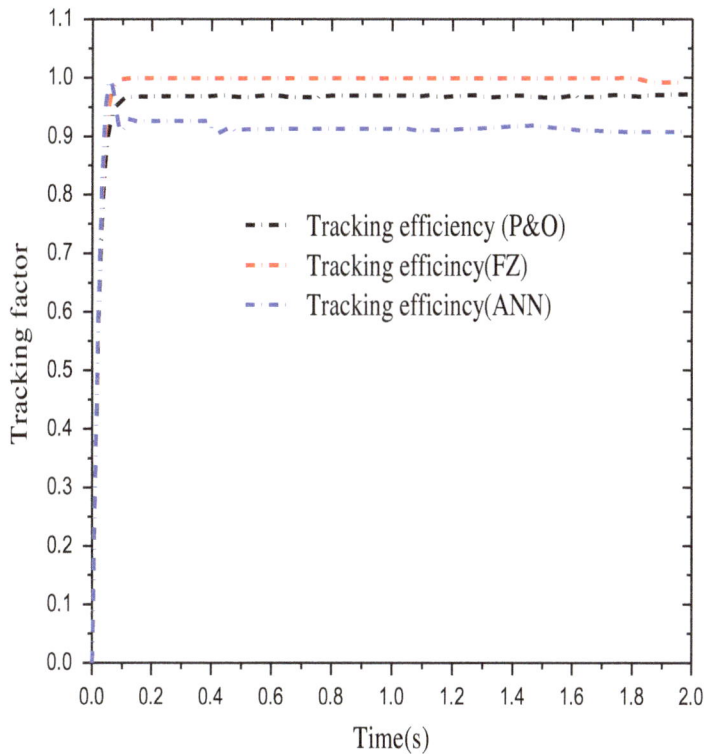

Figure 16. Tracking factor for varying temperature at fixed insolation.

Figure 15 shows the variation in duty ratio as a result of the individual tracking algorithms to attain MPP operation for a changing temperature at fixed insolation. It is observed that the neural network-based tracker has the highest duty ratio, followed by fuzzy logic-controlled duty ratio and P&O duty ratio.

Table 5. Tracking efficiency of the three algorithms for temperature change

MPPT Technique	Tracking factor
Perturb and observe	0.95603
Fuzzy logic	0.98645
Neural Network	0.90655

Table 6. Comparison among the tracking algorithms

Algorithm	Case I			Case II		
	Speed of response to external perturbation	Change of duty ratio	Tracking efficiency	Speed of response to external perturbation	Change of duty ratio	Tracking efficiency
Fuzzy logic	Fastest	Stable and low change	Highest	Fastest	Lower than neural network	Highest
Neural network	Slower than P&O	Highest change	Least efficient	Slower than P&O	Highest value of duty ratio	Least efficient
Perturb and observe	Fast	Fluctuation is less compared to neural networks	Lower than fuzzy logic	Fast	Fluctuation is less compared to neural networks	Lower than fuzzy logic

From the point of view of tracking efficiency of the three algorithms, we observe that the fuzzy controller is most effective in tracking the MPP for the change in temperature, followed closely by the P&O algorithm as shown in Figure 16 and Table 5. The neural network algorithm is seen to be having the least efficient tracking, similar to case I where the results are also in the same corresponding order.

The speed of convergence at the perturbation is also seen to vary with fuzzy logic controller that converges the fastest. This is followed closely by P&O technique and the neural network-based tracking algorithm. The average tracking efficiency of the three algorithms to the change in temperature is recorded in Table 5.

6. Discussion

Table 6 shows a comparative analysis of the three algorithms. The comparison is made on the basis of the speed of response of the algorithm to the change in external perturbation due to variation in the environmental conditions, change of duty ratio and the tracking efficiency. The fast change in response to the change of weather parameters namely insolation and temperature indicates the high sensitivity to external perturbations. This also indicates that an algorithm is able to respond quickly, which is a sought after quality of the MPPT algorithm to make it implementable in systems where fast and swift varying controller response is a desired requirement. The degree fluctuations in duty ratio indicate that the algorithm is unstable at the equilibrium, which causes the converter to oscillate thereby causing reduced efficiency due to system losses. Therefore a stable and low changing algorithm with reduced oscillations at steady state is a sought after characteristic in such algorithms. From Table 6 it is seen that the fuzzy logic-based controller leads the other two techniques in this regard and emerges as the best algorithm for tracking of PV systems.

7. Conclusion

In view of the unavailability of a common bench test on the three most commonly used tracking algorithms, this paper presents a common simulation platform for testing and assessment of P&O,

fuzzy logic, and neural network-based tracking algorithms. The biggest hindrance for implementation of a neural network-based tracking algorithm is the training requirement. For P&O technique, the results are optimal only if there is a proper knowledge of the range of duty ratio variation for a system, corresponding to the changes in the atmospheric conditions. From the simulation results, it is observed that the tracking effectiveness decreases in the order fuzzy logic controller, P&O algorithm, and neural network-based tracking. The results stand for changing insolation at fixed temperature as well as for a change in temperature at a constant insolation level. The fuzzy logic control is better suited for tracking as it gives the highest performance for the changes in the external conditions. From the view point of convergence speed as well, the fuzzy logic controller performs the best among the three techniques achieving a stable state after a change in either the insolation profile or the temperature change. The results can be used as a reference for implementation of fuzzy logic MPP trackers in hardware making use of digital signal processor controller (DSP) or field programmable gate array (FPGA). Moreover possible application of hybrid algorithms like Genetic algorithm (GA) -Fuzzy, Particle swarm optimization (PSO)-Fuzzy, Differential Evolution (DE) -Fuzzy logic-based MPPT controllers can also be investigated to check for better response in the overall system performance.

Funding
The authors received no direct funding for this research.

Author details
Barnam Jyoti Saharia[1]
E-mail: bjsece@tezu.ernet.in
Munish Manas[1]
E-mail: munish@tezu.ernet.in
Bani Kanta Talukdar[2]
E-mail: bktalukdar@yahoo.com
[1] Department of Electronics and Communication Engineering, Tezpur University, Assam 784028, India.
[2] Department of Electrical Engineering, Assam Engineering College, Assam 781013, India.

References
Ahmed, N. A., & Miyatake, M. (2008). A novel maximum power point tracking for photovoltaic applications under partially shaded insolation conditions. *Electric Power Systems Research, 78*, 777–784. http://dx.doi.org/10.1016/j.epsr.2007.05.026
Chih-Lyang, H., Li-Jui, C., & Yuan-Sheng, Y. (2007). Network-based fuzzy decentralized sliding-mode control for car like mobile robots. *IEEE Transactions on Industrial Electronics, 54*, 574–585.
Eltawil, M. A., & Zhao, Z. (2013). MPPT techniques for photovoltaic applications. *Renewable and Sustainable Energy Reviews, 25*, 793–813. http://dx.doi.org/10.1016/j.rser.2013.05.022
Esram, T., & Chapman, P. L. (2007). Comparison of photovoltaic array maximum power point tracking techniques. *IEEE Transactions on Energy Conversion, 22*, 439–449.
Fathabadi, H. (2015). Lambert W function-based technique for tracking the maximum power point of PV modules connected in various configurations. *Renewable Energy, 74*, 214–226. http://dx.doi.org/10.1016/j.renene.2014.07.059
Haykin, S. (1999). *Neural Networks A Comprehensive Foundation* (2nd ed.). Pearson Education.
Hohm, D. P., & Ropp, M. E. (2003). Comparative study of maximum power point tracking algorithms. *Progress in Photovoltaics: Research and Applications, 11*, 47–62. doi:10.1002/pip.459

Iqbal, A., Ahmed, S. K. M., Abu-Rub, H., & Sinan, S. (2010). Adaptive neuro-fuzzy inference system based maximum power point tracking of a solar PV module. *2010 IEEE International Energy Conference & Exhibition, Energycon* (pp. 51–56). Manama. doi:10.1109/ENERGYCON.2010.5771737
Jain, S., & Agarwal, V. (2007). New current control based MPPT technique for single stage grid connected PV systems. *Energy Conversion and Management, 48*, 625–644. http://dx.doi.org/10.1016/j.enconman.2006.05.018
Kim, S. (2007). Robust maximum power point tracker using sliding mode controller for the three- phase-grid connected photovoltaic system. *Solar Energy, 81*, 405–414. http://dx.doi.org/10.1016/j.solener.2006.04.005
Kottas, T. L., Boutalis, Y. S., & Karlis, A. D. (2006). New maximum power point tracker for PV arrays using fuzzy controller in Close cooperation with fuzzy cognitive networks. *IEEE Transactions on Energy Conversion, 21*, 793–803. doi:10.1109/TEC.2006.875430
Kyocera KC120-1 multi-crystalline photovoltaic module datasheet. (2001). Retrieved September 1, 2015 from (www.kyocerasolar.com/assets/001/5180.pdf
Larbes, C., Cheikh, S. M. A., Obeidi, T., & Zerguerras, A. (2009). Genetic algorithms optimized fuzzy logic control for the maximum power point tracking in photovoltaic system. *Renewable Energy, 34*, 2093–2100. http://dx.doi.org/10.1016/j.renene.2009.01.006
Li, S. (2015). A maximum power point tracking method with variable weather parameters based on input resistance for photovoltaic system. *Energy Conversion and Management, 106*, 290–299.
Liu, X., & Lopes, L. A. C. (2004). An improved perturbation and observation maximum power point tracking algorithm for PV arrays. *Power electronics specialists conference, 2004, IEEE 35th annual Power electronics specialist conference 2004, PESC 04* (pp. 2005–2010).
Liu, L., Liu, C., & Gao, H. (2013). A novel improved particle swarm optimization maximum power point tracking control method for photovoltaic array by using current calculated predicted arithmetic under partially shaded conditions. *Journal of Renewable and Sustainable Energy, 5*. doi:10.1063/1.4858615
Manas, M.. (2015, February 13–14). Development of self-sustainable technologies for smart grid in India. *2015 IEEE International Conference on Computational Intelligence & Communication Technology (CICT)* (pp. 563–568).
Manas, M., Zaheeruddin, M., & Sharma, B. B. (2015, March 3–4). Development of a benchmark for the interpretation of

transformer frequency response analysis results. *2015 International Conference on Cognitive Computing and Information Processing (CCIP)* (pp. 1–5).

de Medeiros Torres, A., Antunes, F. L. M., & dos Reis, F. S.. (1998). An artificial neural network-based real time maximum power tracking controller for connecting a PV system to the grid. *Proceeding of IEEE the 24th annual conference on industrial electronics society* (Vol. 1, pp 554–558. Aachen. doi:10.1109/IECON.1998.724303

Mellit, A., & Kalogirou, S. A. (2008). Artificial intelligence techniques for photovoltaic applications: A review. *Progress in Energy and Combustion Science, 34*, 574–632. doi:10.1016/j.pecs.2008.01.001

Mellit, A., Kalogirou, S. A., Hontoria, L., & Shaari, S. (2008). Artificial intelligence techniques for sizing photovoltaic systems: A review. *Renewable and Sustainable Energy Reviews, 13*, 406–419. doi:10.1016/j.rser.2008.01.00

Muralidhar, K. & Susovon, S. (2015). Modified perturb and observe MPPT algorithm for drift avoidance in photovoltaic systems. *IEEE Transactions On Industrial Electronics, 62*, 5549–5559.

Saharia, B. J. (2014). A theoretical study of performance and design constraints of non-isolated Dc-Dc converters. *International Journal Of Innovative Research In Electrical, Electronics, Instrumentation And Control Engineering, 2*, 1920–1925.

Salah, C. B., & Ouali, M. (2011). Comparison of fuzzy logic and neural network in maximum power point tracker for PV systems. *Electric Power Systems Research, 81*, 43–50.

Saleh, C. B., Chaaben, M., & Ammar, M. B. (2008). Multi-criteria fuzzy algorithm for energy management of a domestic photovoltaic panel. *Renewable Energy, 33*, 993–1001. http://dx.doi.org/10.1016/j.renene.2007.05.036

Sher, H. A., Murtaza, A. F., Noman, A., Addoweesh, K. E., Al-Haddad, K., & Chiaberge, M. (2015). A New sensorless hybrid MPPT algorithm based on fractional short-circuit current measurement and P&O MPPT. *IEEE Transactions on Sustainable Energy, 6*, 1426–1434.

Sivasubramaniam, S., Faramus, A., Tilley, R. D., & Alkaisi, M. M. (2014). Performance enhancement in silicon solar cell by inverted nanopyramid texturing and silicon quantum dots coating. *Journal of Renewable and Sustainable Energy, 6*. doi:10.1063/1.4828364

Subhadeep, B., & Saharia, B. J. (2014). A comparative study on converter topologies for maximum power point tracking application in photovoltaic generation. *Journal of Renewable and Sustainable Energy, 6*, 053140. doi:10.1063/1.4900579

Tsang, K. M., & Chan, W. L. (2015). Maximum power point tracking for PV systems under partial shading conditions using current sweeping. *Energy Conversion and Management, 93*, 249–258. http://dx.doi.org/10.1016/j.enconman.2015.01.029

Venelinov, T. A., Leonardo, C. G., Vincenzo, G., Francesco, C., & Okyay, K. (2007). Sliding mode neuro-adaptive control of electrical drives. *IEEE Transactions on Industrial Electronics, 54*, 671–679.

Zaheeruddin, & Manas, M. (2015a). Analysis of Design of technologies, tariff Structures and regulatory policies for sustainable growth of the Smart grid. *Energy Technology and Policy Journal, 2*, 28–38.

Zaheeruddin, & Manas, M.. (2015b). *A new approach for the design and development of renewable energy management system through microgrid central controller. Energy Reports, 1*, 156–163). Elsevier. http://dx.doi.org/10.1016/j.egyr.2015.06.003

13

Optimal power allocation scheme for plug-in hybrid electric vehicles using swarm intelligence techniques

Pandian M. Vasant[1]*, Imran Rahman[1], Balbir Singh Mahinder Singh[1] and M. Abdullah-Al-Wadud[2]

*Corresponding author: Pandian M. Vasant, Department of Fundamental and Applied Sciences, Universiti Teknologi PETRONAS 32610 Seri Iskandar, Perak, Malaysia
E-mails: pandian_m@utp.edu.my, pvasant@gmail.com
Reviewing editor: Kun Chen, Wuhan University of Technology, China

Abstract: Green technologies gain popularity to reduce the pollution and give higher penetration of renewable energy source in the transportation. This research induce that the extensive involvement of plug-in hybrid electric vehicles (PHEVs) requires adequate charging allocation strategy using a combination of smart grid systems and smart charging infrastructures. It is also noticed that daytime charging station are necessary for daily usage of PHEVs due to the limited all-electric-range. Most of the researches in the past have been stated that only proper charging control and infrastructure management can assure the larger participation of PHEVs. Therefore, researchers are trying to develop efficient control mechanism for charging infrastructure in order to facilitate upcoming PHEVs penetration in highway. Nevertheless, most of the past researcher already aware with the issue related to intelligent energy management. Yet, these studies could not fill the gap of the problem associated with intelligent energy management and require formulation of mathematical models with extensive use of computational intelligence-based optimization techniques to solve many technical problems. The outcome of this research study provides four optimization techniques that include Hybrid method

ABOUT THE AUTHOR

Pandian M. Vasant is a senior lecturer at Department of Fundamental and Applied Sciences, Universiti Teknologi PETRONAS in Malaysia. His research interests include Soft Computing, Hybrid Optimization, Holistic Optimization and Applications. He has co-authored research papers and articles in national journals, international journals, conference proceedings, conference paper presentation, and special issues lead guest editor, lead guest editor for book chapters' project, conference abstracts, short communication, editorial note, edited books and book chapters. In the year 2009, Pandian M. Vasant was awarded top reviewer for the journal Applied Soft Computing (Elsevier). He has given keynote lecture at ICTCC 2016 international conference. Currently he is an editor-in-chief of IJCO, IJIEM, IJSIEC, IJEOE, and GJTO.
H-Index SCOPUS Citation = 31, H-Index Google Scholar Citation = 21.
Professional Weblink Reference:
https://scholar.google.com.my/citations?hl=en&user=cOKj9goAAAAJ
https://my.linkedin.com/in/vasant

PUBLIC INTEREST STATEMENT

In the modern era, swarm intelligence has been widely applied to a variety of fields in engineering due to its outstanding features for solving optimization problems with complex fitness function with constraints. This particular research study focuses on swarm intelligent-based approach for the optimal power allocation scheme for Plug-in Hybrid Electric Vehicles (PHEVs). The main objective is to allocate power intelligently for each PHEV coming to the charging station. The State-of-Charge (SoC) is the main parameter which needs to be maximized in order to allocate power efficiently. For this, the fitness function considered in this research is the maximization of average SoC hence allocate power to PHEVs at the next time step. According to the simulation results, the hybrid method, PSOGSA shows best fitness values and takes the longest time to complete 100 iterations. Moreover, both PSO and APSO optimization techniques show better result in terms of computation time.

within swarm intelligence group for the State-of-Charge (SoC) optimization of PHEVs. The finding of this research simulation results obtained for maximizing the highly nonlinear objective function evaluate the comparative performance of all four techniques in terms of best fitness, convergence speed, and computation time. Finally, the hybridization method (PSOGSA) presented in this dissertation uses the advantages of both PSO and GSA optimization and thus produce higher best fitness values. This study evaluates the performance of standard PSO, then Accelerated version of PSO (APSO), GSA algorithm and then Hybrid of PSO and GSA. The hybridization method (PSOGSA) uses the advantages of both PSO and GSA optimization and thus produce higher best fitness values. However, PSOGSA method takes much longer computational time than single methods because of incorporating two single methods in one algorithm. This research study suggests that PSOGSA method is a great promise for SoC optimization but it takes much longer computational time.

Subjects: Artificial Intelligence; Engineering Management; Power & Energy; Transport & Vehicle Engineering

Keywords: APSO; electric vehicle; gravitational search algorithm; optimization; PSO; PSOGSA; swarm intelligence

1. Introduction

Recent researches on sustainable technologies for transportation sector are gaining popularity among the research communities from different areas. In this wake, Plug-in hybrid electric vehicles (PHEVs) have great future because of their charge storage system and charging facilities from traditional grid system. Several researchers have proved that a great amount of reductions in greenhouse gas emissions and the increasing dependence on oil could be accomplished by electrification of transport sector (Rahman et al., 2016a). Future transportation sector will depend much on the advancement of this emerging field of vehicle optimization (Tie & Tan, 2013). PHEVs which is very recently introduced promise to boost up the overall fuel efficiency by holding a higher capacity battery system, which can be directly charged from conventional power grid system, that helps the vehicles to operate continuously in "all-electric-range" (AER). All-electric vehicles or AEVs is a kind of transport that uses electric power as only sources to run the system. PHEVs with a connection to the smart grid can own all of these strategies. Hence, the widely extended adoption of PHEVs might play a significant role in the alternative energy integration into traditional grid systems (Lund & Kempton, 2008). There is a need of efficient mechanisms and algorithms for smart grid technologies in order to solve highly diverse problems like energy management, cost reduction, efficient charging station, etc. with different objectives and system constraints (Hota, Juvvanapudi, & Bajpai, 2014).

According to a statistics of Electric Power Research Institute (EPRI), about 62% of the entire United States (US) vehicle will comprise PHEVs within the year 2050 (Soares et al., 2013). Moreover, there is an increasing demand to implement this technology on the electric grid system. Large numbers of PHEVs have the capability to make threats to the stability of the power system. For example, in order to avoid disturbance when several thousand PHEVs are introduced into the system over a small period, the load on the power grid will need to be managed very carefully. One of the main targets is to facilitate the proper communication between the power grid and the PHEV. For the maximization of customer contentment and minimization of burdens on the grid, a complicated control appliance will need to be addressed in order to govern multiple battery loads from a numbers of PHEVs properly (Su & Chow, 2012). The total demand pattern will also have an important impact on the electricity production due to differences in the needs of the PHEVs parked in the deck at certain time (Su & Chow, 2011a). Proper management can ensure strain minimization of the grid and enhance the transmission and generation of electric power supply. The control of PHEV charging depending on the locations can be classified into two groups; household charging and public charging. The proposed optimization focuses on the public charging station for plug-in vehicles because most of PHEV charging is expected to take place in public charging location (Su, Eichi, Zeng, & Chow, 2012). Wide penetration of PHEVs in the market depends on a well-organized charging infrastructure. The power demand from this new load will put extra stress on the traditional power grid (Morrow, Karner, & Francfort, 2008). In (Amini & Parsa Moghaddam, 2013), a probabilistic approach is utilized to estimate the expected EV charging demand based on the historical data. To this end, Amini et al. used the expected arrival and departure time, and daily driven distance to obtain a realistic model for EV parking lots. Charging stations are needed to be built at workplaces, markets/shopping malls, and home. Boyle (2007) proposed the necessity of building new smart charging station with effective communication among utilities along with sub-station control infrastructure in view of grid stability and proper energy utilization. Furthermore, sizeable energy storage, cost minimization, Quality of Services (QoS), and intelligent charging station for optimal power are underway (Hess et al., 2012). In this wake, numerous techniques and methods were proposed for deployment of PHEV charging stations (Li et al., 2010).

One of the main targets is to facilitate the proper interaction between the power grid and the PHEV. For the maximization of customer satisfaction and minimization of burdens on the grid, a complicated control mechanism will need to be addressed in order to govern multiple battery loads from a numbers of PHEVs appropriately (Su et al., 2012). Charging infrastructures are essential in order to facilitate the large-scale penetration of PHEVs. Some researchers for charging station optimization of PHEV have used different computational intelligence-based methods. Most of them applied traditional methods that needed to be improved furthermore.

Swarm intelligence came from the mimic of the living colony such as ant, bird, and fish in nature, which shows unparalleled excellence in swarm than in single in food seeking or nest building. Drawing inspiration from this, researches design many algorithms simulating colony living, such as Ant Colony Optimization (ACO) algorithm (Dorigo, 2006), Particle Swarm Optimization (PSO) algorithm (Eberhart & Yuhui, 2001), Artificial Bee Colony (ABC) algorithm (Karaboga & Basturk, 2007), and Gravitational Search Algorithm (GSA) (Rashedi, Nezamabadi-pour, & Saryazdi, 2009), which shows excellent performance in dealing with complex optimization problems (Martens, Baesens, & Fawcett, 2011). The intrinsic characteristics of all the population-based metaheuristic algorithms like PSO and GSA are to maintain a good compromise between exploration and exploitation in order to solve the complex optimization problems (Rashedi et al., 2009).

The performance of PHEV depends upon proper utilization of electric power which is solely affected by the battery state-of-charge (SoC). In PHEVs, a key parameter is the SoC of the battery as it is a measure of the amount of electrical energy stored in it. It is analogous to fuel gauge on a conventional internal combustion (IC) car (Chiasson & Vairamohan, 2005). State-of-charge determination becomes an increasingly vital issue in all the areas that include a battery. Previous operation policies made use of voltage limits only to guard the battery against deep discharge and overcharge. Currently, battery operation is changing to what could rather be called battery management than simply protection. For this improved battery control, the battery SoC is a key factor indeed (Piller, Perrin, & Jossen, 2001).

A charging station is one way that the operator of an electrical power grid can adapt energy production to energy consumption, both of which can vary randomly over time. Generally, PHEVs in a charging station are charged during times when production exceeds consumption and are discharged at times when consumption exceeds production (Yang, He, & Fu, 2014). There is a need of in-depth study on maximization of average SoC in order to facilitate intelligent energy allocation for PHEVs in a charging station.

The purpose of this research is to optimize state-of-charge, with respect to charging time, present SoC. For this, swarm intelligence-based methods, PSO and GSA, Accelerated Particle Swarm

Optimization (APSO), and Hybrid version of PSO and GSA (PSOGSA) were applied for solving the particular optimization problem hence presents comparative study on these four techniques.

2. Problem statement

The anticipation of a large penetration of PHEVs and Plug-in Electric Vehicles (PEVs) into the market brings up many technical problems that need to be addressed within the next 10 years (Hannan, Azidin, & Mohamed, 2014). In the future, electric-powered vehicles would be plugged into the grid, and their onboard energy storage systems would be recharged using clean, renewable electricity. One of the key missions is to facilitate the smooth interaction between the plug-in hybrid electric vehicle and the power system (Mwasilu, Justo, Kim, Do, & Jung, 2014). Therefore, there is a need of optimum power allocation scheme in order to facilitate the SoC of PHEVs as well as the overall performance of charging station for large-scale PHEVs in smart grid environment. Lack of proper and adequate charging infrastructure is another issue regarding PHEV compared to typical fuel station. As a result, uncertainty in energy management for the driver named as 'Range anxiety' (Contestabile, Offer, Slade, Jaeger, & Thoennes, 2011) develops which restricts the conventional vehicle owner to switch their vehicle to PHEV. Selecting appropriate place for charging station can decrease this range anxiety. There remain much scientific studies to be performed on how to manage PHEV storage energy, SoC effectively for real-time traffic situations. One of the important constraints for accurate charging is SoC. Charging algorithm can precisely be managed by the precise state of charge evaluation (Shafiei & Williamson, 2010). An approximate graph of a typical Lithium-ion cell voltage vs. SoC shown in Figure 1 indicates that the slope of the curve below 20% and above 90% is high enough to result in a significant voltage difference to be depended on by measurement circuits and charge balancing control. There is a need of in-depth study on maximization of average SoC in order to facilitate intelligent energy allocation for PHEVs in a charging station.

The idea behind smart charging is to charge the vehicle when it is most favorable, which could be when electricity price, demand is lowest, when there is excess capacity (Chang, 2013). When a vehicle is plugged in into a smart charging station a request for energy demand is sent to Substation Control Center (SCC), which decides based on the available energy from utility and either accepts the request or rejects it. Performance of this kind of load management is measured in terms of delay, delivery ration, and jitter. As a matter of fact EVs may be charged at any time of a day depending on requirement to top their batteries even during peak demand hours. In (Amini & Arif, 2014), an optimal reliability-constraint allocation of EVs' parking lots is proposed based on the probabilistic charging demand model of EVs.

3. Scope of study

This research mainly focuses on swarm intelligence-based optimization methods for solving PHEVs charging problem. Here, a single parking lot with the aggregation of distribution network-connected PHEVs is considered. We make use of historical data for office parking from the city of Livermore, CA (Rahman et al., 2016b). There are many swarm intelligence-based method applied for solving real-world problems. Among them four methods have been applied to maximize the state-of-charge. Authors in (Su & Chow, 2012) first applied Genetic Algorithm (GA) and PSO methods to solve this particular problem. According to the results, the authors claim that PSO algorithm converges to a solution in a reasonable amount of time, faster than GA. They also suggest applying other optimization techniques for comparison purposes and selecting the best-suited technique to maximize the SoC of PHEV. From this future research direction, we have come up with swarm intelligence-based methods with comparative study in order to find out the suitable techniques among swarm intelligence-based optimization domain. According to Rahman et al. (2016c), many papers do not document experimental settings in sufficient detail, and hence replication of experiments is almost impossible. We have found the same problem in the research of PSO-based optimization for charging PHEV. Moreover, they used swarm size (population size) 40 which is not sufficient to describe the performance of PSO algorithm properly. Therefore, we have replicated the PSO optimization in our own computer configuration and parameter settings for the fair comparison with other three techniques.

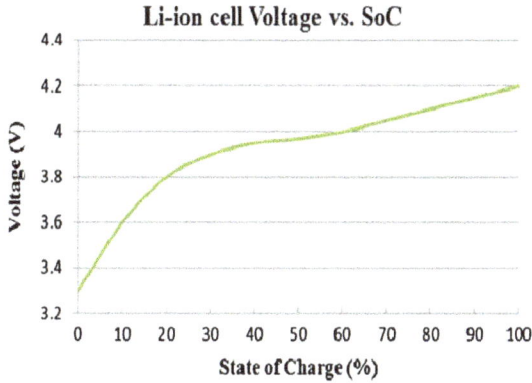

Figure 1. Li-ion cell voltage vs. State-of-Charge.

4. Methodology

Suppose there is a charging station with the capacity of total power P, total N numbers of PHEVs need to be served in a day (24 h). The proposed system should allow PHEVs to leave the charging station before their expected leaving time for making the system more effective. It is worth to mention that each PHEV is regarded to be plugged-in to the charging station once. The main aim is to allocate power intelligently for each PHEV coming to the charging station. The State-of-Charge is the main parameter which needs to be maximized in order to allocate power efficiently. For this, the fitness function considered in this research is the maximization of average SoC and thus allocate energy for PHEVs at the next time step. The constraints considered are: charging time, present SoC, and price of the energy.

The fitness function is defined as:

$$\text{Max } J(k) = \sum_i w_i(k) SoC_i(k+1) \tag{1}$$

$$w_i(k) = f(C_{r,i}(k), T_{r,i}(k), D_i(k)) \tag{2}$$

$$C_{r,i}(k) = (1 - SoC_i(k)) \cdot C_i \tag{3}$$

where $C_{r,i}(k)$ is the battery capacity (remaining) needed to be filled for i No. of PHEV at time step k; C_i is the battery capacity (rated) of the i no. of PHEV; remaining time for charging a particular PHEV at time step k is expressed as $T_{r,i}(k)$; the price difference between the real-time energy price and the price that a specific customer at the i no. of PHEV charger is willing to pay at time step k is presented by $D_i(k)$; $w_i(k)$ is the charging weighting term of the i no. of PHEV at time step k (a function of charging time, present SoC, and price of the energy); $SoC_i(k+1)$ is the state of charge of the i no. of PHEV at time step $k + 1$.

The weighting term gives a reward proportional to the attributes of a specific vehicle. For example, if a vehicle has a lower initial SoC and less remaining charging time, but the driver is willing to pay a higher price, the controller allocates more power to this vehicle battery charger:

$$w_i(k)\alpha[C_{r,i}(k) + D_i(k) + 1/T_{r,i}(k)] \tag{4}$$

The charging current is also assumed to be constant over Δt.

$$[SoC_i(k+1) - SoC_i(k)] \cdot Cap_i = Q_i = I_i(k)\Delta t \tag{5}$$

$$SoC_i(k+1) = SoC_i(k) + I_i(k)\Delta t/Cap_i \tag{6}$$

where the sample time Δt is defined by the charging station operators, and $I_i(k)$ is the charging current over Δt. The battery model is regarded as a capacitor circuit, where C_i is the capacitance of battery (Farad). The model is defined as:

$$C_i\frac{dV_i}{dt} = I_i \tag{7}$$

Therefore, over a small time interval, one can assume the change of voltage to be linear,

$$C_i[V_i(k+1) - V_i(k)]/\Delta t = I_i \tag{8}$$

$$V_i(k+1) - V_i(k) = I_i\Delta t/C_i \tag{9}$$

Since the decision variable is the power allocated to the vehicles, replacing $I_i(k)$ with $P_i(k)$:

$$I_i(k) = P_i(k)/0.5 \times [V_i(k+1) - V_i(k)] \tag{10}$$

$$V_i(k+1) = \sqrt{\frac{2P_i(k)\Delta t}{C_i} + V_i^2(k)} \tag{11}$$

Substituting (10) into (6) yields:

$$SoC_i(k+1) = SoC_i(k) + \frac{P_i(k)\Delta t}{0.5.C_i\left[\sqrt{\frac{2P_i(k)\Delta t}{C_i}+V_i^2(k)} + V_i(k)\right]} \tag{12}$$

Finally, the objective function becomes:

$$J(k) = \sum w_i\left[SoC_i(k)\frac{P_i(k)\Delta t}{0.5.C_i\left[\sqrt{\frac{2P_i(k)\Delta t}{C_i} + V_i^2(k)}+V_i(k)\right]}\right] \tag{13}$$

There are two kinds of inequality constraints used here to optimize the fitness function—(i) Power from the charging station operator and (ii) individual PHEV's SoC. Power obtained from the utility ($P_{utility}$) and the maximum power ($P_{i,max}$) absorbed by a specific PHEV are the primary energy constraints being considered in this chapter. The overall charging efficiency of a particular charging station is described by η. From the system point of view, charging efficiency is supposed to be constant at any given time step. Maximum battery SoC limit for the i No. of PHEV is $SoC_{i,max}$. When SoC_i reaches the values close to $SoC_{i,max}$, the i No. of battery charger shifts to a standby mode. The state of charge ramp rate is confined within limits by the constraint ΔSoC_{max}. The overall control system changes the state when (i) system utility data updates; (ii) a new PHEV is plugged-in; (iii) time period Δt has periodically passed. Obviously, SoC maximization method being considered in this chapter can provide a uniformly higher SoC for all PHEVs/PEVs at plug-out as compared with the alternative schemes. It also proves that the proposed function aims at ensuring some fairness in the SoC distribution at each time step. This will help to ensure that a reasonable level of battery power is attained, even in the event of an early departure. Table 1 shows all the fitness function parameters that were tuned for performing the optimization. There are total three (03) kinds of parameters: fixed, variables, and constraints. Total charging time is fixed to 20 min and charging station efficiency is assumed to be 0.9. The values are retrieved from various literatures (Hota et al., 2014; Kulshrestha, Wang, Chow, & Lukic, 2009). Moreover, SoC is in the range of 0.2–0.8 (Su & Chow, 2011b).

Table 1. Parameter settings of the objective function	
Parameter	**Values**
Fixed parameters	Maximum power, $P_{i,max} = 6.7$ kWh
	Charging station efficiency, $\eta = 0.9$
	Total charging time, $\Delta t = 20$ Min
	Power to each PHEV: 30 W
Variables	$0.2 \leq$ State-of-Charge (SoC) ≤ 0.8
	Waiting time ≤ 30 Min (1,800 S)
	16 kWh $\leq$ Battery capacity ( ) ≤ 40 kWh
Constraints	$\sum_i P_i(k) \leq P_{utility}(k) \times \eta$
	$0 \leq P_i(k) \leq P_{i,max}(k)$
	$0 \leq SoC_i(k) \leq SoC_{i,max}$
	$0 \leq SoC_i(k+1) - SoC_i(k) \leq \Delta SoC_{max}$

The objective function and parameter settings have already been discussed earlier. After that, we set the parameters of each swarm intelligence-based optimization technique. If the experiment is not replicated with sufficient care, any performance measures and statistical approaches cannot remedy the problems introduced by inexact experiment replication. In other words, if collected data are gathered from experiments that exhibit large deviations the comparison is meaningless despite statistical test being applied. Hence, it is crucial that experiment replications are properly conducted. For proper comparison, number of runs as well as iterations should be strictly maintained for all four techniques. After MATLAB simulation with exactly same computer configuration and simulation environment, we analyzed the convergence behavior of each swarm intelligence technique for optimizing the objective function. Finally, the performance evaluation and comparison of each technique was carried out in terms of "Fitness value," "Computational time," and "Robustness". Figure 2 shows the flow chart of our methodology.

4.1. Particle swarm optimization (PSO)

PSO is an evolutionary computation technique that is proposed by Eberhart and Yuhui (2001). The PSO was inspired from social behavior of bird flocking. It uses a number of particles (candidate solutions) which fly around in the search space to find best solution. Meanwhile, they all look at the best particle (best solution) in their paths. In other words, particles consider their own best solutions as well as the best solution has found so far. Each particle in PSO should consider the current position, the current velocity, the distance to pbest, and the distance to gbest in order to modify its position. PSO is initialized with a group of random particles (solutions) and then searches for optima by updating generations. In every iteration, each particle is updated by following two "best" values. The first one is the best solution (fitness) it has achieved so far (The fitness value is also stored). This value is called "pbest". Another "best" value that is tracked by the particle swarm optimizer is the best value, obtained so far by any particle in the population. This best value is a global best and called "gbest".

PSO was mathematically modeled as follows:

$$V_i^{t+1} = wv_i^t + c_1 \text{ rand} \left(pbest_i - x_i^t \right) + c_2 \text{rand} \left(gbest - x_i^t \right) \tag{14}$$

$$x_i^{t+1} = x_i^t + V_i^{t+1} \tag{15}$$

where v_i^t is the velocity of particle i at iteration t, w is a weighting function usually used as follows:

$$w = \omega_{max} - \frac{w_{max} - \omega_{min}}{Itre_{max}} Itre \tag{16}$$

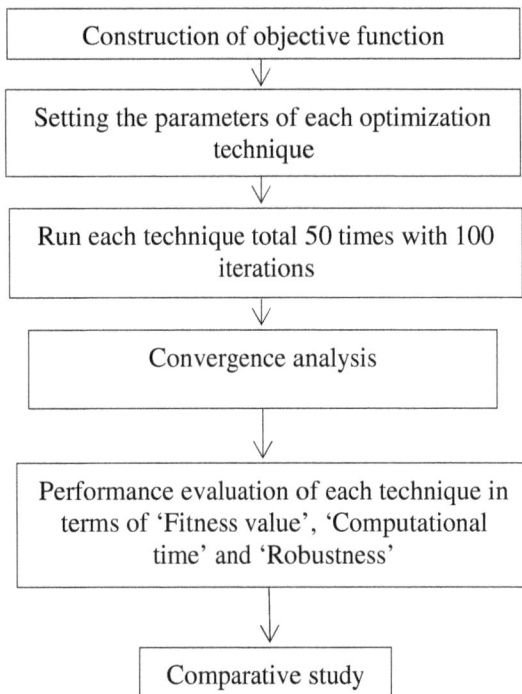

Figure 2. Process flow diagram of methodology.

There are updates of velocities as well as positions of the particles. The algorithm repeats until the maximum number of iterations or the minimum error criteria is met. Figure 3 shows the flowchart for PSO.

Appropriate values for ω_{min} and ω_{max} are 0.4 and 0.9 (Wu, Cao, Wen, & Bian, 2008). Suitable value ranges for c_1 and c_2 1 to 2 (Su & Chow, 2011a), but 2 is most appropriate in many cases (Soares, Morais, & Vale, 2012). rand is a random number between 0 and 1 (Su & Chow, 2011a), x_i^t is the current position of particle i at iteration t, $pbest_i$ is the pbest of agent i at iteration t and gbest is the best solution so far. PSO algorithm works by simultaneously maintaining several particles or potential solutions in the search space. For each iteration of the algorithm, each particle is evaluated by the fitness function being optimized, based on the fitness of that solution. Table 2 shows the parameter settings of PSO method.

4.2. Accelerated particle swarm optimization (APSO)

Accelerated PSO was developed by Xin-She Yang (Martens et al., 2011) at Cambridge University in 2007 in order to accelerate the convergence of the algorithm is to use the global best only. PSO- and APSO-based optimizations have already been studied by the researchers for optimal design of sub-station grounding grid (El_Fergany, 2013), performance analysis of MIMO radar waveform (Reddy, 2012), design of frame structures (Talatahari et al., 2013), dual channel speech enhancement (Prajna et al., 2014), a faster path planner (Mohamed et al., 2012), etc.

In APSO, each member of the population is called a particle and the population is called a swarm. Starting with a randomly initialized population and moving in randomly chosen directions, each particle moves through the searching space and remembers the best earlier positions, velocity and accelerations of itself and its neighbors. Particles of a swarm communicate good position, velocity and acceleration to each other as well as dynamically adjust their own position, velocity and acceleration derived from the best position of all particles. The next step starts when all particles have been shifted. Finally, all particles inclined to fly towards better positions over the searching process until the swarm move close to an optimum of the fitness function.

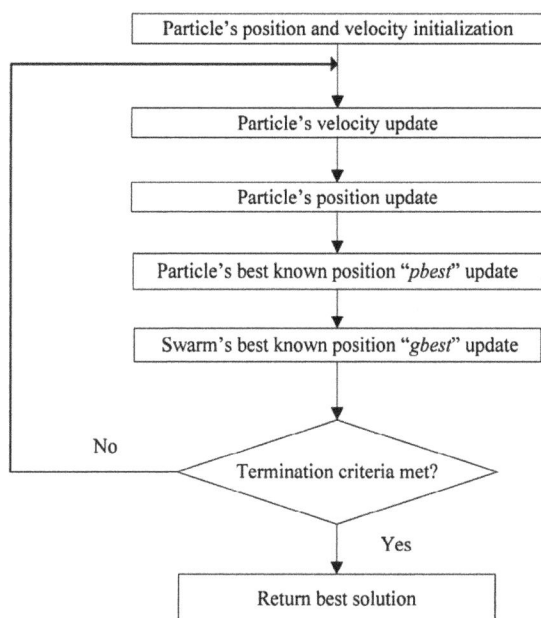

Figure 3. Flowchart of Particle Swarm Optimization (PSO).

A simplified version that could accelerate the convergence of the algorithm is to use the global best only. Thus, in the APSO (Talatahari et al., 2013), the velocity vector is generated by a simpler formula as where *randn* is drawn from (0, 1) to replace the second term. The update of the position is simply like-

$$V_i^{t+1} = V_i^t + \alpha randn(t) + \beta(g^* - x_i^t) \tag{17}$$

where *randn* is drawn from N (0, 1) and the update of the position is like the standard PSO method. In order to increase the convergence even further, the update of the position can be written in a single step as

$$x_i^{t+1} = (1 - \beta)x_i^t + \beta g^* + \alpha r \tag{18}$$

In our simulation, we use (Gandomi et al., 2013):

$$\alpha = 0.7^t \tag{19}$$

Figure 4 shows the flowchart of APSO method.

The typical values for this APSO are $\alpha \approx 0.1–0.4$ and $\beta \approx 0.1–0.7$; however, $\alpha \approx 0.2$ and $\beta \approx 0.5$ are recommended (El_Fergany, 2013). In general, any evolutionary search algorithm shows improved performance with a relatively larger population. However, a very large population will cost more in terms of fitness function evaluations without producing significant improvements. In this simulation, the population size is set to 100. The parameter settings for APSO are demonstrated in Table 3.

4.3. Gravitational search algorithm (GSA)

GSA is an optimization method which has been introduced by Rashedi et al. in the year of 2009 (Rashedi et al., 2009). In GSA, the specifications of each mass (or agent) are total four, which is mass (inertial), position, mass (active gravitational), and mass (passive gravitational). The position of the mass presents a solution of a particular problem, and masses (gravitational and inertial) are obtained by using a fitness function. GSA can be considered as a collection of agents (candidate solutions), whose masses are proportional to their value of fitness function.

Table 2. PSO parameter settings

Parameter	Values
Size of the swarm	100
Maximum no. of steps	100
PSO parameter, c_1	1.4
PSO parameter, c_2	1.4
PSO inertia (w)	0.9
Maximum iteration	100
Number of runs	50

GSA-based optimization has already been used by the researchers for post-outage bus voltage magnitude calculations, economic dispatch with valve-point effects, optimal sizing and suitable placement for distributed generation (DG) in distribution system, optimization of synthesis gas production (Ganesan et al., 2013), solving thermal unit commitment (UC) problem (Roy, 2013) and finding out optimal solution for optimal power flow (OPF) problem in a power system (Duman, Güvenç, Sönmez, & Yörükeren, 2012), etc. Specifically, we are investigating the use of the GSA method for developing real-time and large-scale optimizations for allocating power.

The gravitational force is expressed as follows:

$$F_{ij}^d(t) = G(t)\frac{M_{pi}(t) \times M_{aj}(t)}{R_{ij}(t) + e}\left(x_j^d(t) - x_i^d(t)\right) \tag{20}$$

where M_{aj} is the active gravitational mass related to agent j, M_{pi} is the passive gravitational mass related to agent i, $G(t)$ is gravitational constant at time t, ε is a small constant, and $R_{ij}(t)$ is the Euclidian distance between two agents i and j. The $G(t)$ is calculated as:

$$G(t) = G_0 \times exp\left(-a \times iter/maxiter\right) \tag{21}$$

where a and G_0 are descending coefficient and primary value, respectively, current iteration and maximum number of iterations are expressed as $iter$ and $maxiter$. In a problem space with the dimension d, the overall force acting on agent i is estimated as following equation:

$$F_i^d(t) = \sum_{j=1,j\neq i}^{N} rand_j F_{ij}^d(t) \tag{22}$$

where $rand_j$ is a random number with interval [0, 1]. From law of motion we know that an agent's acceleration is directly proportional to the resultant force and inverse of its mass, so the acceleration of all agents should be calculated as follows:

$$ac_i^d(t) = \frac{F_i^d(t)}{M_{ii}(t)} \tag{23}$$

where t is a specific time and M_{ii} is the mass of the object i. The velocity and position of agents are calculated as follows:

$$vel_i^d(t+1) = rand_i \times vel_i^d(t) + ac_i^d(t) \tag{24}$$

$$x_i^d(t+1) = x_i^d(t) + vel_i^d(t+1) \tag{25}$$

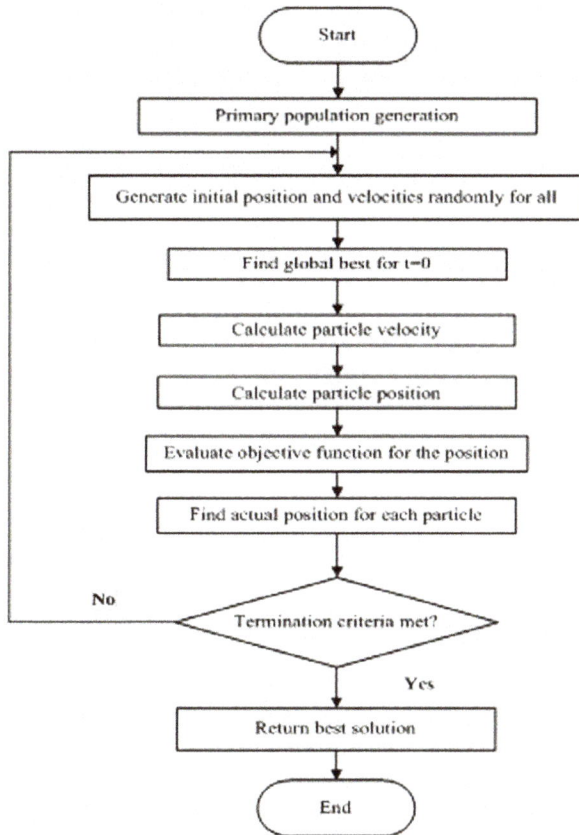

Figure 4. Flowchart of Accelerated Particle Swarm Optimization (APSO).

Table 3. APSO parameter settings	
Parameter	Values
Size of the swarm	100
Maximum no. of steps	100
Alpha, α	0.2
Beta, β	0.5
Maximum iteration	100
Number of runs	50

where $rand_i$ is a random number with interval [0, 1]. Moreover, the step involved in optimization using GSA is shown Figure 5.

In GSA, all agents are initialized first with random values. Each of the agents is a candidate solution. After initialization, velocities for all agents are defined using (24). Moreover, the gravitational constant, overall forces, and accelerations are determined by equations (21), (22), and (23), respectively. The positions of agents are calculated using (25). At the end, GSA will be terminated by meeting the stopping criterion of maximum 100 iterations. The parameter settings for GSA are demonstrated in Table 4.

4.4. Hybrid particle swarm optimization and gravitational search algorithm (PSOGSA)
Hybrid PSOGSA was introduced by Seyedali Mirjalili (Mirjalili & Hashim, 2010) at soft computing research lab of Universiti Teknologi Malaysia (UTM) in 2010 in order to integrate the ability of exploitation in PSO with the ability of exploration in GSA. PSOGSA-based optimization has already been used

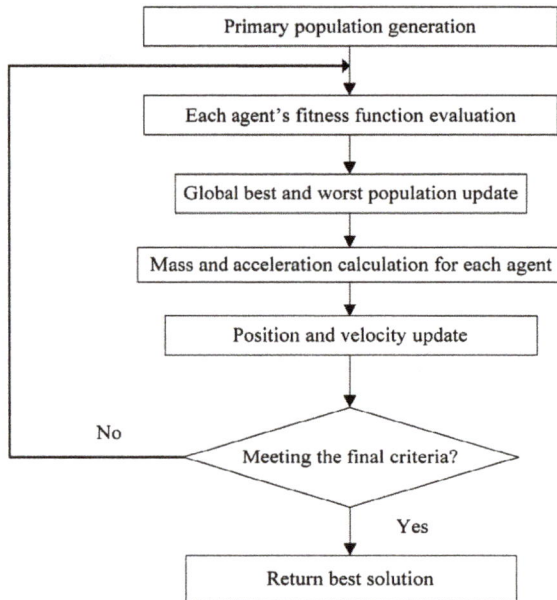

Figure 5. Flowchart of Gravitational Search Algorithm (GSA).

by the researchers for economic load dispatch (Dubey, Pandit, Panigrahi, & Udgir, 2013), optimal static state estimation (Mallick, Ghoshal, Acharjee, & Thakur, 2013), dual channel speech enhancement (Kunche, Rao, Reddy, & Maheswari, 2015), training feed-forward neural networks (Mirjalili, Mohd Hashim, & Moradian Sardroudi, 2012), multi distributed generation planning (Tan, Hassan, Rahman, Abdullah, & Hussin, 2013), etc.

The basic idea is to fit in the exploitation capability in PSO with the exploration capability in GSA to combine both algorithms' strength. In order to combine these two algorithms, velocity update is proposed as

$$v_i(t+1) = w \times v_i(t) + \alpha' \times rand \times ac_i(t) + \beta' \times rand \times (\text{gbest} - x_i(t)) \qquad (26)$$

where $v_i(t)$ is the velocity of agent i at iteration t, w is a weighting factor, $rand$ is a random number between 0 and 1, $ac_i(t)$ is the acceleration of agent at iteration t, and gbest is the best solution so far. Here, α' and β' are the weighting factors. With adjusting α' and β', the abilities of global search and local search can be balanced. The position of the particle $x_i(t+1)$ in each iteration is updated using the equation

$$x_i^d(t+1) = x_i^d(t) + vel_i^d(t+1) \qquad (27)$$

The flowchart of hybrid PSOGSA method is shown in Figure 6.

Table 4. GSA parameter settings	
Parameter	**Values**
Primary parameter	100
No. of mass agents	100
Acceleration coefficient	20
Constant parameter, β	0.01
Power of "R"	1
Maximum iteration	100
Number of runs	50

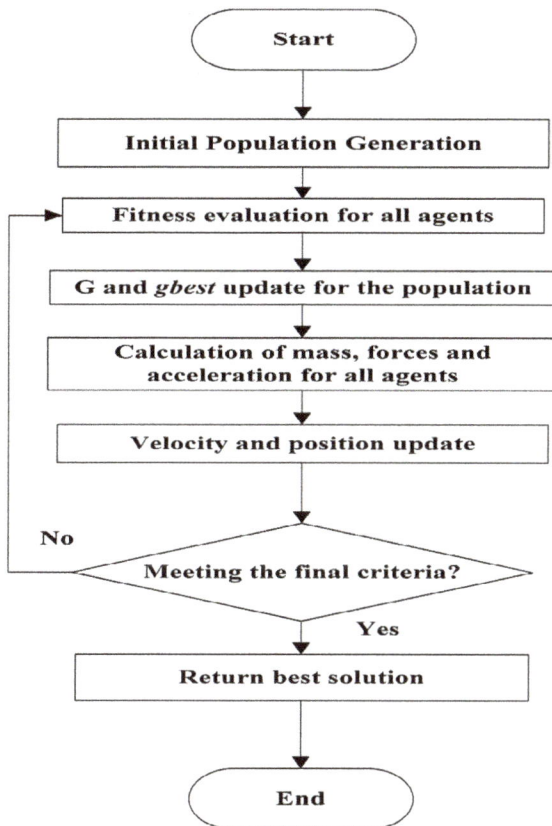

Figure 6. Hybrid Particle Swarm Optimization and Gravitational Search Algorithm (PSOGSA).

Table 5. Parameter settings of PSOGSA	
Parameter	**Values**
Size of the swarm	100
GSA Constant parameter	23
PSO parameter, c_1	0.5
PSO parameter, c_2	1.5
Gravitational constant	1
Maximum iteration	100
Number of runs	50

PSOGSA with the parameter settings stated in Table 5 was also performed for the same fitness function and compared with the performance of GSA in terms of average best fitness. The swarm size and maximum iterations were set exactly the same to that of GSA and PSO techniques for the comparison purpose. The values of parameters c_1, c_2, and alpha were set as standard values, 0.5, 1.5, and 23, respectively.

5. Simulation results of applied swarm intelligence techniques

In order to optimize state-of-charge, with respect to charging time, present SoC. For this, swarm in-telligence-based methods, PSO and GSA, APSO and Hybrid version of PSO and GSA (PSOGSA) were applied. All the optimization techniques were simulated to achieve the best fitness values of objec-tive function stated at equation number 13. All the simulations were run on the following computer configuration stated below:

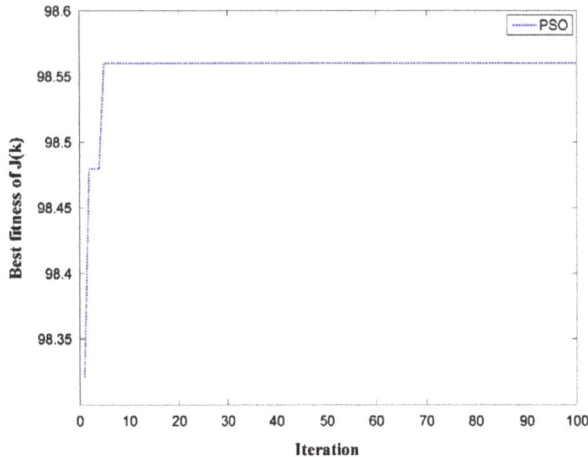

Figure 7. Iteration vs. fitness value, J (k) for PSO (100 PHEVs).

CPU: Core™ i5–3470 M

Processor: 3.20 GHz

RAM: 4.00 GB and

Software: MATLAB version- R2013a.

5.1. Particle swarm optimization (PSO)

Figure 7 shows the convergence behavior (iteration vs. fitness value) of PSO technique for 100 PHEVs. It can be apparently seen from the simulation study that although the algorithm has been set to run maximum 100 iterations, the fitness value converges before five (05) iterations for all five scenarios and become stable. Therefore, an early convergence may cause the fitness function to trap into local minima. This can be avoided by increasing the size of swarm hence the computational time will also be increased as well. As a result, a trade-off should be taken into consideration between the proper convergence and computational time.

Figure 8 depicts the best fitness value for 100 PHEVs. In this case, the maximum best fitness and minimum best fitness were 767.8722 and 9.5076, respectively. The average best fitness increases up to 182.9313.

As PSO is a population-based optimization technique and the fitness function is nonlinear, so the fitness values fluctuate for each iteration (El-Fergany, 2013; Rahman et al., 2016c; Soares et al., 2012;

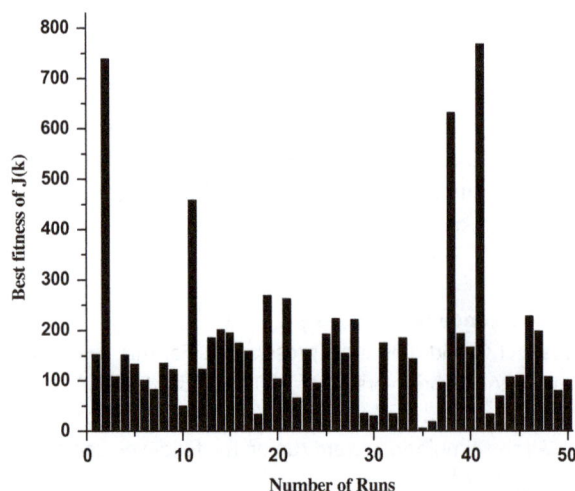

Figure 8. Fitness value vs. number of runs for PSO (100 PHEVs).

Table 6. Fitness evaluation of PSO					
J(k)	50 PHEVs	100 PHEVs	30 PHEVs	500 PHEVs	1,000 PHEVs
Max. best fitness	910.75	767.87	793.09	774.56	697.11
Average best fitness	142.84	171.10	169.31	144.80	156.80
Min. best fitness	4.84	5.38	5.22	7.18	0.73

Table 7. Average computational time for PSO	
Number of PHEVs	Computational Time (s)
50 PHEVs	1.62
100 PHEVs	1.67
300 PHEVs	1.76
500 PHEVs	1.95
1,000 PHEVs	2.33

Wu et al., 2008). However, the maximum best fitness remains in the range of 650 to 950 and the minimum best fitness remains in the range of 0.70–8. Table 6 summarizes the result. From that it can be concluded that average best fitness remain almost in similar pattern for four (05) different scenarios.

Table 7 shows the average computational time requirement for PSO method. The average computational time for 50 PHEVs is 1.620 s while for 1,000 PHEVs it increases up to 2.328 s.

5.2. Accelerated particle swarm optimization (APSO)

Figure 9 shows the convergence behavior (iteration vs. fitness value) of APSO technique for 100 PHEVs. Although the algorithm has been set to run for maximum 100 iterations, the fitness value converges after 10 iterations and becomes stable. Consequently, there is an early convergence that may cause the fitness function to trap into local minima. This can be avoided by increasing the size of swarm hence the computational time will also be increased as well. For instant, a trade-off should be taken into consideration between the proper convergence and computational time.

Figure 10 depicts the best fitness value for 100 PHEVs. In this case, the maximum best fitness and minimum best fitness were 679.7151 and 9.5076, respectively. The average best fitness increases up to 182.9313. In order to evaluate the performance and show the efficiency and superiority of the proposed algorithm, we ran each scenario a total of 50 times.

As APSO is a population-based optimization technique and the fitness function is nonlinear, so the fitness values fluctuate for each iteration (El-Fergany, 2013; Rahman et al., 2016c; Soares et al., 2012; Wu et al., 2008). However, the maximum best fitness remains in the range of 450–700 and the minimum best fitness remains in the range of 0.5–10. Table 8 summarizes the result. From that it can be concluded that average best fitness remains almost in similar pattern for five different scenarios.

Table 9 shows the average computational time requirement for APSO method. The average computational time for 50 PHEVs is 1.696 s while for 1,000 PHEVs it increases up to 2.092 s.

5.3. Gravitational search algorithm (GSA)

Many optimization algorithms involve local search techniques which can get stuck on local maxima. Most search techniques strive to find a global maximum in the presence of local maxima (Amini & Arif, 2014). One of the most important characteristic of GSA is its significant performance during

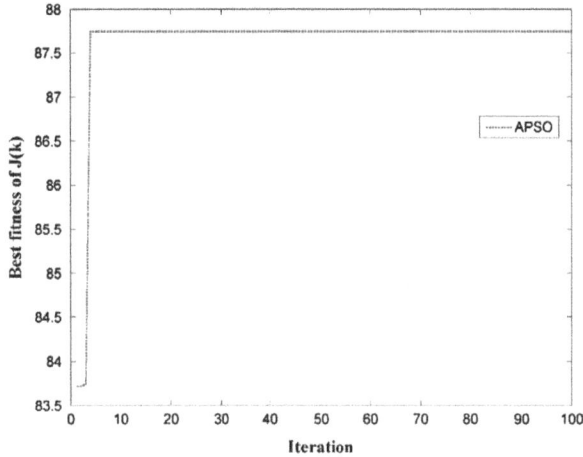

Figure 9. Iteration vs. fitness value, J (k) for APSO (100 PHEVs).

exploration process. The capability of an algorithm to extend the problem in search gap is known as exploration while the ability of an algorithm to recognize optimal solution near a favorable one is exploitation (Kulshrestha et al., 2009; Su & Chow, 2011b).

Figure 11 shows the convergence behavior of GSA for 100 PHEVs. From the simulation study we know that the best fitness function convergences after same iterations (35 iterations) for both 50 and 100 numbers of PHEVs while for 500 and 1,000 numbers of PHEVs, it shows early convergence (converges before 20 iterations).

Figure 12 illustrates the best fitness value for 100 PHEVs by using GSA method. In this case, the maximum best fitness and minimum best fitness were 872.648 and 1.005. The average best fitness decreases up to 182.309.

Finally, Table 10 sums up the results for GSA.

Table 11 shows the average computational time requirement for GSA method.

5.4. Hybrid particle swarm optimization and gravitational search algorithm (PSOGSA)

Figure 13 shows the convergence behavior (iteration vs. fitness value) of Hybrid PSOGSA technique for 100 PHEVs. Here also the algorithm has been set to run for maximum 100 iterations, the fitness

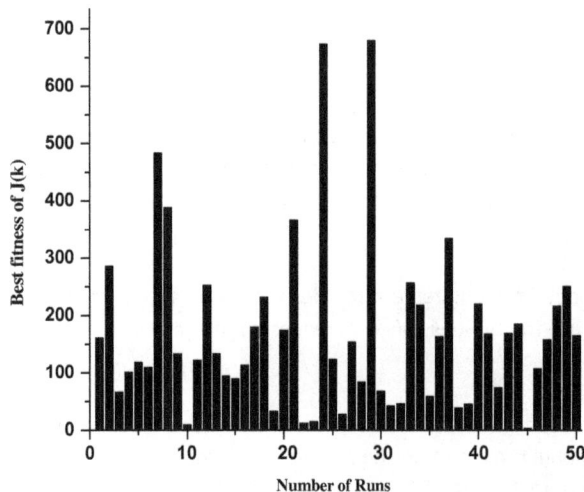

Figure 10. Fitness value vs. number of runs for APSO (100 PHEVs).

Table 8. Fitness evaluation of APSO					
J(k)	50 PHEVs	100 PHEVs	30 PHEVs	500 PHEVs	1,000 PHEVs
Max. best fitness	469.75	679.71	679.55	615.83	678.92
Average best fitness	162.70	168.23	147.42	184.15	171.16
Min. best fitness	7.65	3.46	3.54	5.96	0.99

Table 9. Average computational time for APSO	
Number of PHEVs	Computational time (s)
50 PHEVs	1.69
100 PHEVs	1.71
300 PHEVs	1.76
500 PHEVs	1.83
1,000 PHEVs	2.09

value converges after five (05) iterations and becomes stable. So, there is an early convergence which may cause the fitness function to trap into local minima. This can be avoided by increasing the size of swarm hence the computational time will also be increased as well. In order to evaluate the performance and show the efficiency and superiority of the proposed algorithm, we ran each scenario a total of 50 times.

Figure 14 depicts the best fitness value for 100 PHEVs. In this case, the maximum best fitness and minimum best fitness were 625.82 and 3.39, respectively. The average best fitness decreases up to 184.36.

As PSOGSA is a population-based optimization technique and the fitness function is nonlinear, so the fitness values fluctuate for each iteration (Abro & Mohamad-Saleh, 2012; Ganesan, Vasant, & Elamvazuthi, 2014; Jiménez, Sánchez, & Vasant, 2013). However, the maximum best fitness remains in the range of 400–950 and the minimum best fitness remains in the range of 0.1–8.

Table 12 summarizes the result. From that it can be concluded that average best fitness remains almost in similar pattern for four (05) different scenarios.

Table 13 shows the average computational time requirement for APSO method. The average computational time for 50 PHEVs is 4.228 s while for 1,000 PHEVs it increases up to 72.408 s.

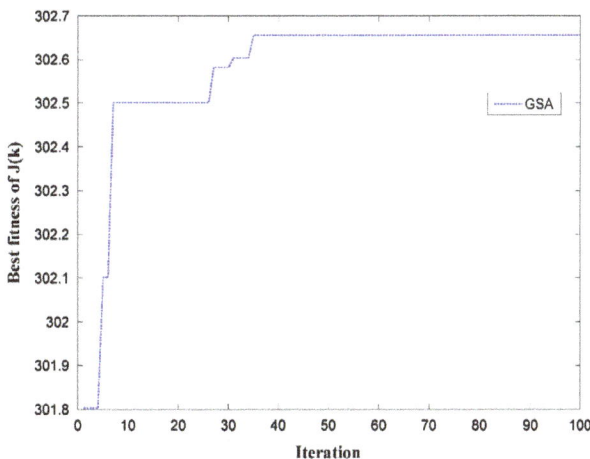

Figure 11. Iteration vs. fitness value, J (k) for GSA (100 PHEVs).

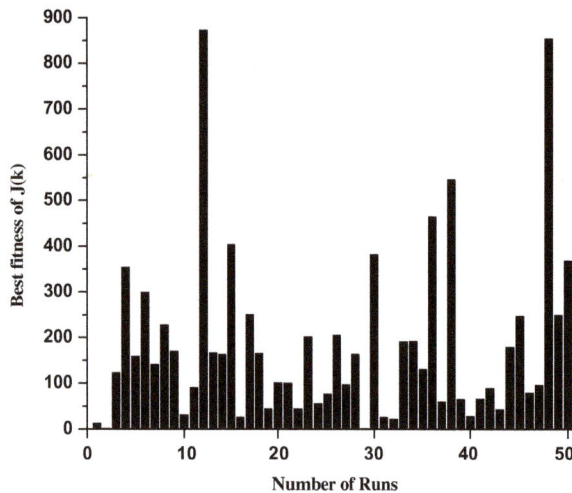

Figure 12. Fitness value vs. number of runs for GSA (100 PHEVs).

Table 10. Fitness evaluation of GSA					
J(k)	50 PHEVs	100 PHEVs	30 PHEVs	500 PHEVs	1,000 PHEVs
Max. best fitness	781.13	872.65	743.13	836.27	968.77
Average best fitness	158.83	182.31	172.43	152.36	161.52
Min. best fitness	0.22	1.00	2.33	0.98	7.27

6. Comparisons among swarm intelligence techniques

This section deals with the comparisons among the applied swarm intelligence-based optimization techniques. All four techniques were run on same computer along with same iterations (100) and total 50 independent runs in order to ensure the fare comparison (Rahman et al., 2016c); (Derrac, García, Molina, & Herrera, 2011). The comparisons among applied swarm intelligence-based techniques are given below.

6.1. Stopping criteria

In any swarm intelligence algorithm there are some initial solutions from which candidate solutions are created. After that, each solution is evaluated and the algorithm chooses the best solution. If the stopping criteria is met, then the algorithm will produce final solution otherwise it will again search for best solutions from the initial step. Proper balance between exploration and the exploitation is the basic criteria to analyze the performance of an algorithm. Proper exploration will diversify the search space of an optimization technique whereas the exploitation ensures the high-quality solutions.

Since an iterative method computes successive approximations to the solution of a system, stopping criteria is needed to determine when to stop the iteration. The maximum number of iteration

Table 11. Average computational time for GSA	
Number of PHEVs	Computational time (s)
50 PHEVs	2.72
100 PHEVs	4.44
300 PHEVs	11.28
500 PHEVs	18.17
1,000 PHEVs	36.28

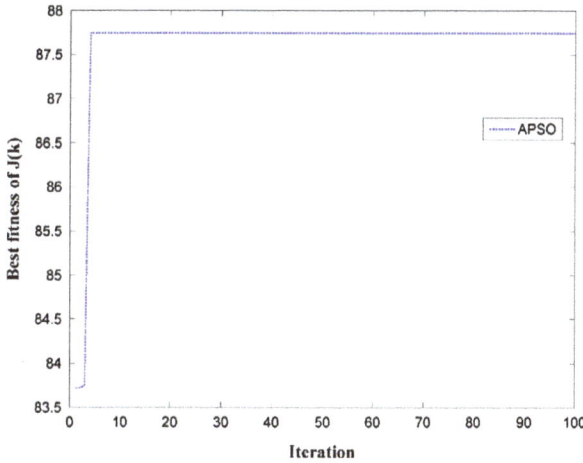

Figure 13. Iteration vs. fitness value, *J* (k) for Hybrid PSOGSA (100 PHEVs).

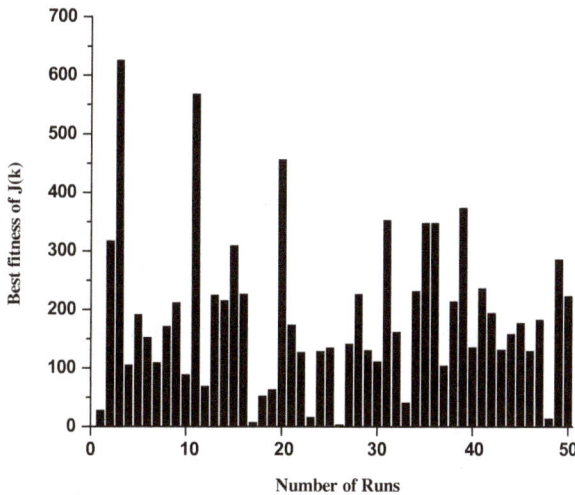

Figure 14. Fitness value vs. number of runs for PSOGSA (100 PHEVs).

was set to 100 for all four optimization techniques. Previous researchers use 100 iterations for their simulation study (Hendtlass, 2007; Martens et al., 2011).

6.2. Convergence analysis

When an algorithm finds an optimal solution to a given problem, one of the important factors is speed and rate of convergence to the optimal solution (Martens et al., 2011). Among the four techniques, the convergence of PSO and APSO techniques is of same pattern while GSA takes higher number of iterations to be converged. Among all four techniques, the hybrid method: PSOGSA takes the least number of iterations for convergence. So, there is an early convergence which may cause the fitness function to trap into local minima. This can be avoided by increasing the size of swarm hence the computational time will also be increased as well. As a result, a trade-off should be taken into consideration between the proper convergence and computational time. Table 14 shows the number of iterations needed to be converged for each algorithm for five different cases.

Table 12. Fitness evaluation of PSOGSA					
J(k)	50 PHEVs	100 PHEVs	30 PHEVs	500 PHEVs	1,000 PHEVs
Max. best fitness	931.03	625.82	434.16	454.04	740.40
Average best fitness	184.36	188.67	181.03	186.70	185.16
Min. best fitness	3.39	3.71	7.43	7.23	0.17

Table 13. Average computational time for GSA

Number of PHEVs	Computational time (s)
50 PHEVs	4.228
100 PHEVs	7.902
300 PHEVs	22.326
500 PHEVs	36.824
1,000 PHEVs	72.408

Table 14. Convergence iteration

Number of PHEVs	Number of iterations taken to be converged			
	PSO	APSO	GSA	PSOGSA
50	5	2	30	2
100	6	3	35	2
300	5	3	15	2
500	7	5	40	2
1,000	7	6	5	3

6.3. Best fitness value

Best fitness value presents the solutions for optimization technique applying to a particular objective function upon given constraints and set of parameters. The best fitness value represents the strength of any optimization technique. For the maximization problems, higher best fitness values indicate the effectiveness of particular techniques whereas the lower fitness values show the effectiveness of any minimization problems. As our optimization goal is maximization, so the higher best fitness values are the best solution.

Maximum best fitness, average best fitness, and minimum best fitness are presented in order to evaluate the performance of applied optimization techniques. According to the simulation result, the hybrid method, PSOGSA shows best fitness values for all five cases (50, 100, 300, 500, and 1,000 PHEVs). The fitness value comparison among all techniques are shown in Figure 15. Moreover single techniques like GSA and APSO show overall better result compared to PSO technique.

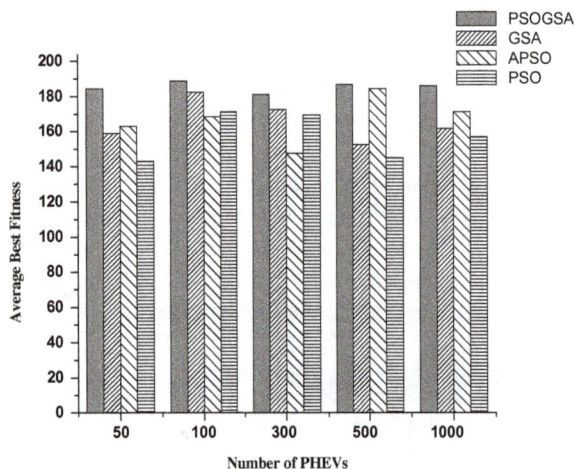

Figure 15. Average best fitness vs. number of PHEVs.

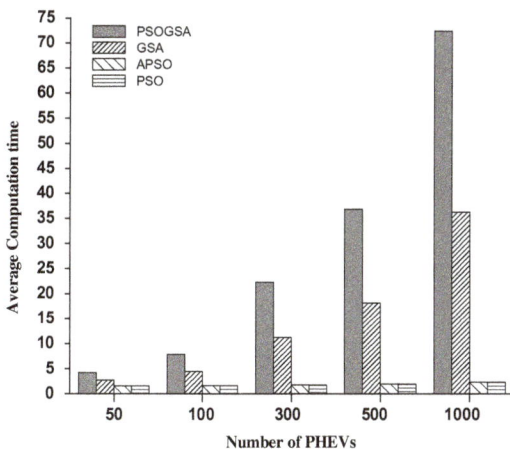

Figure 16. Average best fitness vs. number of PHEVs.

6.4. Computational time

In order to maintain fair comparison, all the simulation runs on same computer as well as same swarm size (population = 100) and iteration (100). Total running time of each optimization technique is also same (total 50 runs). If the number of runs is less for example, 10 or 20 runs only then the purpose of fair comparison will be hampered. Optimizing the design is done using an advanced optimization algorithm, which requires running a large number of simulations. This can be much more efficient than running a parameter sweep, particularly if there is more than one parameter to optimize (Martens et al., 2011). Figure 16 shows the average computational time comparison of all four optimization techniques considering five cases.

PSOGSA-the hybrid technique takes the highest time to complete 100 iterations whereas both PSO and APSO techniques show better result in terms of computation time. The GSA method takes higher time than other single methods, PSO and APSO in order to find best fitness for objective function J (k). Based on the previous literatures, the original GSA has some weaknesses such as using complex operators and long computational time. GSA also suffers from slow searching speed in the last iterations (Mirjalili & Hashim, 2010). Another problem is the difficulty for the appropriate selection of gravitational constant parameter, G. The parameter controls the search accuracy and does not guarantee a global solution at all time.

6.5. Robustness

Robustness is based on the ability of an optimization problem to perform well over a wide range of population (Shaikh, Nor, Nallagownden, & Elamvazuthi, 2013). Furthermore, optimization strategies and parameters must remain either constant over the set of problems or should be automatically set using individual test problems attributes.

Table 15. Paired *t*-test

Paired *t*-test		Paired differences					*t* Value	Sig. (2-tailed)
		Mean	Standard deviation	Standard error mean	95% Confidence interval of the difference			
					Lower	Upper		
Pair 1	PSO-PSOGSA	−28.21	13.69	6.12	−45.21	−11.21	−4.60	0.010
Pair 2	APSO-PSOGSA	−18.45	11.36	5.08	−32.56	−4.33	−3.63	0.022
Pair 3	GSA-PSOGSA	−19.69	11.88	5.31	−34.45	−4.93	−3.70	0.021

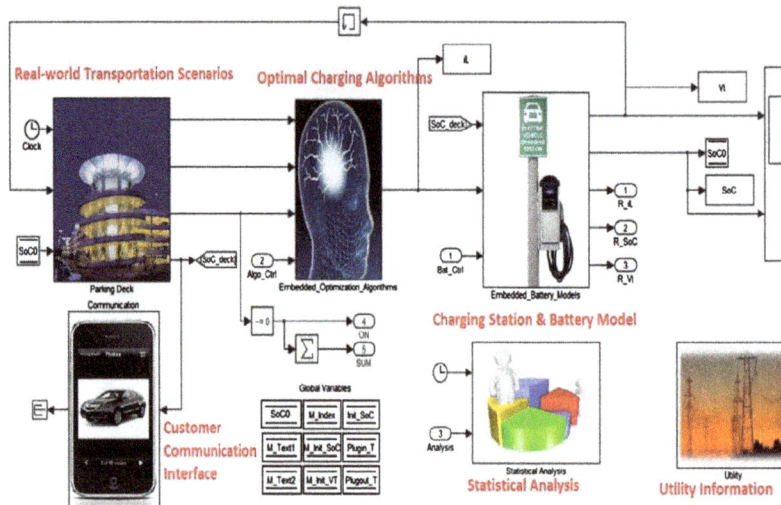

Figure 17. Large-scale PHEV/PEV Charging Infrastructure Digital Testbed (Simulink-based Energy Management Module) [121].

A paired t-test was performed on the simulation results in order to determine which method outperforms others. In statistical significance testing, the p-value is the probability of obtaining a test statistic at least as extreme as the one that was actually observed, assuming that the null hypothesis is true. A small p-value indicates that the null hypothesis is less likely to be true. Let J_{PSOGSA} is the average best fitness value achieved by the PSOGSA, while J_E is the average best fitness value achieved by the others methods (PSO, APSO, and GSA). First, the comparison was performed between PSOGSA's performance based on average best fitness values with PSO for a total of five different scenarios (starting from 50 PHEVs up to 1,000 PHEVs). Then, the null hypotheses is defined as

$$H_0: J_{PSOGSA} \leq J_{PSO} \tag{28}$$

This was tested against the alternative hypotheses

$$H_1: J_{PSOGSA} > J_{PSO} \tag{29}$$

Here, the comparisons were performed between PSOGSA with other three optimization methods (PSO, APSO, and GSA) and shown in Table 15. From the table it is clear that the significant level p-value is 0.01 (less than 0.05). The lower the p-value is, the less likely the null hypothesis is true. Thus we reject the null hypothesis and draw the conclusion that there is significant evidence that PSOGSA can achieve better results in terms of average best fitness.

By following a similar procedure as described before, p-values for the other cases (APSO-PSOGSA and GSA-PSOGSA) were obtained less than 0.05. Therefore, the null hypotheses were rejected and concluded that PSOGSA method is robust than other methods in solving this specific problem.

7. Suggestions for future work
This section suggests future directions of optimization techniques and procedures. The specific research field is relatively new and possible future perspectives have to be emphasized, so that new techniques can be realized.

The concepts proposed here may be utilized and tested for a wider range of multi-objective problems with a variety of problems characteristics in the future (Jiménez et al., 2013). Besides, objective function, $J(k)$ should also be tested with other swarm intelligence algorithms like ACO, ABC optimization, Firefly Algorithm (FA). Although swarm intelligence-based methods have established their capability to explore large search spaces, they are comparatively incompetent in fine-tuning the solution. This weakness is usually avoided by means of local search method that is applied to the individuals of the population (Derrac et al., 2011). In our case, further studies can be carried out by hybridizing PSO or GSA (swarm intelligence-based algorithm) with local search method. The future optimization tools should be capable of performing parallel processing evaluations on the same computer by using modern multi-core processor technology or to distribute the calculations to a cluster of computers. Such ability will substantially improve the simulation runtime. Advanced controlling mechanisms (Su, Wang, & Hu, 2015) are necessary for allocating sufficient energy to a particular charging station in order to facilitate large-scale PHEV penetration in upcoming years (Su, Zeng, & Chow, 2012). The future optimization tools should have the capability of stable convergence and thus provide good solution to the desired fitness functions. Exploration and exploitation of the search space is essential in order to get desired solution within acceptable computation time. Finally, optimization of charging station needs proper assortment of available resources as well as efficient available technique implementation. Proper charging infrastructure management can assist the larger participation of PHEVs.

At the same time, researchers should try to improve available device mechanism for the infrastructure with a view to simplify future PHEVs dispersion in roads and highways. In future, more vehicles should be considered for intelligent power allocation strategy as well as other single and hybrid techniques should be applied to ensure higher fitness value and low computational time.

8. Prototype model

Researchers are trying to design efficient controller for charging station and several literatures on optimization-based methods were published in this wake. These vehicles will help the government in its role of promoting energy security and environmental protection, when successfully marketed to consumers [123]. Efforts are also to be taken for provision of affordable and accessible infrastructure for recharging (Su et al., 2015). Hence, thrust in research and development on the aforementioned design considerations and technological challenges coupled with government support in terms of incentives to the automobile owners and to the manufacturers will go a long way in accelerating the deployment of large-scale PHEVs. The Figure 17 shows the prototype of digital testbed for Large-scale PHEV/PEV Charging Infrastructure from Future Renewable Electric Energy Delivery and Management (FREEDM) Systems Center with Advanced Diagnosis Automation and Control (ADAC) Lab at North Carolina State University and Advanced Transportation Energy Center (ATEC) (Su et al., 2012). The applied swarm intelligence-based algorithms are a step towards real-life implementation of such controller for PHEV charging stations.

Proper charging infrastructure management can assist the larger participation of PHEVs. At the same time, researchers should try to improve available device mechanism for the infrastructure with a view to simplify future PHEVs dispersion in roads and highways. In future, more vehicles should be considered for intelligent power allocation strategy as well as other single and hybrid techniques should be applied to ensure higher fitness value and low computational time.

9. Conclusion

SoC is needed to be optimized in order to develop the future charging infrastructures for PHEVs. The objective function is highly nonlinear which makes the optimization problem as a complex one. Simple Linear programming (LP) is not useful for solving this kind of problem. Swarm intelligence methods are within the group of metaheuristic algorithms. A metaheuristic is high-level problem-independent algorithmic framework that provides a set of guidelines or strategies to develop heuristic optimization algorithms. The aim of this work was to a comprehensive framework to solve single-objective optimization problems by analyzing the effects of best fitness and computational time using four swarm intelligence techniques: PSO, APSO, GSA, and Hybrid PSOGSA. In addition, this work was also targeted to analyze and compare the performance of swarm intelligence algorithms for PHEV charging station. The effects of convergence and robustness on the performance of applied techniques have been studied rigorously for solving charging problems.

Among the all four swarm intelligence-based optimization methods, three are single optimization techniques (PSO, APSO, and GSA) and one is hybrid version (PSOGSA). The convergence speed is the fastest in PSOGSA whereas GSA takes more iteration to be converged. In order to optimize the objective function $J(k)$, a hybrid technique (PSOGSA) has been introduced for the first time along with other three single techniques such as PSO, APSO, and GSA for comparative study. One of the comparison scales is best fitness value. Maximum best fitness, average best fitness, and minimum best fitness are shown in order to evaluate the performance of applied optimization techniques. Total five scenarios in terms of PHEVs number in the charging station are taken into consideration for the simulation starting from 50 PHEVs up to 1,000. These five scenarios show how the best fitness value and computational time changes with the increase of PHEVs in a charging station per day.

Acknowledgment
The authors would like to thank Universiti Teknologi PETRONAS (UTP) for supporting the research under UTP Graduate Assistantship (GA) scheme.

Funding
This work was supported by Graduate Assistance Scheme of Universiti Teknologi PETRONAS.

Author details
Pandian M. Vasant[1]
E-mails: pandian_m@utp.edu.my, pvasant@gmail.com
Imran Rahman[1]
E-mail: imran.iutoic@gmail.com
ORCID ID: http://orcid.org/0000-0002-6732-8751
Balbir Singh Mahinder Singh[1]
E-mail: balbir@petronas.com.my
M. Abdullah-Al-Wadud[2]
E-mail: mwadud@ksu.edu.sa
[1] Department of Fundamental and Applied Sciences, Universiti Teknologi PETRONAS 32610 Seri Iskandar, Perak, Malaysia.
[2] Department of Software Engineering, College of Computer and Information Sciences, King Saud University, Riyadh, Saudi Arabia.

References
Abro, A. G., & Mohamad-Saleh, J. (2012). An enhanced artificial bee colony optimization algorithm. *Recent Advances in Systems Science and Mathematical Modelling* (pp. 222–227).

Amini, M. H., & Arif, I. (2014). Allocation of electric vehicles' parking lots in distribution network. IEEE PES 5th Innovative Smart Grid Technologies Conference (ISGT 2014), February 19–22. Washington, DC.

Amini, M. H., & Parsa Moghaddam, M. (2013). Probabilistic modelling of electric vehicles' parking lots charging demand. Electrical Engineering (ICEE), 2013 21st Iranian Conference on. IEEE.

Boyle, G. (2007). *Renewable electricity and the grid: the challenge of variability.* Earthscan.

Chang, W.-Y. (2013). The state of charge estimating methods for battery: A review. *ISRN Applied Mathematics, 2013.*

Chiasson, J., & Vairamohan, B. (2005). Estimating the state of charge of a battery. *IEEE Transactions on Control Systems Technology, 13,* 465–470. http://dx.doi.org/10.1109/TCST.2004.839571

Contestabile, M., Offer, G. J., Slade, R., Jaeger, F., & Thoennes, M. (2011). Battery electric vehicles, hydrogen fuel cells and biofuels. Which will be the winner? *Energy & Environmental Science, 4,* 3754–3772.

Derrac, J., García, S., Molina, D., & Herrera, F. (2011). A practical tutorial on the use of nonparametric statistical tests as a methodology for comparing evolutionary and swarm intelligence algorithms. *Swarm and Evolutionary Computation, 1,* 3–18. http://dx.doi.org/10.1016/j.swevo.2011.02.002

Dorigo, M. (2006). *Ant colony optimization and swarm intelligence.* 5th International Workshop ANTS 2006, Brussels, Belgium, September 4–7, 2006, Proceedings vol. 4150, Springer.

Dubey, H. M., Pandit, M., Panigrahi, B., & Udgir, M. (2013, July). Economic load dispatch by hybrid swarm intelligence based gravitational search algorithm. *International Journal of Intelligent Systems & Applications, 5,* 21–32, doi:10.5815/ijisa.2013.08.03.

Duman, S., Güvenç, U., Sönmez, Y., & Yörükeren, N. (2012). Optimal power flow using gravitational search algorithm. *Energy Conversion and Management, 59,* 86–95. http://dx.doi.org/10.1016/j.enconman.2012.02.024

Eberhart, R. C., & Yuhui, S. (2001). Particle swarm optimization: Developments, applications and resources. In *Evolutionary Computation , 2001. Proceedings of the 2001 Congress on,* (Vol. 1, pp. 81–86). http://dx.doi.org/10.1109/CEC.2001.934374

El-Fergany, A. (2013, July). Accelerated particle swarm optimization-based approach to the optimal design of substation grounding grid. *Przeglad Elektrotechniczny, 89,* 30-34.

Gandomi, A. H., Yun, G. J., Yang, X.-S., & Talatahari, S. (2013). Chaos-enhanced accelerated particle swarm optimization. *Communications in Nonlinear Science and Numerical Simulation, 18,* 327–340. http://dx.doi.org/10.1016/j.cnsns.2012.07.017

Ganesan, T., Elamvazuthi, I., Ku Shaari, K. Z., & Vasant, P. (2013). Swarm intelligence and gravitational search algorithm for multi-objective optimization of synthesis gas production. *Applied Energy, 103,* 368–374, 3//2013. http://dx.doi.org/10.1016/j.apenergy.2012.09.059

Ganesan, T., Vasant, P., & Elamvazuthi, I. (2014). Hopfield neural networks approach for design optimization of hybrid power systems with multiple renewable energy sources in a fuzzy environment. *Journal of Intelligent and Fuzzy Systems, 26,* 2143–2154.

Hannan, M. A., Azidin, F. A., & Mohamed, A. (2014). Hybrid electric vehicles and their challenges: A review *Renewable and Sustainable Energy Reviews,* (Vol. *29,* pp. 135–150), 1//2014. http://dx.doi.org/10.1016/j.rser.2013.08.097

Hendtlass, T. (2007). Fitness estimation and the particle swarm optimisation algorithm In *Evolutionary Computation, CEC 2007, IEEE Congress on* (pp. 4266–4272). Singapore: IEEE. http://dx.doi.org/10.1109/CEC.2007.4425028

Hess, A., Malandrino, F., Reinhardt, M. B., Casetti, C., Hummel, K. A., & Barceló-Ordinas, J. M. (2012, December). Optimal deployment of charging stations for electric vehicular networks. In *Proceedings of the first workshop on Urban networking* (pp. 10–13). Nice, France. http://dx.doi.org/10.1145/2413236

Hota, A. R., Juvvanapudi, M., & Bajpai, P. (2014). Issues and solution approaches in PHEV integration to smart grid *Renewable and Sustainable Energy Reviews, 30,* 217–229, 2//2014.
http://dx.doi.org/10.1016/j.rser.2013.10.008

Jiménez, F., Sánchez, G., & Vasant, P. (2013). A multi-objective evolutionary approach for fuzzy optimization in production planning. *Journal of Intelligent & Fuzzy Systems: Applications in Engineering and Technology, 25,* 441–455.

Karaboga, D., & Basturk, B. (2007). A powerful and efficient algorithm for numerical function optimization: Artificial bee colony (ABC) algorithm. *Journal of Global Optimization, 39,* 459–471.
http://dx.doi.org/10.1007/s10898-007-9149-x

Kulshrestha, P., Wang, L., Chow, M.-Y., & Lukic, S. (2009). Intelligent energy management system simulator for PHEVs at municipal parking deck in a smart grid environment. In *Power & Energy Society General Meeting. PES'09 IEEE, 2009* (pp 1–6). Calgary: IEEE.

Kunche, P., Rao, G. S. B., Reddy, K., & Maheswari, R. U. (2015, March). A new approach to dual channel speech enhancement based on hybrid PSOGSA. *International Journal of Speech Technology, 18,* 45–56.

Li, Z., Sahinoglu, Z., Tao, Z., & Teo, K. H. (2010). Electric vehicles network with nomadic portable charging stations. In *Vehicular Technology Conference Fall (VTC 2010-Fall), 2010 IEEE 72nd* (pp. 1–5). Ottawa: IEEE.

Lund, H., & Kempton, W. (2008). Integration of renewable energy into the transport and electricity sectors through V2G. *Energy Policy, 36,* 3578–3587.
http://dx.doi.org/10.1016/j.enpol.2008.06.007

Mallick, S., Ghoshal, S. P., Acharjee, P., & Thakur, S. S. (2013). Optimal static state estimation using improved particle swarm optimization and gravitational search algorithm. *International Journal of Electrical Power & Energy Systems, 52,* 254–265.

Martens, D., Baesens, B., & Fawcett, T. (2011). Editorial survey: swarm intelligence for data mining. *Machine Learning, 82,* 1–42.
http://dx.doi.org/10.1007/s10994-010-5216-5

Mirjalili, S., & Hashim, S. Z. M. (2010). A new hybrid PSOGSA algorithm for function optimization. In *Computer and Information Application (ICCIA), 2010 International Conference on* (pp. 374–377). Tianjin: IEEE.

Mirjalili, S., Mohd Hashim, S. Z., & Moradian Sardroudi, H. (2012).Training feedforward neural networks using hybrid particle swarm optimization and gravitational search algorithm. *Applied Mathematics and Computation, 218,* 11125–11137.
http://dx.doi.org/10.1016/j.amc.2012.04.069

Mohamed, A. Z., Lee, S. H., Hsu, H. Y., & Nath, N. (2012). A faster path planner using accelerated particle swarm optimization. *Artificial Life and Robotics, 17,* 233–240.
http://dx.doi.org/10.1007/s10015-012-0051-3

Morrow, K., Karner, D., & Francfort, J. (2008). *Plug-in hybrid electric vehicle charging infrastructure review* (Final Report Battelle Energy Alliance Contract No. 58517). The Idaho National Laboratory is a U.S. Department of Energy National Laboratory Operated by Battelle Energy Alliance.

Mwasilu, F., Justo, J. J., Kim, E.-K., Do, T. D., & Jung, J.-W. (2014). Electric vehicles and smart grid interaction: A review on vehicle to grid and renewable energy sources integration *Renewable and Sustainable Energy Reviews, 34,* 501–516, 6//2014. http://dx.doi.org/10.1016/j.rser.2014.03.031

Piller, S., Perrin, M., & Jossen, A. (2001). Methods for state-of-charge determination and their applications. *Journal of Power Sources, 96,* 113–120.
http://dx.doi.org/10.1016/S0378-7753(01)00560-2

Prajna, K., Rao, G. S. B., Reddy, K., & Maheswari, R. U. (2014). A new dual channel speech enhancement approach

based on accelerated particle swarm optimization (APSO). *International Journal of Intelligent Systems and Applications (IJISA), 6,* 1.
http://dx.doi.org/10.5815/ijisa

Rahman, I., Vasant, P. M., Singh, B. S. M., Abdullah-Al-Wadud, M., & Adnan, N. (2016a). Review of recent trends in optimization techniques for plug-in hybrid, and electric vehicle charging infrastructures. *Renewable and Sustainable Energy Reviews, 58,* 1039–1047.
http://dx.doi.org/10.1016/j.rser.2015.12.353

Rahman, I., Vasant, P. M., Singh, B. S. M., & Abdullah-Al-Wadud, M. (2016b). On the performance of accelerated particle swarm optimization for charging plug-in hybrid electric vehicles. *Alexandria Engineering Journal, 55,* 419–426.
http://dx.doi.org/10.1016/j.aej.2015.11.002

Rahman, I., Vasant, P. M., Singh, B. S. M., & Abdullah-Al-Wadud, M. (2016c). Novel metaheuristic optimization strategies for plug-in hybrid electric vehicles: A holistic review *Intelligent Decision Technologies, 10,* 149–163.
http://dx.doi.org/10.3233/IDT-150245

Rashedi, E., Nezamabadi-pour, H., & Saryazdi, S. (2009). GSA: A Gravitational Search Algorithm. *Information Sciences, 179,* 2232–2248.
http://dx.doi.org/10.1016/j.ins.2009.03.004

Reddy, B. R. (2012, August). Performance Analysis of MIMO Radar Waveform using Accelerated Particle Swarm Optimization Algorithm. *Signal & Image Processing : An International Journal (SIPIJ). arXiv preprint arXiv:1209.4015, 3,* 193–202.

Roy, P. K. (2013). Solution of unit commitment problem using gravitational search algorithm. *International Journal of Electrical Power & Energy Systems, 53,* 85–94.

Shafiei, A., & Williamson, S. S. (2010). Plug-in hybrid electric vehicle charging: Current issues and future challenges. In Vehicle Power and Propulsion Conference (VPPC), 2010 IEEE (pp. 1–8).

Shaikh, P. H., Nor, N. M., Nallagownden, P., & Elamvazuthi, I. (2013). Intelligent Optimized Control System for Energy and Comfort Management in Efficient and Sustainable Buildings. *Procedia Technology, 11,* 99–106.
http://dx.doi.org/10.1016/j.protcy.2013.12.167

Soares, J., Morais, H., & Vale, Z. (2012). Particle swarm optimization based approaches to vehicle-to-grid scheduling. In *Power and Energy Society General Meeting, 2012 IEEE* (pp. 1–8). San Diego, CA: IEEE.

Soares, J., Sousa, T., Morais, H., Vale, Z., Canizes, B., & Silva, A. (2013). Application specific modified particle swarm optimization for energy resource scheduling considering vehicle-to-grid. *Applied Soft Computing, 13,* 4264–4280.

Su, W., & Chow, M.-Y. (2011a). Performance evaluation of a PHEV parking station using particle swarm optimization. In *Power and Energy Society General Meeting, 2011 IEEE* (pp. 1–6). San Diego, CA: IEEE.

Su, W., & Chow, M.-Y. (2011b). Sensitivity analysis on battery modeling to large-scale PHEV/PEV charging algorithms. In *IECON 2011-37th Annual Conference on IEEE Industrial Electronics Society* (pp. 3248–3253). Melbourne, VIC: IEEE.
http://dx.doi.org/10.1109/IECON.2011.6119831

Su, W., & Chow, M. Y. (2012). Computational intelligence-based energy management for a large-scale PHEV/PEV enabled municipal parking deck. *Applied Energy, 96,* 171–182.
http://dx.doi.org/10.1016/j.apenergy.2011.11.088

Su, W., Eichi, H., Zeng, W., & Chow, M.-Y. (2012). A survey on the electrification of transportation in a smart grid environment. *IEEE Transactions on Industrial Informatics, 8,* 1–10.
http://dx.doi.org/10.1109/TII.2011.2172454

Su, W., Zeng, W., & Chow, M.-Y. (2012). A digital testbed for a PHEV/PEV enabled parking lot in a smart grid environment. In *Innovative Smart Grid Technologies (ISGT), IEEE PES, 2012* (pp. 1–7).

Su, W., Wang, J., & Hu, Z. (2015). Planning, Control, and Management Strategies for Parking Lots for PEVs. In S. Rajakaruna, F. Shahnia, A. Ghosh (Eds.), *Plug In Electric Vehicles in Smart Grids* (pp. 61–98). Singapore: Springer. doi:10.1007/978-981-287-299-9_3

Talatahari, S., Khalili, E., & Alavizadeh, S. (2013). Accelerated particle swarm for optimum design of frame structures. *Mathematical Problems in Engineering, 2013*. Article ID 649857, 6p. http://dx.doi.org/10.1155/2013/649857

Tan, W. S., Hassan, M. Y., Rahman, H. A., Abdullah, M. P., & Hussin, F. (2013). Multi-distributed generation planning using hybrid particle swarm optimisation-gravitational search algorithm including voltage rise issue. *IET*

Generation, Transmission & Distribution, 7, 929–942.

Tie, S. F., & Tan, C. W. (2013). A review of energy sources and energy management system in electric vehicles. *Renewable and Sustainable Energy Reviews, 20*, 82–102. http://dx.doi.org/10.1016/j.rser.2012.11.077

Wu, X., Cao, B., Wen, J., & Bian, Y. (2008). Particle swarm optimization for plug-in hybrid electric vehicle control strategy parameter. In *Vehicle Power and Propulsion Conference, VPPC'08 IEEE, 2008*(pp. 1–5). Harbin: IEEE.

Yang, J., He, L., & Fu, S. (2014). An improved PSO-based charging strategy of electric vehicles in electrical distribution grid *Applied Energy, 128*, 82–92. 9/1/2014. http://dx.doi.org/10.1016/j.apenergy.2014.04.047

14

PV-wind hybrid system: A review with case study

Yashwant Sawle[1]*, S.C. Gupta[1] and Aashish Kumar Bohre[1]

*Corresponding author: Yashwant Sawle, Department Electrical Engineering, Maulana Azad Institute of Technology, Bhopal, India
E-mail: yashsawle@gmail.com
Reviewing editor: Wei Meng, Wuhan University of Technology, China

Abstract: Renewable energy systems are likely to become widespread in the future due to adverse environmental impacts and escalation in energy costs linked with the exercise of established energy sources. Solar and wind energy resources are alternative to each other which will have the actual potential to satisfy the load dilemma to some degree. However, such solutions any time researched independently are not entirely trustworthy because of their effect of unstable nature. In this context, autonomous photovoltaic and wind hybrid energy systems have been found to be more economically viable alternative to fulfill the energy demands of numerous isolated consumers worldwide. The aim of this paper is to give the idea of the hybrid system configuration, modeling, renewable energy sources, criteria for hybrid system optimization and control strategies, and software used for optimal sizing. A case study of comparative various standalone hybrid combinations for remote area Barwani, India also discussed and found PV–Wind–Battery–DG hybrid system is the most optimal solution regarding cost and emission among all various hybrid system combinations. This paper also features some of the near future improvements, which actually has the possibility to improve the actual monetary attraction connected with this sort of techniques and their endorsement by the consumer.

Subjects: Computer Science; Engineering & Technology; Urban Studies

Keywords: PV–wind-based hybrid systems; photovoltaic; wind turbine; modeling; optimization techniques

ABOUT THE AUTHORS

The key research area of authors is optimal sizing of renewable energy system.

This paper information is very helpful for pre-analysis of hybrid renewable energy system design. This work analyzed the different combinations of hybrid renewable energy source model and compared each other on the basis of emission, fuel consumption, cost, and component used in the system. This study gives the hybrid system consisting of PV/Wind/Battery/Generator which is a feasible solution. The total net present cost, cost of energy, operating cost, and emission are very less for the presented hybrid renewable energy combination compared to the other. This paper addresses the issues related to the feasibility of the system, combination of renewable source and cost function for pre-analysis of any hybrid practical system and wider projects.

PUBLIC INTEREST STATEMENT

The aim of the paper is to electrify those remote locations where the utility supply is not available. In all over the world many remote location areas where the electricity supply is so costly due to the higher transportation cost, transmission losses, etc. to sort out all these problems renewable energy is the better option. Solar and wind energy resources are freely available in atmosphere thus utilizing these renewable energy sources to power generation is easy and economic. This type of hybrid system can be modeled near to the consumer, which reduces the transmission cost, losses, and transportation cost. Hybrid renewable energy system is environment friendly because it does not produce harmful gasses such as carbon dioxide, unburned hydrocarbons, sulfur dioxide, and nitrogen oxides.

1. Introduction

Many remote communities around the world cannot be physically or economically connected to an electric power grid. The electricity demand in these areas is conventionally supplied by small isolated diesel generators. The operating costs associated with these diesel generators may be unacceptably high due to discounted fossil fuel costs together with difficulties in fuel delivery and maintenance of generators. In such situations, renewable energy sources, such as solar photovoltaic (PV) and wind turbine generator provide a realistic alternative to supplement engine-driven generators for electricity generation in off-grid areas. It has been demonstrated that hybrid energy systems can significantly reduce the total life cycle cost of standalone power supplies in many off-grid situations, while at the same time providing a reliable supply of electricity using a combination of energy sources. Numerous hybrid systems have been installed across the world, and the expanding renewable energy industry has now developed reliable and cost competitive systems using a variety of technologies. In a report, India's gross renewable energy potential (up to 2032) is estimated at 220 GW. It is likewise noted in the report that, with a renewable energy capacity of 14.8 GW (i.e. 9.7% of the total installed generation capacities of 150 GW as on 30 June 2009), India has barely scratched the surface of a huge opportunity. However, in the last couple of years itself, the share of renewable energy in installed capacity has grown from 5 to 9.7% (The Economic Times, 2009). This implies an enormous potential in energy generation, which can achieve several hundred GW with current renewable energy technologies. As the cost of building solar PV–wind capacity continues to fall over the next five to ten years; a significant scale-up of renewable generation is a very realistic possibility in the developing world. Thousands of villages across the globe are still being exiled from electricity and energizing these villages by extended grids or by diesel generators alone will be uneconomical. Moreover, with the current resource crunch with government, these villages receive low priority for grid extension because of lower economic return potential. Standalone solar PV–wind hybrid energy systems can provide economically viable and reliable electricity to such local needs. Solar and wind energy are non-depletable, site dependent, non-polluting, and possible sources of alternative energy choices. Many countries with an average wind speed in the range of 5–10 m/s and average solar insolation level in the range of 3–6 KWh/m^2 are pursuing the option of wind and PV system to minimize their dependence on fossil-based non-renewable fuels (Bellarmine & Urquhart, 1996; Nayar, Thomas, Phillips, & James, 1991). Autonomous wind systems (in spite of the maturity of state-of-the-art) do not produce usable energy for a considerable portion of time during the year. This is primarily due to relatively high cut-in wind speeds (the velocity at which wind turbine starts produces usable energy) which ranges from 3.5 to 4.5 m/s. In decree to overcome this downtime, the utilization of solar PV and wind hybrid system is urged. Such systems are usually equipped with diesel generators to meet the peak load during the short periods when there is a deficit of available energy to cover the load demand. Diesel generator sets, while being relatively inexpensive to purchase, are generally expensive to operate and maintain, especially at low load levels (Nayar, Phillips, James, Pryor, & Remmer, 1993). In general, the variation of solar and wind energy does not match the time distribution of the demand. Thus, power generation system dictates the association of battery bank storage facilities to overcome/smoothen the time distribution-mismatch between the load and renewable (solar PV and wind) energy generation (Borowy & Salameh, 1996). A drawback common to wind and solar system is their unpredictable nature and dependence on weather and climatic change. Both of these (if used independently) would have to be oversized to make them completely reliable, resulting in an even higher total cost. However, a merging of solar and wind energy into a hybrid generating system can attenuate their individual fluctuations, increase overall energy output, and reduce energy storage requirement significantly. It has been shown that because of this arrangement, the overall expense for the autonomous renewable system may be reduced drastically (Bagul & Salameh, 1996). Nowadays, the integration of PV and wind system with battery storage and diesel backup system is becoming a viable, cost-effective approach for remote area electrification. Wind and solar systems are expandable, additional capacity may be added as the need arises. Moreover, the combination of wind and solar PV system shrinks the battery bank requirement and further reduces diesel consumption. The prospects of derivation of power from hybrid energy systems are proving to be very promising worldwide (Beyer & Langer, 1996; Erhard & Dieter, 1991; Seeling-Hochmuth, 1997). The use of hybrid energy systems also reduces combustion of fossil fuels and consequent CO_2 emission which

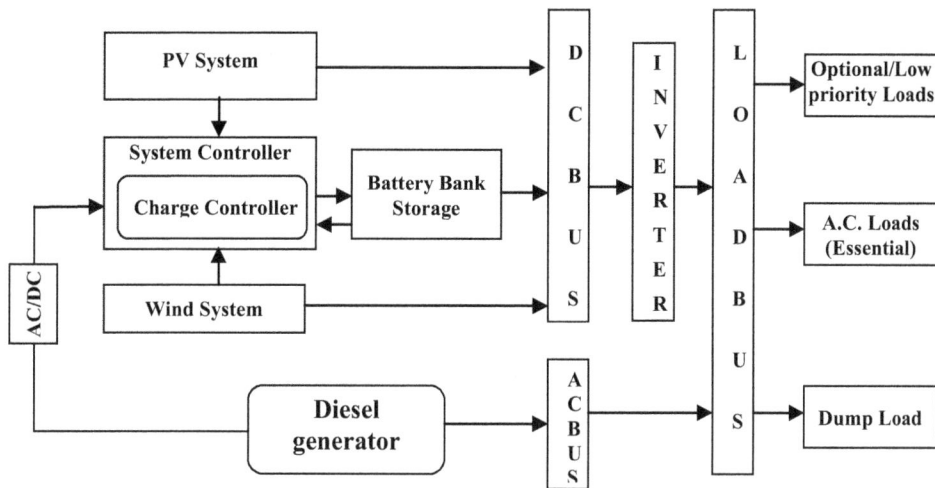

Figure 1. Block diagram of a typical PV–wind hybrid system is depicted.

is the principle cause of greenhouse effect/global warming. The global warming is an international environmental concern which has become a decisive factor in energy planning. In wake of this problem and as a remedial measure, strong support is expected from renewables such as solar and winds (Diaf, Notton, Belhamel, Haddadi, & Louche, 2008). The smart grid readying is associate optimum resolution to the present-day power sector issues like environmental pollution caused by typical power generation, grid losses, as well as poor reliableness and accessibility of power in rural areas (Zaheeruddin & Manas, 2015). The PV–wind hybrid energy system using battery bank and a diesel generator as a back-up can be provided to electrify the remotely located communities (that need an independent source of electrical power) where it is uneconomical to extend the conventional utility grid. All possible advantages of a hybrid energy system can be achieved only when the system is designed and operated appropriately (Gupta, Kumar, & Agnihotri, 2011). In these systems, sizing, control setting, and operating strategies are interdependent. In addition, some of the system components have non-trivial behavior characteristics. Thus, the task of assessing different design possibilities to plan a hybrid system for a specific location becomes quite difficult. The block diagram of a typical PV–wind hybrid system is depicted in Figure 1

The paper is organized as follows: Section 2 description of hybrid renewable energy systems; Section 3 depicts a discussion on hybrid PV/wind energy system modeling; Section 4 provides criteria for PV–wind hybrid system optimization; Section 5 discusses control strategies; Section 6 provides an overview of software tool used for optimal sizing; Section 7 case study of standalone hybrid system; and Section 8 highlights the challenges and future scope and also discussed with a conclusion.

2. Description of hybrid renewable energy schemes
A hybrid renewable PV–wind energy system is a combination of solar PV, wind turbine, inverter, battery, and other addition components. A number of models are available in the literature of PV–wind combination as a PV hybrid system, wind hybrid system, and PV–wind hybrid system, which are employed to satisfy the load demand. Once the power resources (solar and wind flow energy) are sufficient excess generated power is fed to the battery until it is fully charged. Thus, the battery comes into play when the renewable energy sources (PV–wind) power is not able to satisfy the load demand until the storage is depleted. The operation of hybrid PV–wind system depends on the individual element. In order to evaluate the maximum output from each component, first the single component is modeled, thereafter which their combination can be evaluated to meet the require dependability. If the electric power production, though this type of individual element, is satisfactory the actual hybrid system will offer electrical power at the very least charge.

2.1. Hybrid photovoltaic system

Solar energy is one of the non-depletable, site-dependent, non-polluting energy sources, and is available in abundance. It is a potential source of alternative/renewable energy and utilization of solar radiation for power generation reduces the dependence on fossil fuel (Douglas, 1997; Erhard & Dieter, 1991; Mahmoud, 1990; Post & Thomas, 1988; Richard, 1989; Traca De Almeida, Martins, & Jesus, 1983). Solar PV power generation unit consists of PV generator, diesel generator, and inverter and battery system shown in Figure 2. For improved performance and better control, the role of battery storage is very important (Shaahid & Elhadidy, 2003, 2004a). The necessary condition for the design of the hybrid PV systems for maximum output power is hot climate. This type of system is cost effective and reliable, especially for those locations where the power supplies though the grid is not suitable and the cost of the transmission line is very high such as remote and isolated areas (Valente & de Almeida, 1998). Table 1 shows the summary of subjects based on PV hybrid system. In literature a number of methods are used to evaluate performance of the hybrid PV system as a combination of PV with battery, diesel generator, and PV without battery. Muselli, Notton, Poggi, and Louche (2000) in the hybrid system modeling of battery with respect to the state of charge and best possible sizing of the system can also be achieved. El-Hefnawi (1998) developed a technique for minimizing the PV area and evaluate of least number of storage days in a PV hybrid system. Syafaruddin, Narimatsu, and Miyauchi (2015) designed the real-time output power, PV system for calculating the accumulative energy and capacity factor. This information is used for evaluating the energy production model based on the capacity factor. Designed a system for computing production cost associated with hybrid PV battery method in which the size associated with PV method is calculated on such basis as electrical requirements not attained (Abouzahr & Ramakumar, 1991). For standalone hybrid PV system, analysis of reliability is determined in the term of loss of load (LOL) probability. A number of numerical and analytical models are employed for measuring the LOL probability (Egido & Lorenzo, 1992). Execution of hybrid PV system is assessed on the premise of the reliability of the power supply under broadly differing conditions (Marwali, Shahidehpour, & Daneshdoost, 1997).

2.2. Hybrid wind energy system

For the design of a reliable and economical hybrid wind system a location with a better wind energy potential must be chosen (Mathew, Pandey, & Anil Kumar, 2002). In addition, analysis has to be conducted for the feasibility, economic viability, and capacity meeting of the demands (Elhadidy & Shaahid, 2004; Nfaoui, Buret, & Sayigh, 1996; Nfaoui, Buret, Sayigh, & Dunn, 1994; Papadopoulos & Dermentzoglou, 2002; Rehman, Halawani, & Mohandes, 2003). The algorithm for calculating the size of wind turbines and optimal location of distributed energy system has to be developed by using a hybrid configuration of ant colony optimization (ACO), artificial bee colony (ABC) (Kefayat, Lashkar Ara, & Nabavi Niaki, 2015). Optimal sizing of a hybrid wind system and forecasting of a hybrid system based on regression analysis, neural network, Monte Carlo simulation technique, and genetic algorithm were described in the literature (Feijoo, Cidras, & Dornelas, 1999; Li, Wunsch, O'Hair, &

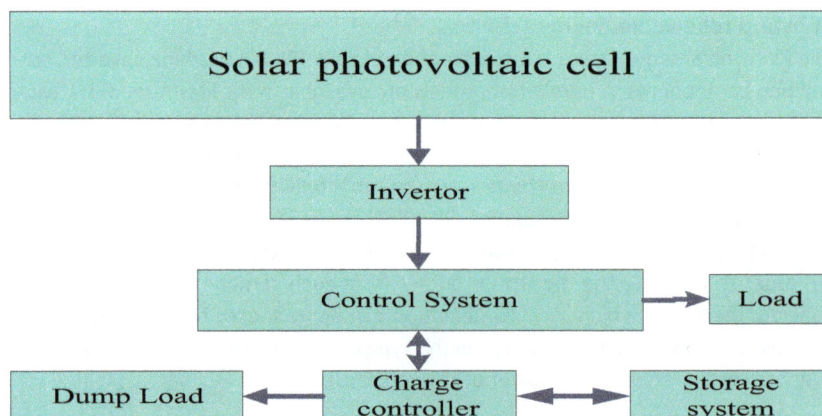

Figure 2. Architecture of PV hybrid system.

Table 1. Summary of studies based on PV hybrid system

Author	Indicator optimized	SA/GC	Location	Load type	Outcome	Algorithm used
Mahmoud (1990)	Reliability and economic feasibility study	SA	Jordan	Water pumping motor	The report exemplifies the invention and testing of water pumping systems powered by PV generators	Matlab
Post and Thomas (1988)	Cost	SA/GC		–	Study on PV systems for present and upcoming relevancies	–
Richard (1989)	Systems with a fixed tilt array, product or energy storage	SA	US	–	Built up the correlations for optimal sizing which give storage capacity and array size as a function of horizontal insolation and the long-term loss-of-load probability, respectively	LLP
Traca De Almeida et al. (1983)	Reliability	GC	Portugal	Grid connected	Design an optimal hybrid system which reduce system cost and give higher reliability	Monte Carlo simulation
Shaahid and Elhadidy (2003)	Potential of utilizing hybrid system	SA	Dhahran, Saudi Arabia	Residential buildings	An attempt has been made to address monthly average daily energy generated by the PV systems or different situations while meeting the load allocation	Matlab
S. M. Shaahid et al. (2004)	Feasibility of hybrid system	SA	Saudi Arabia	Commercial	1. System load can be satisfied in the optimal way	Matlab
					2. Diesel efficiency can be maximized	
					3. Diesel maintenance can be minimized	
					4. A reduction in the capacities of diesel and battery can occur	
L. Carlos et al. (1998)	Costs and the reliability	SA	Northern, Brazil	Residential	1. Software has been developed to optimize the generation cost starting from a given load curve	Matlab
					2. The PV/diesel option is more reliable and economical than the diesel system	
M. Muselli et al. (2000)	Lower kilowatt-hour cost	SA	Corsica island	Residential	The design hybrid system for remote location to fulfill load requirement	–
Tahrir Street et al. (1998)	Minimum number of storage days and the minimum PV array area	SA	Egyptian Eastern Desert	Farm	1. The sized hybrid system is reliable and can absorb any load disturbances	A program has been designed using FORTRAN language
					2. The hybrid system is more economic than the standalone system	
R. Ramakumar et al. (1991)	Energy storage and the loss of power supply probability	SA	–	Residential	Evaluate relationships between the amount of energy storage and the loss of power supply probability under various operating conditions can be investigated using the results	LPSP
M. Egido et. al. (1992)	Reliability	SA	Spain	Residential	Developed a new model which is more accurate and simple as compare to analytical and numerical models	LLP
M. K. C. Marwali et al. (1997)	Production cost	GC	–	Utility systems	Examine valuable method for generation expectation, production assessment and EENS in a PV-utility with battery storage	Probabilistic approach

Giesselmann, 2001; Papaefthymiou & Stavros, 2014). Salameh and Safari (1995) propose a methodology for identifying the wind turbine generator parameters as capacity factor which relates to identically rated available wind turbine and capacity factor calculated on the basis of wind speed data at different hours of the day of many years. Hybrid wind system installation planning for a particular site and system control strategies have also been reported by researchers (Chedid, Karaki, & El-Chamali, 2000; Jangamshetti & Ran, 2001). For calculating the monthly performance of wind energy system without hourly wind data, a Weibull function is needed (Celik, 2003a). Hybrid wind system performance, reliability, and reduction in the cost of energy (COE) can be obtained by using a

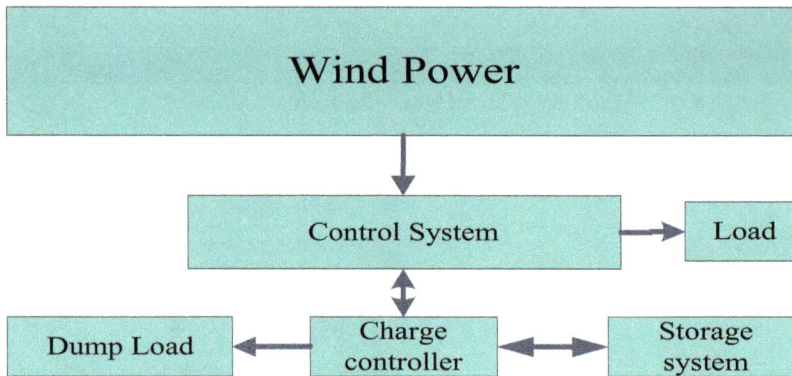

Figure 3. Architecture of wind hybrid system.

battery backup system. When the hybrid system generated power is in surplus, this power is used for loading the batteries for backup security and this charge battery power is used when the load requirement is not supplied by design hybrid system (Elhadidy & Shaahid, 2000). Figure 3 shows the architecture of wind hybrid energy system and Table 2 shows the summary of studies based on wind hybrid system.

2.3. Hybrid photovoltaic/wind energy system

PV and wind system, both depending on weather condition, individual hybrid PV and hybrid wind system does not produce usable energy throughout the year. For better performance of the standalone individual PV combination or wind combination need battery backup unit and diesel generator set, which increase the hybrid system cost (Elhadidy & Shaahid, 2004; Giraud & Salameh, 2001; McGowan, Manwell, Avelar, & Warner, 1996) for proper operation and better reliability, and lower cost of the system, studies are reported by researchers regarding the combination of hybrid PV–wind system. The current report offers a new strategy determined by the iterative approach, to accomplish the suitable sizing of any standalone hybrid PV/wind/hydrogen method, supplying a desalination unit which feeds the area's inhabitants with fresh water (Smaoui, Abdelkafi, & Krichen, 2015). Gupta, Kumar, and Agnihotri (2011) designed a Matlab software tool for evaluating the economic cost and loss of power supply probability (LPSP) technique is used as a key system constraint to assess the reliability and net present cost (NPC) of the system. González, Riba, Rius, and Puig (2015) suggested a system which is able to seek the sizing leading into a minimum life cycle cost of the system while matching the electrical supply with the local requirement. In the present post, the system is examined through a case study that precise by the hour electrical energy store and also current market rates are actually implemented for getting practical estimations of life cycle costs and also benefits. Design an off-grid hybrid PV–wind battery system with high reliability and minimum production cost of the system. The main objective of the design is to obtain a cost-effective solution (Cano, Jurado, Sánchez, Fernández, & Castañeda, 2014; Sawle & Gupta, 2014, 2015). Maleki and Askarzadeh (2014) use different artificial techniques for the optimal size of the hybrid system to minimize total annual cost. For this aspire sizing is formulated in four different techniques such as particle swarm optimization (PSO), tabu search (TS), simulated annealing (SA), and harmony search (HS). Shang, Srinivasan, and Reindl (2016) this specific paper will take the actual dispatch-coupled sizing approach through adding the actual battery to the procedure on the generation unit inside a process, and formulates this particular program issue employing optimum control. A couple of renewable energy sources––PV panels and wind turbines––are viewed as, together with traditional diesel generators. Shin, Koo, Kim, Jung, and Kim (2015) in order to optimally design ability as well as functioning, preparing of the hybrid system, per hour electricity demand data should be applied more than 8,760 h of 12 months. An optimization that matches hourly supply and demand problem had been resolved to have sparse matrices and also the linear programming algorithm. Lingfeng Wang and Singh (2009) study on techno-economic and environmental for hybrid system PV–wind, and battery banks and optimized for total cost, energy index of reliability, and pollutant emissions (PEs) and evaluate. A set of trade-off solution is obtained using multi-criteria meta-heuristic method

Table 2. Summary of studies based on wind hybrid system						
Author	Indicator optimized	SA/GC	Location	Load typ0065	Outcome	Algorithm used
Mathew et al. (2002)	Distribution of wind velocity	SA	Kerala, India	Water pumping	A method to calculate the energy potential of a wind regime is suggested	Matlab
Rehman et al. (2003)	Cost calculation of three different wind turbine capacities	SA	Saudi Arabia	Residential	The wind duration curves have been formulated in addition to employed to estimate the cost every kWh involving power created coming from several decided on the wind machines	Matlab
Nfaoui et al. (1996)	Cost of electricity generated and fuel saving	SA	Morocco	Residential	Develop an optimum hybrid system which reduces the cost energy generation	–
Elhadidy and Shaahid (2004b)	Role of hybrid power systems		Saudi Arabia	Commercial	Design an optimal system which capable to minimize maintains, the cost of generation and maximized the efficiency	Matlab
Papadopoulos and Dermentzoglou (2002)	Economic viability	GC	Greece	Utility system	Developed software which analysis the economic viability for two different cases in this study	Software developed
Kefayat et al. (2015)	Optimal location and sizing of distributed energy resources (DERs) on distribution systems	GC	–	Utility system	In this study found to minimize power loss, emission, cost of energy and increase the voltage stability	Hybrid ACO-ABC
Papaefthymiou and Stavros (2014)	Two alternative perspectives regarding the optimization targets:	SA	Greece	Residential	Enhance the penetration of green power technique plus decrease in levelized price of energy	Genetic algorithms
	1. The investor's perspective, profit					
	2. The system perspective					
Li et al. (2001)	Compares regression and artificial neural network models	GC	Fort Davis, Texas	Utility system	The neural network model is found to own far better effectiveness than the regression model pertaining to wind generator energy curve evaluation within challenging have an effect on components.	Regression and artificial neural
Feijoo et al. (1999)	Optimization based on two methods	GC	–	–	Two methods have been proposed to calculate the probability of occurrence of wind speed in several wind farms simultaneously	Monte Carlo simulation
	1. Wind speed distribution, assumed to be Rayleigh					
	2. Application of the simulation to the wind speed series					
Salameh and Safari (1995)	Finding the capacity factors (CF)	–	Irbid-Jordan	–	Time calculation and selection of windmill is done capacity factor.	–
Chedid et al. (2000)	To generate fuzzy membership functions and control rules for the controller.	GC	–	Motor load	Develop a adaptive fuzzy control for wind-diesel weak power systems	Fuzzy logy
Celik (2003a)	Estimate the monthly performance of autonomous hybrid system with battery storage	SA	Athens		Design a model for estimating the monthly performance of autonomous wind energy systems	ARES
Elhadidy and Shaahid (2000)	Identified the viability of hybrid system in Dhahran	SA	Saudi Arabia	Commercial	Parametric study of hybrid generating systems	Matlab

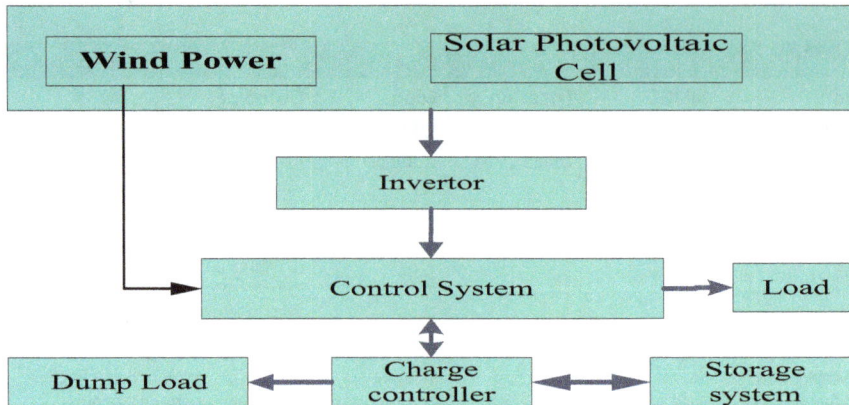

Figure 4. Architecture of PV–wind hybrid system.

that offers many design alternatives to decision-maker. Bilal, Sambou, Ndiaye, Kébé, and Ndongo (2013) developed a methodology to size and to optimize a hybrid PV–wind system minimizing the levelized COE and the carbon emission using a multi-objective genetic algorithm approach. Kamel and Dahl (2005) study on standalone hybrid PV–wind–diesel generate-battery system for economic analysis and evaluated annualized cost LPSP optimization results show that hybrid systems are less costly than diesel generation from a NPC perspective. Dufo-López, Bernal-Agustín, and Mendoza (2009) design a grid connected hybrid PV–wind system, taking constraints of land surface acquired by system and initial installation cost and evaluated that system is economical if the selling price of the electric energy is roughly 10 €/kg. Katsigiannis, Georgilakis, and Karapidakis (2010) work on economic and environmental study of a standalone hybrid system, the main aim of the work is to calculate the greenhouse gas emission based on life cycle cost of each component of the hybrid system. Bernal-Agustín, Dufo-López, and Rivas-Ascaso (2006) design is posed as a possible optimization problem whose solution allows having the configuration of the system as well as the control strategy that simultaneously minimizes the total cost through the particular useful life of system plus the PEs. Tina and Gagliano (2010) study on the probabilistic method for standalone hybrid system on the basis of the energy index of the reliability, internal rate of return and expected energy not supplied, evaluate the inform at the design of a pre-processing stage for the input of an algorithm that probabilistically optimized the design of hybrid power systems. In literature various types of method are used for most feasible solution, high reliability, and minimizing the COE such as (Yang, Lu, & Burnett, 2003) probabilistic method (Diaf, Notton, et al., 2008; Dufo-López et al., 2009; Kamel & Dahl, 2005), analytical method (Khatod, Pant, & Sharma, 2010), iterative method (Ekren & Ekren, 2009; Yang, Wei, & Chengzhi, 2009), hybrid method (Bernal-Agustín et al., 2006; Katsigiannis et al., 2010; Lingfeng Wang & Singh, 2009). Figure 4 shows the architecture of PV–wind hybrid energy system and Table 3 shows the summary of studies based on PV–wind hybrid system.

3. Modeling of hybrid renewable energy system components
Different modeling techniques are suggested by researches for modeling the component of a hybrid renewable energy system. The modeling of hybrid system component is discussed below.

3.1. Modeling of photovoltaic system
The outputs of the PV fully depend on solar radiation. Hourly solar radiation on a fixed inclined surface (I_T) can be evaluated as (Onar, Uzunoglu, & Alam, 2006).

$$I_T = I_b R_b + I_d R_d + (I_b + I_d)R_r \tag{1}$$

where I_T = solar radiation on an incident surface; I_b = direct normal and diffuse; I_d = solar radiations; R_b = the tilt factors for the beam; R_d = the tilt factors for the diffuse; and R_r = reflected part of the solar radiations.

PV power output with respect to area is calculated by

Table 3. Summary of studies based on PV–wind hybrid system

Author	Indicator optimized	SA/GC	Location	Load type	Outcome	Algorithm used
McGowan et al. (1996)	Life cycle cost	SA	Brazil	Telecom	The major performance parameters for the design and sizing of renewable energy systems can be set up	HYBRID 2 and SOME
Francois Giraud et al. (2001)	Reliability, power quality, loss of supply	GC	England	Residential	Evaluate performance of hybrid system regarding cost, reliability	LPSP
Elhadidy et al. (2004)	Load distribution and power generation	GC	Saudi Arabia	Commercial	Investigate the potential of utilizing hybrid energy conversion systems to meet the load requirements	Matlab
Mariem Smaoui et al. (2015)	Economic	SA	South of Tunisia	Residential	Evaluate a hybrid system, which is designed to supply sea water desalination	Iterative technique
Arnau González et al. (2015)	Minimum life cycle cost	GC	Catalonia Spain	Residential	Design a hybrid system to meet the load demand at minimum life cycle cost on the basis of net present cost	GA and PSO
Antonio Cano et al. (2014)	Unit-sizing and the total net present cost	SA	Malaga Spain.	Residential	Investigate a hybrid system by dissimilar methods and examine hybrid off-grid system is more reliable and cost effective	HOMER, HOGA, MATLAB
Akbar Maleki et al. (2014)	Annual cost	SA	-	Residential	Evaluate an optimal system by using PSO tool which result at minimum cost while comparing to other artificial intelligence techniques	PSO, HS, TS, SA
Ce Shang et al. (2016)	Economic, levelized cost	SA	Singapore	Residential	The author describes the sizing optimization in the dispatch-coupled way, and derives the optimal size of battery for systems with different penetration levels of renewable	PSO
Younggy Shin et al. (2015)	Capacity design and operation planning	SA	South Korea	Building load	It observes the hybrid renewable energy system is more reliable as compare to diesel generator system for island location.	Pareto optimal front
Lingfeng Wang et al. (2009)	Cost, reliability, and emissions	GC	-	Utility system	A set of trade-off clarifications is obtained by way of the multi-criteria meta-heuristic scheme that provides numerous design substitutes to the decision-maker	PSO
Ould. Bilal et al. (2013)	Levelized cost of energy (LCE) and the CO_2 emission	SA	North-western of Senegal	Three different loads	Author takes variation of three dissimilar load profile for hybrid system and minimized LCE and the CO_2 emission	Genetic Algorithm
Sami Kamel et al. (2005)	Economic	SA	Egypt	Agricultural load	Hybrid renewable energy system is more cost valuable and environmentally pleasant as compare to diesel generator scheme	HOMER
Rodolfo Dufo-López et al. (2009)	Net present value	GC	Spain	Utility system	Design in addition to cost-effective analysis connected with hybrid techniques connected to the grid for the irregular generation connected with hydrogen	GRHYSO
Banu Y. Ekren et al. (2009)	Economic	SA	Urla, Turkey	Institute load	Evaluate optimal sizing at different loads and auxiliary energy positions and output shown by loss of load probability and autonomy analysis	ARENA
Katsigiannis et al. (2009)	Cost of energy and greenhouse gas (GHG) emissions	SA	Crete, Greece	Residential	The main uniqueness of the anticipated methodology is the thought of LCA results for the estimate of CO_2 emissions	Genetic Algorithm
Yang Hongxing et al. (2010)	Economic	SA	China	Telecommunication station	Design an optimal hybrid system which annualized cost is least while load demand is satisfied on the basis of loss of power supply probability	Genetic algorithm
Bernal-Agustin et al. (2006)	Pollutant emissions, cost	SA	-	Farm Load	Developed a software tool which objective is to reduce cost of energy and co_2 emission	Pareto Evolutionary
Yang et al. (2003)	Reliability, and probability of power supply	SA	Hong Kong	Telecommunication	Study on weather data and probability analysis of hybrid power generation systems	Matlab
Khatod et al. (2010)	Well-being assessment and production cost	SA	Gujarat, India	-	Design a technique which has high accuracy and taking less calculating time as compare to Monte Carlo method	Monte Carlo simulation
Tina et al. (2010)	Probability distribution function	-	Italy	-	Developed a algorithm which results gives information about more reliable and optimal configurations for design of hybrid system	Matlab

$$P = I_T A_{PV} \eta_{PV}$$
(2)

A_{PV} and η_{PV} are PV system area and PV system efficiency, respectively.

The PV system efficiency is defined as

$$\eta_{PV} = \eta_M \, \eta_{PC} \left[1 - \beta \left(T_C - T_R \right) \right]$$
(3)

where η_M = module efficiency; η_{PC} = power conditioning efficiency; T_C = monthly average cell temperature; T_R = reference temperature; and β = array efficiency temperature coefficient.

In the ideal equivalent circuit of PV cell a current source is connected in parallel with diode. Onar et al. (2006) connected PV cell with load, voltage, and current equation of cell which is calculated by

$$I_{PV} = I_{PH} - I \left(e^{QV_{PV}/KT} - 1 \right)$$
(4)

where I_{PV} = is the PV current (A); I = the diode reverse saturation current (A); Q = the electron charge = 1.6 _ 10_19 (C); k = the Boltzman Constant = 1.38 _ 10_23 (J/K); and T = the cell temperature (K).

3.2. Modeling of wind energy system

The actual mathematical modeling of wind energy conversion process comprises wind turbine dynamics as well as generator modeling. Borowy and Salameh (1997) took a three blade, horizontal axis and repair free wind generator is installed for modeling. Power generation through the wind turbine can be calculated by wind power equation. The turbine is characterized by non-dimensional performance as a function of tip the speed quantitative relation. Bhave (1999) estimates the generated output power and torque by the wind turbine by giving the formula.

$$P_T = \left(\frac{C_p \lambda_\rho A V^3}{2} \right)$$
(5)

Torque developed by wind turbine given as

$$T_T = \frac{P_T}{\omega M}$$
(6)

$$\lambda = \frac{\omega R}{V}$$
(7)

where P_T = output power; T_T = the torque developed by wind turbine; C_p = the power co-efficient; λ = the tip speed ratio; ρ = the air density in kg/mg^3; A = the frontal area of wind turbine; and V = the wind speed.

Many researchers work on different mathematical modeling for wind energy conversion. Arifujjaman, Iqbal, Quaicoe, and Khan (2005) has worked on small wind turbine by controlling horizontal furling scheme. This furling scheme is used to control aerodynamic, power extraction through the wind. The system is designed in Matlab/Simulink for evaluating appropriate control approach. Two controllers are designed and simulated. For the first scheme, a controller uses rotor speed and wind speed information and controls the load in order to operate the wind turbine at optimal tip speed ratio. In the second scheme, controller compares the output power of the turbine with the previous power and based on this comparison it controls the load.

3.3. Modeling of diesel generator

Hybrid PV–wind system's operation and power generation depends on weather conditions. If poor sunshine and low wind speeds then hybrid PV–wind system's operation and efficiency are affected and the load requirement is not satisfied by either hybrid system or by batteries. All this issue can be resolved by using a diesel generator in hybrid PV–wind system. The application of diesel generator depends on the type and nature of load demand. Notton, Muselli, and Louche (1996) present two essential conditions for calculating the rated capacity of the generator to be installed. The first condition, if the diesel generator is directly connected to the load then the rated capacity of the generator must be at least equal to the maximum load. Second condition, if the diesel generator is used as a battery charger then the current produced by the generator should not be greater than CAh/5 A, where CAh is the ampere-hour capacity of the battery. The efficiency of a diesel generator is specified by the formula (Kaldellis & Th, 2005; Nag, 2001).

$$\eta_T = \eta_B + \eta_G \tag{8}$$

where η_T total efficiency and η_B, η_G are the thermal and generator efficiency. In hybrid system, a generator is used to maintain the reliability and load requirement. To obtain the lowest cost of system generator should work between the ranges of 70–90% of full load (El-Hefnawi, 1998; Valenciaga & Puleston, 2005). Generator fulfills the load demand and battery charging if peak load is not available.

3.4. Modeling of battery system

Sinha (2015) mentions that battery is used to store surplus generated energy, regulate system voltage and supply load in case of insufficient power generation from the hybrid system. Battery sizing depends on the maximum depth of discharge (DD), temperature, and battery life. A battery's state of charge (S_C) is expressed as follows:

During charging process

$$S_C(t + 1) = S_C(t)\big[1 - \sigma(t)\big] + \big[I_B(t)\Delta t.\eta_C(t)/C_B\big] \tag{9}$$

During discharging process

$$S_C(t + 1) = S_C(t)\big[1 - \sigma(t)\big] - \big[I_B(t).\Delta t.\eta_D(t)/C_B\big] \tag{10}$$

where S_C = state of charge; $\sigma(t)$ = hourly self-discharge rate depending on the battery; I_B = battery current; C_B = nominal capacity of the battery (Ah); η_C = charge efficiency (depends on the S_C and the charging current and has a value between 0.65 and 0.85); and η_D = discharge efficiency (generally taken equal to one)

and

$$\big[1 - DD\big] \leq S_C(t) \leq 1 \tag{11}$$

where DD = depth of discharge.

4. Criteria for PV–wind hybrid system optimization

In literature, optimal and reliable solutions of hybrid PV–wind system, different techniques are employed such as battery to load ratio, non-availability of energy, and energy to load ratio. The two main criteria for any hybrid system design are reliability and cost of the system. The different methods used for these criteria are given below.

4.1. Reliability analysis

Hybrid PV–wind system performance, production, and reliability depend on weather conditions. Hybrid system is said to be reliable if it fulfills the electrical load demand. A power reliability study is important for hybrid system design and optimization process. In literature, several methods are used to determine the reliability of the hybrid system. Al-Ashwal (1997) has developed, LOL risk method for reliability analysis. LOL risk evaluation is performed using a probabilistic model. LOLR is defined as the probability of the generating system failure to meet the daily electrical energy demand due to the deficient energy of the renewable energy sources used (Planning & installing PV system: A guide for installers, architects & engineers, 2005) LOLR can be represented as $LOLR = 1-P$ or $LORL = Q$, where P is the cumulative probability of meteorological status which corresponds to electrical energy generation and Q is the probability of failure. Maghraby, Shwehdi, and Al-Bassam (2002) worked on system performance level. System performance level is defined as the probability of unsatisfied load. Shrestha and Goel's Shrestha & Goel, (1998) reliability calculated on the basis of LOL hours. LOL hours is the summation of LOL expectation in hours over a specified time (usually one year) that the power system is unable to meet load requirements due to lack of power at an instant excluding the effects of component breakdown or maintenance time. LA is defined as one minus the ratio between the total number of hours in which LOL occurs and the total hours of operation (Celik, 2003b).

$$LA = 1 - H_{LOL}/H_{TOT} \tag{12}$$

where H_{LOL} hours which LOL occurs (h) and H_{TOT} total hours operation system (h). Kaldellis (2010) uses different methods for analysis of hybrid system reliability as LPSP, LOL probability (LOLP), unmet load (UL). LPSP is the most widely used method on condition when power supplies do not fulfill the required load demand. LPSP is the ratio of power supply deficits to the electric load demand during a certain period. As for the LOL probability (LLP), it is defined as the power failure time period divided by the total working time of the hybrid system. Lastly, UL can be defined as the load which cannot be served divided by a total load of a time period (normally one year).

4.2. Cost analysis

NPC or net present worth (NPW) is defined as the total present cost of a time series of cash flows. It is a standard method for using the time value of money to appraise long-term projects. The basis of NPC analysis is to be an ability to express a series of yearly costs in constant currency taking into account the changing value of money as well as cost escalation due to inflation (Dufo-López et al., 2009; Gupta et al., 2011). Therefore, the NPC means the present value of the cost of installing and operating the hybrid system over the lifetime of the project. It is calculated as follows:

$$NPC = C_{ANN}/CRF(i, T_{PLT}) \tag{12}$$

where C_{ANN} = total annualized cost; CRF = capital recovery factor; i = interest rate; and T_{PLT} = project lifetime. Life cycle costs (LCC) are the sum of all the hybrid system component costs and discounted operational costs arising during the project until the end of the project horizon, which is usually set between 20 and 30 years (Bhuiyan, Asgar, Mazumder, & Hussain, 2000). The component costs are the capital cost incurred at the beginning of hybrid system project; operational costs include system running costs, maintenance, and replacement costs. The COE reflects the cost of energy or electricity generation and is expressed as the ratio of total annualized cost of the system to the annual electricity delivered by the system. Total annualized cost includes all the costs over the system's lifetime from initial investment and capital costs to operations and maintenance (e.g. fuel) and financing costs (Zhou, Lou, Li, Lu, & Yang, 2010).

5. Control strategies

As the hybrid renewable energy system is the combination of different renewable energy sources, diesel generator–conventional sources, and energy storage system it is very difficult to get output at maximum efficiency and reliability without applying any proper control strategy (Dimeas & Hatziargyriou, 2005). In hybrid renewable energy system, for a variable, monitoring and power supply load for the requirement is done by the controller. Controller also keeps the output voltage, frequency and determines the active and reactive power from different energy sources. Different types of controller are applied in a hybrid renewable energy system according to the requirement of different energy sources, output power, and control strategy. Controller, predominantly are of four types as centralized, distributed, hybrid (combination of centralized and distributed) control, and multiple control system. In each one of the cases, every source is expected to have its own controller that can focus on ideal operation of the relating unit taking into account current data. In the centralized control arrangement, the entire energy source's signals and storage system are controlled by centralized (master controller) arrangement. Multi-objective energy unit framework can accomplish global optimization in view of all accessible data (Abido, 2003; Azmy & Erlich, 2005; Lagorse, Simoes, & Miraoui, 2009; Miettinen, 1998; Sawle & Gupta, 2014). The disadvantage of this centralized unit is that it suffers from heavy computation load and is subjected to single-point failures. The second control unit is the distributed control unit; in this, unit single energy source is connected to individual to local control unit and thus control units are connected to each other for communicating measurement signals and take a suitable assessment for global optimization. This control unit more advantageous as compared to the centralized control unit because it calls for a minimum computational load without any failure (Hajizadeh & Golkar, 2009; Huang, Cartes, & Srivastava, 2007; Kelash, Faheem, & Amoon, 2007; Ko & Jatskevich, 2007; Lagorse et al., 2009; Nagata & Sasaki, 2002; Nehrir et al., 2011; Toroczckai & Eubank, 2005; Weiss, 1999; Yang et al., 2006). Withal, this control structure has the shortcoming of multi-faceted communication systems among local controllers. This problem of distributed control unit can be solved by artificial algorithm techniques. Multi-agent system is a standout among the most encouraging methodologies for a distributed control unit. The third control arrangement is a hybrid control unit (Ko & Jatskevich, 2007; Torreglosa, García, Fernández, & Jurado, 2014; Torres-Hernandez, 2007). Hybrid control unit is the arrangement of centralized and distributed control units. In hybrid control unit, renewable sources are assembled within the integrated system. In this hybrid control unit, local optimization in a group and global optimization with different groups are obtained by centralized control unit and distributed control unit, respectively. This hybrid control unit is more advantageous and suitability over other control units because it takes less computation burden which reduces the failure problem of the system. The main drawback of the system is the potential complexity of its communication system. The fourth control is a multi-level control unit. The working operation of this control unit is almost the same as the hybrid control unit but the advantage is it has supervisor control which takes care about real-time operation of each energy unit on the basis of control objective within millisecond range. It also facilitates with the two-way communication existing among diverse levels to execute choices (Torreglosa et al., 2014; Upadhyay & Sharma, 2014). The drawback of this control unit is the potential complexity of its communication system. Figure 5 shows the energy flow and data communication information.

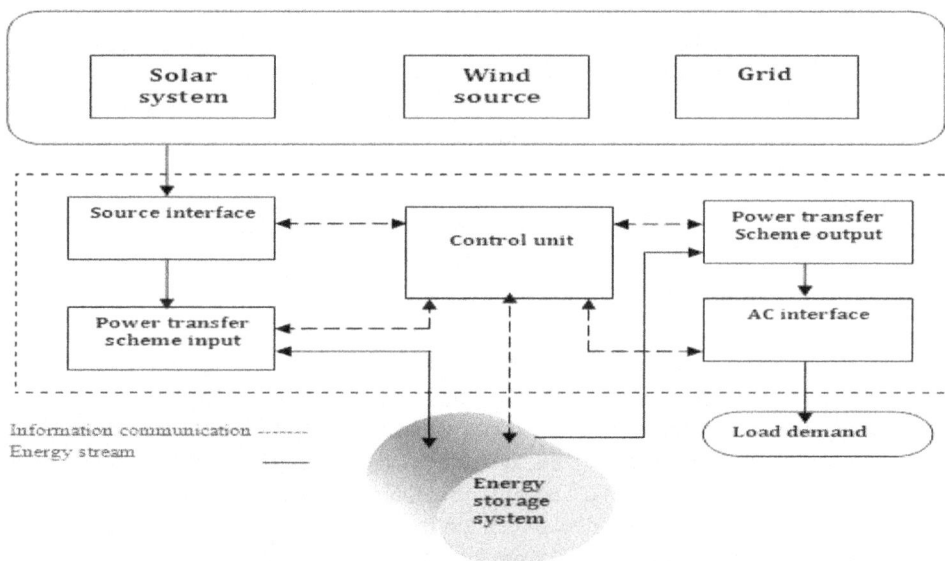

Figure 5. The energy flow and data communication information.

6. Software based on optimization of hybrid system

There are many software tools that are capable to assess the renewable energy system performance for pre-defined system configurations. These include HYBRID 2, PVSYST, INSEL, SOLSIM, WATSUN-PV, PV-DESIGNPRO, RAPSIM, PHOTO, SOMES, HOMER, RAPSYS, RETScreen, ARES, and PVF-chart. Out of all these software tools only two (SOMES and HOMER) are exactly relevant to this investigation, because these two software are capable of providing optimal design of hybrid system. A brief description of each tool is given below sections.

6.1. Software tools for pre-defined system configurations

6.1.1. HYBRID 2

HYBRID 2 (http://www.ceere.org/rerl/projects/software/ hybrid2/index) is a simulation tool that aims to provide a versatile model for the technical and economical analysis of renewable hybrid energy system. The tool was developed in NREL, Canada in the year 1993. This programming model utilizes both the time series and a statistical approach to evaluate the operation of renewable hybrid system. This allows the model to determine long-term performance while still taking into account the effect of short-term variability of solar and wind data. A range of system components, control and dispatch option can be modeled with users specified time steps. HYBRID 2 comprises all kinds of energy dispatch strategies researched by Barley (1995). HYBRID 2 is an extensively validated model. Though the technical accuracy of the model is very high but the model is incapable to optimize the energy system. The HYBRID 2 code employs a user-friendly graphical user interface (GUI) and a glossary of terms commonly associated with hybrid power systems. HYBRID 2 is also packaged with a library of equipment to assist the user in designing hybrid power systems. Each piece of equipment is commercially available and uses the manufacturer's specifications. In addition the library includes sample power systems and projects that the user can use as a template. Two levels of output are provided, a summary and a detailed time step by time step description of power flows. A graphical results interface (GRI) allows for easy and in-depth review of the detailed simulation results.

6.1.2. PVSYST

PVSYS 4.35 (2009) developed by Geneva University in Switzerland is a software package for the study sizing, simulation, and data analysis of complete PV systems. It allows determination of PV size and battery capacity, given a user's load profile and the acceptable duration that load cannot be satisfied. The software offers a large database of PV components, metrological sites, an expert system, and a 3-D tool for near shading detailed studies. This software is oriented toward architects, engineers, and researchers, and holds very helpful tools for education. It includes an extensive contextual help, which explains in detail the procedures and the models used. Tool performs the database meteo and components management. It provides also a wide choice of general solar tools (solar geometry, meteo on tilted planes, etc.), as well as a powerful mean of importing real data measured on existing PV systems for close comparisons with simulated values.

6.1.3. INSEL

Integrated simulation environment and a graphical performing language (INSEL) is software developed by University of Oldenburg, Germany, in which simulation models can be created from existing blocks in the graphic editor HP VEE with a few mouse clicks (Swift & Holder, 1988). The simulation of systems like on-grid PV generators with MPP tracker and inverter, for instance, becomes practically a drawing exercise. This software supports the designer with database for PV modules, inverters, thermal collectors, and meteorological parameters. Even more, INSEL offers a programming interface for the extension of the block library. The main advantage of this model is the flexibility in creating system model and configuration compared to simulation tools with fixed layouts. A disadvantage is that INSEL does not perform system optimization, though it completes or even replaces the experimental laboratory for renewable energy system, since components can be interconnected like in reality.

6.1.4. SOLSIM
Simulation and optimization model for renewable Energy Systems (SOLSIM) (Schaffrin, 1998) is developed at Fachhochschule Konstanz, Germany. SOLSIM is a simulation tool that enables users to design, analyze, and optimize off-grid, grid connected hybrid solar energy systems. It has detailed technical models for PV, wind turbine, diesel generator, and battery components as well as for biogas and biomass modeling. SOLSIM software package consists of different tools: the main simulation program called SOLSIM; the unit to optimize the tilting angle of PV module called SolOpti; the unit to calculate life cycle cost called SolCal; and the unit to simulate wind generators called SolWind. This program is also incapable to find the optimal size of hybrid system for any location on techno-economical ground.

6.1.5. WATSUN-PV
WATSUN-PV 6.0 (Tiba & Barbosa, 2002) developed by University of Waterloo, Canada, is a program intended for hourly simulation of various PV systems: standalone battery back-up, PV/diesel hybrid, utility grid-connected system, and PV water pumping system simulations. The modules standalone battery back-up and PV/diesel hybrid system simulation modules are very complete; on the other hand, the module that deals with PV water pumping systems only allows the analysis of configurations using DC electric motors, which is not a configuration very frequently used nowadays. The modeling systems for solar radiation, PV arrangement, and the battery are quite detailed and updated. The model used for DC motors is a simple relationship between the voltage and current supplied by the array and the torque and angular speed of the motor. WATSUN-PV 6.0 has a library containing information on PV modules, batteries, inverters, and diesel and gasoline generators. The database does not include information on motors or pumps.

6.1.6. PV-DESIGN PRO
The PV-design pro simulation program (Planning & installing PV system: A guide for installers, architects & engineers, 2005) comprises three variants for simulating standalone system, grid-connected system, and PV pump system. For standalone systems, a reserve generator and a wind generator can be integrated into the PV system, and a shading analysis can be carried out. The system can be optimized by varying the individual parameters. Detailed calculations are performed for operating data and characteristics curves. The module and climate database are very comprehensive. This program is recommended for the PV systems that have battery storage. Simulation is carried out on hourly basis. An advantage of PV-design pro is that its database already includes most information needed for PV system design.

6.1.7. RAPSIM
RAPSIM (Pryor, Gray, & Cheok, 1999) or remote area power supply simulator is a computer modeling program developed at the Murdoch University Energy Research Institute, Australia. It is designed to simulate alternative power supply options, including PV, wind turbine, battery, and diesel system. The user selects a system and operating strategy from a few pre-defined options and optimization is sought by varying component sizes and by experimenting with the control variables that determine on-off cycles of the diesel generator. Battery aging effect is not considered in this model.

6.1.8. RETScreen
RETScreen is developed and maintained by the Government of Canada through Natural Resources Canada's Canmet Energy research centre in 1996. RETScreen software is capable to calculate the energy efficiency, renewable energy, and risk for various types of renewable-energy, energy-efficient technologies and also analyze the cost function of the design system and hybrid system feasibility (RETScreen, 2009). RETScreen working is based on Microsoft excel software tool. The main characteristics of this software are to minimize the green house gas emission, life cycle cost, and energy generation (Sinha & Chandel, 2014).

6.1.9. PHOTO

The computer code PHOTO (Manninen, Lund, & Vikkula, 1990) developed at the Helsinki University of Technology in Finland simulates the performance of renewable energy system, including PV–wind hybrid configuration. A back up diesel generator can also be included in the system configuration. The dynamic method developed uses accurate system component models accounting for component interactions and losses in wiring and diodes. The PV array can operate in a maximum power mode with the other subsystems. Various control strategies can also be considered. Individual subsystem models can be verified against real measurements. The model can be used to simulate various system configurations accurately and evaluate system performance, such as energy flows and power losses in PV array, wind generator, backup generator, wiring, diodes, and maximum power point tracking device, inverter, and battery. A cost analysis can be carried out by PHOTO. This code has facility to create a stochastic weather generation database in the cases where hourly data are not available. The simulation results compare well with the measured performance of a PV test plant.

6.1.10. SOMES

The computer model SOMES (simulation and optimization model for renewable energy systems) developed at University of Utrecht Netherlands (RETScreen, 2009) can simulate the performance of renewable energy systems. The energy system can comprise renewable energy sources (PV arrays, wind turbines), diesel generator, a grid, battery storage, and several types of converters. An analysis of the results gives technical and economical performance of the system and the reliability of power supply. The simulation is carried out on hourly basis for the simulation period of, for example, one year. Hourly average electricity produced by solar and wind system is determined. Hourly results are accumulated for simulation period. The accumulated values are used to evaluate technical and economical performance of system. The model contains an optimization routine to search for the system with lowest electricity cost, given the customer's desired reliability level.

6.1.11. HOMER

HOMER (https://analysis.nrel.gov/homer/includes/downloads/HOMERBrochure_English.pdf) is a computer model that simplifies the task of evaluating design options for both off-grid and grid-connected power systems for remote, standalone, and distributed generation (DG) applications. HOMER is developed by the National Renewable Energy Laboratory (NREL, USA), HOMER's optimization and sensitivity analysis algorithms allow us to evaluate the economic and technical feasibility of a large number of technology options and to account for variation in technology costs and energy resource availability. HOMER models both conventional and renewable energy technologies: such as PV, wind turbine, run-of-river hydropower, diesel or biogas generator, fuel cell, utility grid, battery bank, micro turbine, and hydrogen storage. HOMER performs simulation for all of the possible system configurations to determine whether a configuration is feasible. Then, HOMER estimates the cost of installing and operating cost of the system, and displays a list of configurations sorted by their life cycle cost. This tool offers a powerful user interface and accurate sizing with detail analysis of the system.

6.1.12. RAPSYS

RAPSYS (version 1.3) was developed in the University of New South Wales, Australia in the year 1987 (RAPSYS, www.upress.uni-kassel.de/online/frei/978-3-933146-19-9). This software can simulate a wide range of renewable system components that may be included in a hybrid system configuration. The software can be used only by those who are experts in remote area power supply system. RAPSYS does not optimize the size of components. The user is required to pre-define the system configuration. The simulation recommends the switch ON and OFF timings of diesel generator. RAPSYS does not calculate the life cycle COE system, though it is capable to provide detailed information about the operating cost of the system.

6.1.13. ARES

A refined simulation program for sizing and optimization of autonomous hybrid energy systems (ARES) developed at University of Cardiff, UK determines whether a system meets the desired reliability level while meeting the project budget based on user specified cost data (Morgan, Marshall, & Brink worth, 1995; Morgan, Marshall, & Brinkworth, 1997). This program, unlike the majority of other hybrid simulation program, predicts the battery state of voltage (SOV) rather than its state of charge (SOC). LOL occurs when the battery voltage drops below the low voltage cut-off limit. Given the load and weather profiles, ARES is able to predict the occurrence of LOL thus giving a direct measure of the system autonomy. The model predicts electrical quantities measured at the terminals of battery bank and do not describe the electro-chemical phenomena occurring within the individual cells. The simulation code has been validated by comparison with measured data obtained form a 200 W wind and PV system. Accurate prediction of battery voltage requires a fairly extensive knowledge of the descriptive parameters of system components. These details are rarely found in manufacturing data sheets. The lack of data concerning charge characteristics and temperature effects is even more blatant. It would be advantageous if a data bank with such parameters were to be made available. The battery aging and its effects on system performance has not been addressed as part of this program. The precision and reliability of the simulation results obtained by this software depend mostly on the accuracy of the descriptive parameters.

6.1.14. PVF-CHART

The computer program PVF-chart (Klein & Beckman, 1993; Planning & installing PV system: A guide for installers, architects & engineers, 2005) developed by F-chart software is suitable for prediction of long-term average performance of PV utility interface system, battery storage system, and system without interface or battery storage. It is a comprehensive PV system analysis and design program. The program provides monthly-average performance estimates for each hour of the day. The calculations are based upon methods developed at the University of Wisconsin which use solar radiation utilizability to account for statistical variation of radiation and the load. The PVF-Chart method consists of combination of correlation and fundamental expression for hourly calculation of solar radiation at given location.

6.2. Search methodology based on optimization of hybrid system

In addition to the software tool stated above, other search methods for the design of hybrid energy system are described in various technical publications. These methods, which are described below, include amp-hour (AH) method, knowledge-based approach, simulation approach, trade-off method, probability method, analytical method, linear programming, goal programming, dynamic programming, and non-linear programming.

6.2.1. AH method

AH method is the most straightforward method to size PV-battery–diesel hybrid system. This method detailed out in a handbook of PV design practices by Sandia National Laboratory (SANDIA, 1995). The storage capacity is determined by number of autonomous days (number of continuous days that the battery can cover the load without sunshine), which is arbitrary selected by designer (typically 3–7 days). The size of diesel generator is selected to cover peak demand. This method does not take into account the relationship between the output of PV, generator sets, and storage capacity. Unless the very accurate data are used to select the value for autonomous days, this can easily lead to the specification of oversized components and suboptimal results. This method is used in Bhuiyan and Ali Asgar (2003), Ming, Buping, and Zhegen (1995), Protogeropulos, Brinkworth, and Marshall (1997) to size the standalone PV systems.

6.2.2. Trade-off method

The trade-off method is introduced by Gavanidou and Bakirtzis (1993) for multi-objective planning under uncertainty. The idea is intended for use in the design of standalone systems with renewable energy sources. This is done first by developing a database that contains all possible combinations of PV plants, wind generator, and battery, given ranges and steps of component sizes. Next, all possible planes are simulated over all possible futures, i.e. ±1 m/s variation in the wind velocity, ±10% variation in the global solar insolation. The author then creates a trade-off curve by plotting investment cost and LOL probability (LOLP) for all possible scenarios, eliminating options with LOLP greater than 10%, and identifying the knee-sets. Robust plans are then identified by the frequency of the occurrence of discrete option values in the conditional decision set. This method yield a small set of robust designs that are expected to work well under most foreseeable conditions. The final decision for the selection of the unique design is left to the decision-makers.

6.2.3. Probability method using LPSP technique

The concept of LPSP was introduced (Ofry & Braunstein, 1983) to design standalone PV systems. This technique enables the determination of the minimum sizes of the PV system and storage capacity, and yet assures a reliable power supply to load. The reliability of power supply is measured by total number of hours per year for which the consumer's power demand is greater than PV supply. The study is performed during a period of one year to collect the state of charge (SOC) of battery as function of time. The cumulative distribution function of the battery SOC is derived. The LPSP is then determined by calculating the value [1 – {cumulative proportion of the time where battery SOC is higher than the SOC min.}]. Similar work is done including wind generators in Borowy and Salameh (1996), Ghali, El Aziz, & Syam. (1997) and Ali, Yang, Shen, and Liao (2003), then adopted this concept to find the optimum size of the battery bank storage coupled with a hybrid PV–wind autonomous system. Long-term data of wind speed and insolation recorded for every hour of the day are deduced to produce the probability density function of combined generation. For the load distribution being considered, the probability density function of the storage is obtained. Finally, the battery size is calculated to give the relevant level of the system reliability using the LPSP technique.

6.2.4. Analytical method with LPSP technique

A closed form solution approach to the evaluation of LPSP of standalone PV system with energy storage, as well as standalone wind electric conversion system, is presented in Abouzahr and Ramakumar (1990). Similar to Borowy and Salameh (1996), in this paper also, authors have defined the LPSP as probability of encountering the state of charge (SOC) of battery bank falling below a certain specified minimum value. However, instead of using long-term historical data to determine LPSP, LPSP is determined by integrating the probability density function of power input to the storage. In addition to the above publications, there are several other publications that analyze and estimate reliability of a standalone PV system, using LOL probability (LOLP). These publications include Diaf, Belhamel, Haddadi, and Louche (2008), Diaf, Notton, et al. (2008), Klein and Beckman (1987), Yang, Lu, and Zhou (2007), Diaf, Diaf, Belhamel, and Haddadi (2007).

6.2.5. Knowledge-based approach

A knowledge-based design approach that minimizes the total capital cost at a pre-selected reliability level is introduced in Ramkumar, Abouzahr, and Ashenayi (1992) and Ramkumar, Abouzahr, Krishnan, and Ashenayi (1995). The overall design approach is as follows: first, a year is divided into as many times sections as needed. For each section the rating of energy converter and the sizes of energy storage system that satisfy the energy needs at the desired reliability level at the minimum capital cost are determined. Then, a search algorithm is used to search for feasible configurations. Since the final design is selected based on the seasonal designs, the user must decide whether to select the worst or best case designs or the designs in between.

6.2.6. Simulation approach

In this approach, design of hybrid renewable energy system comprising PV/wind/battery systems is carried out using the same concept as used in HOMER (https://analysis.nrel.gov/homer/includes/downloads/HOMERBrochure_English.pdf). Initially, simulation is performed using a time step of usually one hour (though not necessary) to identify all possible combinations that satisfy the desired level of reliability of user. An optimal combination is then extracted from these combinations on the basis of economic parameters. The reliability level is calculated by total number of load unmet hours divided by the total number of hours in simulation period. The similar approach is used by Ali et al.

(2003), Bernal-Agustín et al. (2006), Celik (2002) to find the optimal configurations. The simulations are done by varying fraction of wind and PV energy from zero to one, at the battery-to-load ratio (the number of days that the battery is able to supply the load while fully charged) of 1.25, 1.5, and 2.0, and various energy-to-load ratios (the ratio of the energy produced by renewable component to energy demand).

6.2.7. Linear programming method
This is a well-known popular method used by number of researchers to find the optimum size of renewable energy systems. A very good explanation and insights into how linear programming (LP) method can be applied to find the size of wind turbine and PV system in a PV–wind hybrid energy system is detailed out in Markvast (1997). The method employs a simple graphical construction to determine the optimum configuration of the two renewable energy generators that satisfies the energy demand of the user throughout the year. It is essential to note that method does not include battery bank storage and diesel generator. LP method was used in Swift and Holder (1988) to size PV–wind system, considering reliability of power supply system. The reliability index used is defined as the ratio of total energy deficit to total energy load. Other applications of this method are available in Chedid and Rahman (1997) and Ramakumar, Shetty, and Ashenayi (1986).

6.2.8. Non-linear programming method
The basic approach used in this method aims to take the interdependency between sizing and system operation strategy into account. Thus, it can simultaneously determine the optimal sizing and operation control for renewable hybrid energy system. This method has been used in Seeling-Hochmuth (1997) to determine the optimum size of hybrid system configuration.

6.2.9. Genetic algorithm method
Genetic algorithms are an adequate search technique for solving complex problems when other techniques are not able to obtain an acceptable solution. This method has been applied in Tomonobu, Hayashi, and Urasaki (2006), Dufo-Lopez and Bernal-Augustin (2005), and Shadmand and Balog (2014). The works reported in these papers use the hourly average metrological and load data over a few years for simulation. In reality, the weather conditions are not the same every day and in every hours of the day. Therefore, under varying every hour and every day weather conditions, the optimum number of facilities to use the hourly average data may not be able to be supplied without outages over a year. In such situations, the use of genetic algorithm method has been found most suitable.

6.2.10. Particle swarm optimization
The particle swarm algorithm was first presented by Kennedy and Eberhart (1995) as an optimization method to solve non-linear optimization problems. This procedure is inspired by certain social behavior. For a brief introduction to this method, consider a swarm of p particles, where each particle's position represents a possible solution point in the design problem space D. Every single particle is denoted by its position and speed; in an iterative process, each particle continuously records the best solution thus far during its flight. As an example of optimal sizing of hybrid energy systems by means of PSO, refers to Hakimi and Moghaddas-Tafreshi (2009) and Haghi, Hakimi, and Tafreshi (2010).

6.3. Outcomes
Above literature review leads to following conclusion:
  (1) There are different software packages existing with varying degree in user friendliness, validation of simulation models, accuracy of system models, and possible configuration to simulate.
  (2) Most of these software tools simulate a given and predefined hybrid system based on a mathematical description of component characteristic operation and system energy flow. But, the

Table 4. Software based on optimization of hybrid system						
Software	Developed by	Advantages	Disadvantages	Ref.	Year	Availability
HYBRID 2	NREL; Canada	Technical accuracy of the model is very high	Model is incapable to optimize the energy system	HYBRID 2 (http://www.ceere.org/rerl/projects/software/hybrid2/index), Barley (1995)	1993	http://www.ceere.org/rerl/rerl_hybrid-power.html
PVSYS	Geneva University in Switzerland	It allows determination of PV size and battery capacity	Limitation for renewable energy sources	PVSYST 4.35 (2009)	1992	Not free www.pvsyst.com
INSEL	University of Oldenburg, Germany	Flexibility in creating system model and configuration	Does not perform system optimization	Planning and installing PV system (2005)	1996	Not free www.insel.eu
SOLSIM	Fachhochschule Konstanz, Germany	The unit to calculate life cycle cost	In capable to find the optimal size of hybrid system	Schaffrin (1998)	1987	NOT Free
WATSUN-PV	University of Waterloo, Canada	The model used for DC motors is a simple relationship between the voltage and current supplied by the array and the torque and angular speed of the motor	The database does not include information of motors or pumps	Tiba and Barbosa (2002)	–	NOT FREE
PV-DESIGN PRO	–	Database already includes most information needed for PV system design	The module and climate database are very comprehensive	Planning and installing PV system (2005)	–	–
RAPSIM	Murdoch University Energy Research Institute, Australia	The control variables that determine on-off cycles of the diesel generator	Battery aging effect is not considered in this model	Pryor et al. (1999)	1997	Unknown, after 1997 any change are not accounted
RETScreen	Government of Canada through Natural Resources Canada's Canmet ENERGY research centre	This software is to minimized the green house gas emission, life cycle cost, energy generation	1. Does not support calculation of more advanced statistics/analysis 2. Limited search and retrieval features 3. Limited visualization features 4. Data sharing problems 5. Data validation 6. Difficult to relate different data-sets, hence need for duplication	RETScreen (2009), Sinha and Chandel (2014)	1996	Free http://www.retscreen.net/
PHOTO	The Helsinki University	Various control strategies can also be considered	High computational time	Manninen et al. (1990)	1990	Unknown
SOMES	University of Utrecht Netherlands	The model contains an optimization routine to search for the system with lowest electricity cost	SOMES does not give optimal operating strategy	SOME (http://www.web.co.bw/sib/somes_3_2_description.pdf)	1987	Not free http://www.uu.nl/EN/Pages/default.aspx
HOMER	National Renewable Energy Laboratory (NREL, USA),	This tool offers a powerful user interface and accurate sizing with detail analysis of the system	Technical accuracy of HOMER is low because its components mathematical models are linear and do not include many correction factors	HOMER (https://analysis.nrel.gov/homer/includes/downloads/HOMER-Brochure_English.pdf)	1993	Not Free www.homerenergy.com
RAPSYS	University of New South Wales, Australia	This software can simulate a wide range of renewable system components that may be included in a hybrid system configuration	It does not optimize the size of components	RAPSYS (www.upress.uni-kassel.de/online/frei/978-3-933146-19-9)	1987	www.upress.uni-kassel.de/online/frei/978-3-933146-19-9

(Continued)

Software	Developed by	Advantages	Disadvantages	Ref.	Year	Availability
Table 4. (Continued)						
ARES	University of Cardiff, UK	ARES is able to predict the occurrence of loss of load thus giving a direct measure of the system autonomy	The battery aging and its effects on system performance has not been addressed as part of this program.	Morgan et al. (1995, 1997)	–	Not found
PV F-chart	F-chart software	It suitable for prediction of long-term average performance. Extremely fast execution	Tracking options fixed	Planning and installing PV system (2005), Klein and Beckman (1993)	1993	Not free www. fchart.com

mathematical models used for characterizing system components are unknown due to commercial reasons.

(3) Some of these software tools (such as HYBRID 2, RAPSIM), though incorporate financial costing but incapable of determining optimal hybrid system configuration.

(4) For the optimal hybrid system design problem so far only two software tools (HOMER and SOMES) exist, using simplified linear system components mathematical models but varying the design randomly within a chosen range of component sizes.

(5) Technical accuracy of HOMER is low because its components mathematical models are linear and do not include many correction factors. SOMES does not give optimal operating strategy and not freely available to the designer/users.

(6) Majority of the software packages require the user to come up with a pre-design system. Therefore, a better system performance with lower cost could be achieved in many of these designs only if the system configuration could be optimized.

(7) With reference to search methodology-based optimization of hybrid systems, several previous works have certain limitations: some gives oversized components; some leave many design configurations for user to select; some do not consider the important parameters/correction factors in the design; and some are very lengthy and time consuming.

(8) Many papers are available for sizing by using artificial intelligent techniques, such as GA and PSO, etc. these new artificial intelligent techniques which can also be considered while sizing of hybrid renewable energy system. These artificial intelligent techniques provide best possible solution as compared to other software tools, but they face a crisis in the form of poor performance when a number of hybrid system components are increased such as PV, wind, generator, batteries, etc.

7. Case studies

A routine of software tools are used to design hybrid system which are discussed in Section 6. Among all these software GA, PSO, and HOMER found to be more suitable for evaluating optimal sizing of hybrid renewable solution. In this case study design of optimal sizing of different combinations of PV/wind hybrid energy-based power system for rural electrification in the key area by using HOMER software tool is presented.

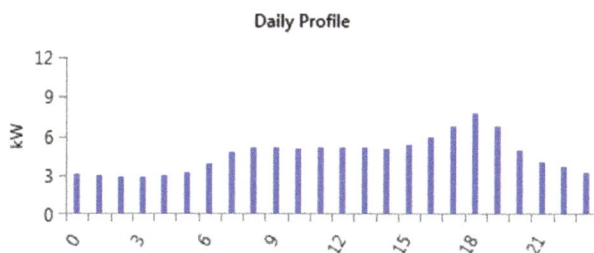

Figure 6. Daily load profile.

Ref.	Software	SA/GC	Parameters optimized	Load type	Highlights
Ghali et al. (1997), Bhuiyan and Ali Asgar (2003), Ming et al. (1995), Proto-geropulos et al. (1997)	AH method	SA	PV, battery	Residential	To operate the estimated load reliably in the month of minimum insolation taking into account different types of power losses
Gavanidou and Bakirtzis (1993)	Trade-off method	SA	PV, wind battery	Residential	Design that is a reasonable compromise between the conflicting design objectives under most foreseeable conditions
Ghali et al. (1997), Borowy and Salameh (1996), Ali et al. (2003), Ofry and Braunstein (1983)	Probability method using LPSP technique	SA	PV, battery	Residential	Determine the minimum (and thus the economical) sizes of the solar cell array and storage system capacity
Abouzahr and Ramakumar (1990), Klein and Beckman (1987), Yang et al. (2007), Diaf et al. (2007), Diaf, Notton, et al. 2008), Diaf, Belhamel, et al., 2008; Borowy and Salameh (1996)	Analytical method with LPSP technique	SA	Wind battery	Industrial	To evaluate the relationship between the amount of energy storage and the loss of power supply probability under various operating conditions
Ramkumar et al. (1992, 1995)	Knowledge-based approach	SA	PV, wind biogas, battery	Residential	A knowledge-based design approach that minimizes the total capital cost at a pre-selected reliability level
Ali et al. (2003), Bernal-Agustin et al. (2006), Celik (2002)	Simulation approach	SA	PV, Wind Battery	Residential	An optimum combination of the hybrid PV–wind energy system provides higher system performance than either of the single systems for the same system cost for every battery storage capacity
Chedid and Rahman (1997), Markvast (1997), Ramakumar et al. (1986), Swift and Holder (1988)	Linear programming (LP) method	GA	PV, wind battery	Residential	1. Linear programming techniques to minimize the average production cost of electricity while meeting the load requirement in a reliable manner
					2. A controller that monitors the operation of the autonomous grid-linked system is designed
8	Non-linear programming method	SA	PV, wind battery	Residential	A general method has been developed to jointly determine the sizing and operation control of hybrid-PV systems
Dufo-Lopez and Bernal-Augustin (2005), Shadmand and Balog 2014), Tomonobu et al. (2006)	Genetic algorithm method	GA	PV, wind battery	Residential	The proposed methodology employs a techno-economic approach to determine the system design optimized by considering multiple criteria including size, cost, and availability
Haghi et al. (2010), Hakimi and Moghaddas-Tafreshi (2009), Kennedy and Eberhart (1995)	Particle swarm optimization	SA	Wind fuel cells, hydrogen tanks	–	Minimize the total costs of the system in view of wind power uncertainty to secure the demand

Table 5. Summary of search methodology for design hybrid system

7.1. Renewable energy resources

A Jamny Ven village Barwani (latitude 22.71 and longitude 75.85) Madhya Pradesh, India site renewable energy resource is an important factor for developing hybrid systems. According to IMD wind and solar energy are available in many parts of India in large quantities (http:/homepage.mac.com/unarte/solar_radiatio n.html). These energy sources are discontinuous and naturally obtainable; because of these issues our primary preference to power the village base power station is renewable energy sources like wind and solar. Climate data for particular site renewable hybrid energy systems are important factors to study the possibility of the former the confidential information, wind and solar energy resources data for the village are taken from NASA (Lilienthal & Flowers, 1995).

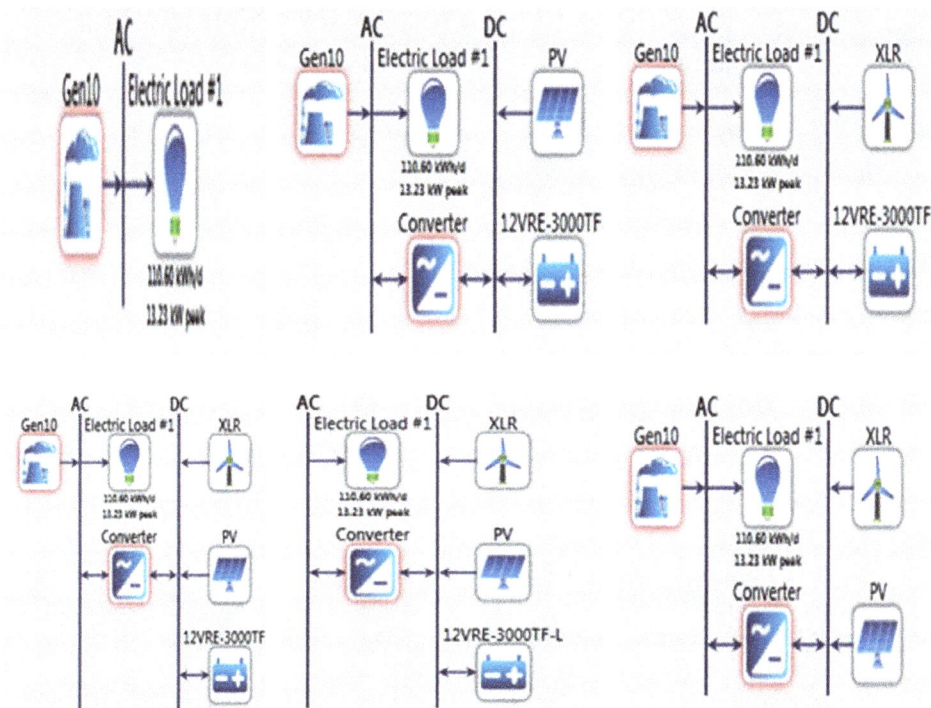

Figure 7. Architecture of hybrid renewable energy system with (i) DG, (ii) PV–Battery–DG, (iii) Wind–Battery–DG, (iv) PV–Wind–DG, (v) PV–Wind–Battery,(vi) PV–Wind–Battery–DG.

7.2. Solar energy resource

Hourly solar emission information was collected from the environment Barwani Jamny village. Long-term average annual resource scaling (5.531). Solar power is higher in summer season when compared to the winter season. Here solar insolation and clearance index data are shown in Table 4.

7.3. Wind energy resource

Confidential information may be an occurrence that is associated with the connection of air, plenty caused mainly by the degree of difference star heating of the Earth's surface. Seasonal and position variations within the energy arriving from the Sun have an effect on the strength and manner of the wind. Power from the wind depends upon the swept space of the rotary engine blades and, therefore,

S. No	Months	Insolation (KWh/m²/d)	Clearance index	Wind speed (m/s)
Table 6. Resource of PV–Wind data				
1	January	4.810	0.684	4.794
2	February	5.650	0.697	5.702
3	March	6.350	0.675	3.338
4	April	6.990	0.668	4.121
5	May	7.210	0.656	4.062
6	June	6.080	0.546	2.664
7	July	4.770	0.432	3.572
8	August	4.170	0.393	3.630
9	September	5.190	0.533	3.594
10	October	5.790	0.684	4.823
11	November	4.900	0.675	6.587
12	December	4.510	0.675	7.195
13	Average	5.531	0.598	4.500

Table 7. Input parameters used hybrid system				
S. No	Items cost($)		Other parameters	Life span
1	Wind turbine		Hub height: 30 m Rotor diameter 1.75 m	20 year
	Initial: 23220	Replacement: 1775	O&M: 480	
2	PV		Derating factor: 80% Ground reflectance: 20%	20 year
	Initial: 1590	Replacement: 750	O&M: 2	
3	Inverter		Efficiency: 90%	15 year
	Initial: 2400	Replacement: 2350	O&M: 1	
4	Battery		Capacity: 240 Ah voltage: 12 V	3550 h
	Initial: 250	Replacement: 250	O&M: 10	
5	Generator		Minimum load ratio: 30	15000 h
	Initial: 15300	Replacement: 1450	O&M: 0.2	

the cube of the wind speed, wind energy has been considered as potential toward meeting the continually increasing demand for energy. The wind sources of energy the alteration processes are pollution-free, and it is freely available. Periodical regular wind information for Jamny Ven village was together beginning environmental of Barwani climate. The scaled annual average wind speed is 4.5 the highest value of monthly average wind speed is observed during the month of December with a maximum of 7.195 m/s and the lowest value is observed during June with 2.664 m/s monthly average wind speed. Resource data are shown in Table 4 (Lilienthal & Flowers, 1995).

7.4. Electrical load data
The average estimation of daily energy consumption is 110.6 (kWh/day), peak load is found to be13.23 KW, and average is 4.61 KW. The information was computed for the entire hour basis daily electrical load condition of a demand for a village of Barwani district. The daily load profile with respective 24 h of day is shown in Figure 6.

7.5. Cost of hybrid system components
The cost of input components which are used to design optimal combination solution is given in Table 5.

7.6. Result and discussion
The study is to design of optimal sizing of different combinations of PV/wind hybrid energy-based power system for rural electrification in the key area (Jamny Ven Barwani) Madhya Pradesh, India where utility supply cost is really high due to limited consumer higher transmission and higher transportation cost. The chosen case study presents a power demand 110.6 kWh/d. The system is designed and optimized as hybrid energy base power system in parliamentary procedure to meet the existing user's power require at a minimum price of energy. The simulation-based optimization generates the best-optimized sizing of different combinations of wind and PV array with diesel generators for a rural hybrid base power system. Optimal sizing of various combinations such as DG (diesel generator), PV–Battery–DG, Wind–Battery–DG and PV–Wind–DG, PV–Wind–Battery and PV–Wind–Battery–DG are shown in Figure 7. Simulation and optimization result calculated by using HOMER software and analysis on the base of sensitive parameters of PV, wind resources data, and variation in diesel price. Among all six hybrid combinations only two hybrid system Wind-DG and PV–Wind–Battery–DG are more cost-effective, reliable and environmentally friendly solution. Emission and levelized COE of the both hybrid systems are nearly equal, but the total NPC and operating cost of the PV–Wind–Battery–DG is less as compared to Wind-DG hybrid system. As the penetration of solar, wind system will increase; the surplus energy is multiplied. It can be saved and used by foreseeable

S.No	Description	DG	PV/Battery/DG	Wind/Battery/DG	PV/Wind/DG	PV/Wind/Battery	PV/Wind/Battery/DG
	Table 8. Comparative examination of different hybrid system configurations						
1	Emission						
	Carbon dioxide (kg/yr)	62,204.00	36,334.00	28,394.00	61,517.00	0	29,201.00
	Carbon monoxide (kg/yr)	153.54	89.69	70.09	151.85	0	72.08
	Unburned hydrocarbons (kg/yr)	17.01	9.93	7.76	16.82	0	7.98
	Particulate matter (kg/yr)	11.58	6.76	5.28	11.45	0	5.43
	Sulfur dioxide (kg/yr)	124.92	72.97	57.02	123.54	0	58.64
	Nitrogen oxides (kg/yr)	1,370.10	800.27	625.39	1,354.90	0	643.17
2	Production						
	Excess electricity (KWh/yr)	5,471.00	0	14,785.00	6,087.40	63,747.00	7,701.90
	Unmet electric load (KWh/yr)	0	11.6	10	0	22.3	9.2
	Capacity shortage (KWh/yr)	0	40.2	38	0	35	37.3
	Renewable fraction	0	33.8	47.2	0	100	46.5
	Max. renew. penetration	0	496.8	2,180.70	49.7	3,634.60	1,453.80
3	Cost						
	Total net present cost ($)	3,20,873	3,44,576	3,30,844	4,08,347	6,27,750	3,24,178
	Levelized cost ($)	0.6149	0.6605	0.6341	0.7825	1.2	0.6213
	Operating cost ($)	24,465.94	21,355	19,297	30,607.03	23,909	19,261
4	Fuel						
	Total fuel consumed (L)	23,622.00	13,798.00	10,783.00	23,361.00		11,089.00
	Avg fuel per day (L/day)	64.73	37.81	29.55	64.01		30.38
	Avg fuel per hour (L/hour)	2.7	1.58	1.23	2.67		1.27
5	Battery						
	Energy input (KWh/yr)		6,953.90	6,897.10		13,415.00	6,433.50
	Energy out (KWh/yr)		5,965.00	5,862.60		11,403.00	5,468.70
	Storage depletion (KWh/yr)		49.8	−8.9		−35.96	−8.7
	Losses (KWh/yr)		939.05	1,043.40		2,048.20	973.47
	Annual throughput (KWh/yr)		6,470.00	6,358.90		12,368.00	5,931.70
	Expected life (yr)		13.72	13.96		18	14.96
6	Efficiency						
	Mean electrical efficiency	19.72	19.69	20.10	19.55		19.81
7	Components						
	Generic flat plate PV	×	✓	×	✓	✓	✓
	BWC Excel-R			✓	✓	✓	✓
	10 kW genset	✓	✓	✓	✓		✓
	Discover 12VRE-3000TF-L	×	✓	✓	×	✓	✓
	Converter	×	✓	✓	✓	✓	✓

future objective by making use of battery bank. The comparative analysis of all optimal combination is shown in Table 6.

8. Conclusion
For hybrid renewable energy system design number of new technologies are discussed in the literature, but due to some new problems like parameters of renewable source material and design, constraints of load, generator, battery, converter, and cost function, the system performance has

Table 9. Suggestion for hybrid system implementation		
Number	Issue	Comments
1	Converters losses	The loss involved with electrical power converters are actually reduced to some sufficient stage; on the other hand, it should be guaranteed that there's minimal quantity of electrical power reduction within these converters
2	Life-cycle	The life-cycle associated with storage units, such as batteries along with UCs, should be improved upon by means of innovative systems
3	Disposal of storage equipment	The convenience connected with storage space products, like power packs and also other storages, is among the significant problems for producers
4	Renewable energy sources	Photovoltaic and other renewable energy options require break-through systems for removing much more quantity of use full strength. The poor effectiveness involving a solar PV is often an important barrier inside stimulating its use
5	Control unit	With the entire supplement associated with unique turbines inside developing a hybrid renewable energy system raises the strain about power alteration devices. Any possible hybrid renewable energy system requires the feasible associated with right keeping track of design system that will record important info to its productive functioning. Each time almost any mismatch inside the power generation in addition to desire exists the system may open the circuit breakers with regard to much better safety in addition to functioning
6	Grid control	For controlling different generators which are linked to the hybrid renewable energy system, to the function of saving power and carry through the load demand a development of a small grid system required
7	Manufacturing cost	This making price tag of renewable energy sources needs a significant lessening considering that the higher capital price tag causes an elevated payback time. cost lessening will supply a motivation on the marketplace to be able to apply like devices
8	Load management	The particular renewable resources tend to be independent of the load variations and as such suitable energy management should be designed, in order that the prolonged existence on the hybrid renewable energy system can be increased. Big deviation inside the load could even result in a whole system fall
9	Stability	Hybrid renewable energy system depends on weather conditions so that theirs is needed to carry out transient analysis of the system for varying constraint like solar radiation, wind velocity, load demand
10	Government support	For reducing the cost of components, production costs of generation and wide deployment of Hybrid renewable energy system network, it is essential to give subsidy on renewable energy goods from central to the state government

decreased. These kinds of issues have to be attended properly, to resolve these shorted out. Table 7 presents some important suggestion and also scope with regard to potential research. This paper explains several hybrid system combinations for PV and wind turbine, modeling parameters of hybrid system component, software tools for sizing, criteria for PV–wind hybrid system optimization, and control schemes for energy flow management. In this paper for the sizing purpose of the hybrid system, 25 different types of computational software tools are discussed. Among all these software tools, HOMER and GA, PSO gives more feasible result for hybrid system design. Another technique for sizing of hybrid scheme which presents more promising result, such as genetic algorithm and PSO. At least to obtain the operational efficiency, highest system reliability and proper energy flow management, control strategies are suggested in this paper. Controller work as monitoring whole hybrid system and maintain the requirement of load demand while keeping system frequency and output voltage. Additionally, it is been located in which wide range of research work in the community associated with hybrid renewable energy system has been completed. A case study of various standalone hybrid system combinations for a remote location in India by using HOMER and evaluate best optimal hybrid system configuration such as PV–Wind–Battery–DG with respective total NPC, operating cost, COE, and also emission. The optimal hybrid system has following advantages (Table s 8 and 9).

- This non-conventional power PV–Wind–Battery–DG hybrid energy method is available to be technically achievable, emission much less along with less expensive with years to come.
- Its environment-friendly dynamics helps it be a nice-looking substitute for complementing the energy present inside countryside regions.
- Load demand is fulfilled in an optimal way.

On the other hand much more research along with the work usually is needed to increase battery's strength along with effectiveness with giving attention to decreasing the cost. Hybrid system performance depends on weather condition so as to minimize the issue related to the system reliability and operation there is a need to carry out transient analysis of the system for varying constraint like solar radiation, wind velocity, load demand. The COE sources used in hybrid system are very high so that there is needed to provide subsidy from central and state government to minimize initial cost of the system and also reduced the COE.

Funding
The authors received no direct funding for this research.

Author details
Yashwant Sawle[1]
E-mail: yashsawle@gmail.com
S.C. Gupta[1]
E-mail: scg.nit.09@gmail.com
Aashish Kumar Bohre[1]
E-mail: aashu371984@gmail.com
[1] Department Electrical Engineering, Maulana Azad Institute of Technology, Bhopal, India.

References
Abido, M. A. (2003). Environmental/economic power dispatch using multiobjective evolutionary algorithms. *IEEE Transactions on Power Systems, 18*, 1529–1537. http://dx.doi.org/10.1109/TPWRS.2003.818693
Abouzahr, I., & Ramakumar, R. (1990). Loss of power supply probability of stand-alone wind electric conversion systems: A closed form solution approach. *IEEE Transactions on Energy Conversion, 5*, 445–452. http://dx.doi.org/10.1109/60.105267
Abouzahr, I., & Ramakumar, R. (1991). Loss of power supply probability of stand-alone photovoltaic systems: A closed form solution approach. *IEEE Transactions on Energy Conversion, 6*, 1–11. http://dx.doi.org/10.1109/60.73783
Al-Ashwal, A. M. (1997). Proportion assessment of combined PV-wind generating systems. *Renewable Energy, 10*, 43–51. http://dx.doi.org/10.1016/0960-1481(96)00011-0
Ali, B., Yang, H., Shen, H., & Liao, X. (2003). Computer aided design of PV/wind hybrid system. *Renewable Energy, 28*, 1491–1512.
Arifujjaman, M., Iqbal, M. T., Quaicoe, J. E., & Khan. M. J. (2005). Modeling and control of a small wind turbine. *IEEE Transactions on Electrical and Computer Engineering,*, 778–781.
Azmy, A. M., & Erlich, I. (2005). Online optimal management of PEM fuel cells using neural networks. *IEEE Transactions on Power Delivery, 20*, 1051–1058. http://dx.doi.org/10.1109/TPWRD.2004.833893
Bagul, A. D., & Salameh, Z. M. (1996). Sizing of a stand-alone hybrid wind-photovoltaic system using a three-event probability density approximation. *Solar Energy,56*, 323–335. http://dx.doi.org/10.1016/0038-092X(95)00116-9

Barley, C. D. (1995). Optimal control of remote hybrid power system, Part I: Simplified model. In *Proceedings of Wind Power 95*. Washington, DC.
Bellarmine, G. T., & Urquhart, J. (1996). Wind energy for the 1990s and beyond. *Energy Conversion Management, 37*, 1741–1752. http://dx.doi.org/10.1016/0196-8904(96)00009-X
Bernal-Agustín, J. L., Dufo-López, R., & Rivas-Ascaso, D. M. (2006). Design of isolated hybrid systems minimizing costs and pollutant emissions. *Renewable Energy, 31*, 2227–2244. http://dx.doi.org/10.1016/j.renene.2005.11.002
Beyer, H. G., & Langer, C. (1996). A method for the identification of configurations of PV/wind hybrid systems for the reliable supply of small loads. *Solar Energy, 57*, 381–391. http://dx.doi.org/10.1016/S0038-092X(96)00118-1
Bhave, A. G. (1999). Hybrid solar–wind domestic power generating system—case study Renew. *Energy, 17*, 355–358.
Bhuiyan, M. M. H., & Ali Asgar, M. (2003). Sizing of a stand-alone photovoltaic power system at Dhaka. *Renewable Energy, 28*, 929–938. http://dx.doi.org/10.1016/S0960-1481(02)00154-4
Bhuiyan, M. M. H., Asgar, M., Mazumder, R. K., & Hussain, M. (2000). Economic evaluation of a stand-alone residential photovoltaic power system in Bangladesh. *Renewable Energy, 21*, 403–410. http://dx.doi.org/10.1016/S0960-1481(00)00041-0
Bilal, B. O., Sambou, V., Ndiaye, P. A., Kébé, C. M. F., & Ndongo, M. (2013). Study of the influence of load profile variation on the optimal sizing of a standalone hybrid PV/Wind/Battery/Diesel system. *Energy Procedia, 36*, 1265–1275. http://dx.doi.org/10.1016/j.egypro.2013.07.143
Borowy, B. S., & Salameh, Z. M. (1996). Methodology for optimally sizing the combination of a battery bank and PV array in a wind/PV hybrid system. *IEEE Transactions on Energy Conversion, 11*, 367–375. http://dx.doi.org/10.1109/60.507648
Borowy, B. S., & Salameh, Z. M. (1997). Dynamic response of a stand-alone wind energy conversion system with battery energy storage to a wind gust. *IEEE Transactions on Energy Conversion, 12*, 73–78. http://dx.doi.org/10.1109/60.577283
Cano, A., Jurado, F., Sánchez, H., Fernández, L. M., & Castañeda, M. (2014). Optimal sizing of stand-alone hybrid systems based on PV/WT/FC by using several methodologies. *Journal of the Energy Institute, 87*, 330–340. http://dx.doi.org/10.1016/j.joei.2014.03.028
Celik, A. N. (2002, January). Optimization and techno-economic analysis of autonomous photovoltaic-wind hybrid energy systems in comparison to single

photovoltaic and wind system. *Energy Conversion and Management, 43*, 2453–2468.

Celik, A. N. (2003a). A simplified model for estimating the monthly performance of autonomous wind energy systems with battery storage. *Renewable Energy, 28*, 561–572. http://dx.doi.org/10.1016/S0960-1481(02)00067-8

Celik, A. N. (2003b). Techno-economic analysis of autonomous PV-wind hybrid energy systems using different sizing methods. *Energy Conversion and Management, 44*, 1951–1968. http://dx.doi.org/10.1016/S0196-8904(02)00223-6

Chedid, R., & Rahman, S. (1997). Unit sizing and control of hybrid wind-solar power systems. *IEEE Transactions on Energy Conversion, 12*, 79–85. http://dx.doi.org/10.1109/60.577284

Chedid, R. B., Karaki, S. H., & El-Chamali, C. (2000). Adaptive fuzzy control for wind-diesel weak power systems. *IEEE Transactions on Energy Conversion, 15*, 71–78. http://dx.doi.org/10.1109/60.849119

Diaf, S., Belhamel, M., Haddadi, M., & Louche, A. (2008). Technical and economic assessment of hybrid photovoltaic/wind system with battery storage in Corsica island. *Energy Policy, 36*, 743–754. http://dx.doi.org/10.1016/j.enpol.2007.10.028

Diaf, S., Diaf, D., Belhamel, M., & Haddadi, M. (2007). A methodology for optimal sizing of autonomous hybrid PV/wind system. *Energy Policy, 35*, 5708–5718. http://dx.doi.org/10.1016/j.enpol.2007.06.020

Diaf, S., Notton, G., Belhamel, M., Haddadi, M., & Louche, A. (2008). Design and techno-economical optimization for hybrid PV/wind system under various meteorological conditions. *Applied Energy, 85*, 968–987. http://dx.doi.org/10.1016/j.apenergy.2008.02.012

Dimeas, A. L., & Hatziargyriou, N. D. (2005). Operation of a multiagent system for microgrid control. *IEEE Transactions on Power Systems, 20*, 1447–1455. http://dx.doi.org/10.1109/TPWRS.2005.852060

Douglas, B. (1997, May/June). The essence way renewable energy system. *Solar Today, 5*, 16–19.

Dufo-Lopez, R., & Bernal-Augustin, J. L. (2005). Design and control strategies of PV-diesel using genetic algorithm. *Solar Energy, 79*, 33–46.

Dufo-López, R., Bernal-Agustín, J. L., & Mendoza, F. (2009). Design and economical analysis of hybrid PV–wind systems connected to the grid for the intermittent production of hydrogen. *Energy Policy, 37*, 3082–3095. http://dx.doi.org/10.1016/j.enpol.2009.03.059

Egido, M., & Lorenzo, E. (1992). The sizing of stand alone PV-system: A review and a proposed new method. *Solar Energy Materials and Solar Cells, 26*, 51–69. http://dx.doi.org/10.1016/0927-0248(92)90125-9

Ekren, B. Y., & Ekren, O. (2009). Simulation based size optimization of a PV/wind hybrid energy conversion system with battery storage under various load and auxiliary energy conditions. *Applied Energy, 86*, 1387–1394. http://dx.doi.org/10.1016/j.apenergy.2008.12.015

Elhadidy, M. A., & Shaahid, S. M. (2000). Parametric study of hybrid (wind+solar+diesel) power generating systems. *Renewable Energy, 21*, 129–139.

Elhadidy, M. A., & Shaahid, S. M. (2004a). Promoting applications of hybrid (wind+photovoltaic+diesel+batte ry) power systems in hot regions. *Renewable Energy, 29*, 517–528. http://dx.doi.org/10.1016/j.renene.2003.08.001

Elhadidy, M. A., & Shaahid, S. M. (2004b). Role of hybrid (wind+diesel) power systems in meeting commercial loads. *Renewable Energy, 29*, 109–118. http://dx.doi.org/10.1016/S0960-1481(03)00067-3

El-Hefnawi, S. H. (1998). Photovoltaic diesel-generator hybrid power system sizing. *Renewable Energy, 13*, 33–40. http://dx.doi.org/10.1016/S0960-1481(97)00074-8

Erhard, K., & Dieter, M. (1991). Sewage plant powered by combination of photovoltaic, wind and bio-gas on the Island of Fehmarl, Germany. *Renewable Energy, 1*, 745–748.

Feijoo, A. E., Cidras, J., & Dornelas, J. L. G. (1999). Wind speed simulation in wind farms for steady-state security assessment of electrical power systems. *IEEE Transactions on Energy Conversion, 14*, 1582–1588. http://dx.doi.org/10.1109/60.815519

Gavanidou, E. S., & Bakirtzis, A. G. (1993). Design of a standalone system with renewable energy sources using trade-off methods. *IEEE Transactions on Energy Conversion, 7*, 42–48.

Ghali, F. M. A., El Aziz, M. M. A., & Syam, F. A (1997, August 3–6). Simulation and analysis of hybrid systems using probabilistic techniques. In *Proceedings of the Power Conversion Conference* (Vol. 2, pp. 831–835). Nagaoka. http://dx.doi.org/10.1109/PCCON.1997.638338

Giraud, F., & Salameh, Z. M. (2001). Steady-state performance of a grid-connected rooftop hybrid wind–photovoltaic power system with battery storage. *IEEE Transactions on Energy Conversion, 16*, 1–7. http://dx.doi.org/10.1109/60.911395

González, A., Riba, J. R., Rius, A., & Puig, R. (2015). Optimal sizing of a hybrid grid-connected photovoltaic and wind power system. *Applied Energy, 154*, 752–762. http://dx.doi.org/10.1016/j.apenergy.2015.04.105

Gupta, S. C., Kumar, Y., & Agnihotri, G. (2011). Design of an autonomous renewable hybrid power system. *International Journal of Renewable Energy Technology, 2*, 86–104. http://dx.doi.org/10.1504/IJRET.2011.037983

Haghi, H. V., Hakimi, S. M., & Tafreshi, S. M. M. (2010, June 14–17). Optimal sizing of a hybrid power system considering wind power uncertainty using PSO-embedded stochastic simulation. In *IEEE 11th International Conference on Probabilistic Methods Applied to Power Systems (PMAPS)* (pp. 722–727). Singapore.

Hajizadeh, A., & Golkar, M. A. (2009). Fuzzy neural control of a hybrid fuel cell/battery distributed power generation system. *IET Renewable Power Generation, 3*, 402–414. http://dx.doi.org/10.1049/iet-rpg.2008.0027

Hakimi, S. M., & Moghaddas-Tafreshi, S. M. (2009). Optimal sizing of a stand-alone hybrid power system via particle swarm optimization for Kahnouj area in south-east of Iran. *Renewable Energy, 34*, 1855–1862. http://dx.doi.org/10.1016/j.renene.2008.11.022

Huang, K., Cartes, D. A., & Srivastava, S. K. (2007). A multiagent-based algorithm for ring-structured shipboard power system reconfiguration. *IEEE Transactions on Systems, Man and Cybernetics, Part C (Applications and Reviews), 37*, 1016–1021. http://dx.doi.org/10.1109/TSMCC.2007.900643

Jangamshetti, S. H., & Ran, V. G. (2001). Optimum siting of wind turbine generators. *IEEE Transactions on Energy Conversion, 16*, 8–13. http://dx.doi.org/10.1109/60.911396

Kaldellis, J. K. (2010). *Stand-alone and hybrid wind energy systems*. Tigaday of island of El Hierro: Woodhead Publishing Limited and CRC Press LLC. http://dx.doi.org/10.1533/9781845699628

Kaldellis, J. K., & Th, G. (2005). Vlachos Optimum sizing of an autonomous wind–diesel hybrid system for various representative wind-potential cases. *Applied Energy, 83*, 113–132.

Kamel, S., & Dahl, C. (2005). The economics of hybrid power systems for sustainable desert agriculture in Egypt. *Energy, 30*, 1271–1281. http://dx.doi.org/10.1016/j.energy.2004.02.004

Katsigiannis, Y. A., Georgilakis, P. S., & Karapidakis, E. S. (2010). Multiobjective genetic algorithm solution to the

optimum economic and environmental performance problem of small autonomous hybrid power systems with renewables. *IET Renewable Power Generation, 4*, 404–419.
http://dx.doi.org/10.1049/iet-rpg.2009.0076

Kefayat, M., Lashkar Ara, A., & Nabavi Niaki, S. A. (2015). A hybrid of ant colony optimization and artificial bee colony algorithm for probabilistic optimal placement and sizing of distributed energy resources. *Energy Conversion and Management, 92*, 149–161.
http://dx.doi.org/10.1016/j.enconman.2014.12.037

Kelash, H. M., Faheem, H. M., & Amoon, M. (2007). It takes a multiagent system to manage distributed systems. *IEEE Potentials, 26*, 39–45.
http://dx.doi.org/10.1109/MP.2007.343026

Kennedy, J., & Eberhart, R. C. (1995). Particle swarm optimization. In *Proceedings of IEEE International Conference on Neural Network IV*. Piscataway, NJ.
http://dx.doi.org/10.1109/ICNN.1995.488968

Khatod, D. Kumar, Pant, V., & Sharma, J. (2010). Analytical approach for well-being assessment of small autonomous power systems with solar and wind energy sources. *IEEE Transactions on Energy Conversion, 25*, 535–545.
http://dx.doi.org/10.1109/TEC.2009.2033881

Klein, S. A., & Beckman, W. A. (1987). Loss-of-load probability for stand-alone photovoltaic systems. *Solar Energy, 39*, 449–512.

Klein, S. A., & Beckman, W. A. (1993). *PVF-chart user's manual window version*. Middleton, WI. Retrieved from F-Chart Software: www.fchart.com

Ko, H., & Jatskevich, J. (2007). Power quality control of wind-hybrid power generation system using fuzzy-LQR controller. *IEEE Transactions on Energy Conversion, 22*, 516–527.
http://dx.doi.org/10.1109/TEC.2005.858092

Lagorse, J., Simoes, M. G., & Miraoui, A. (2009). A multiagent fuzzy-logic-based energy management of hybrid systems. *IEEE Transactions on Industry Applications, 45*, 2123–2129.
http://dx.doi.org/10.1109/TIA.2009.2031786

Li, S., Wunsch, D. C., O'Hair, E., & Giesselmann, M. G. (2001). Comparative analysis of regression and artificial neural network models for wind turbine power curve estimation. *Journal of Solar Energy Engineering, 123*, 327–332.
http://dx.doi.org/10.1115/1.1413216

Lilienthal, P., & Flowers, L. (1995). HOMER: The hybrid optimization model for electrical renewable. In *Proceedings of WindPower 95* (pp. 475–480). Washington, DC: American Wind Energy Associations.

Lingfeng Wang, L., & Singh, C. (2009). Multicriteria design of hybrid power generation systems based on a modified particle swarm optimization algorithm. *IEEE Transactions on Energy Conversion, 24*, 163–172.
http://dx.doi.org/10.1109/TEC.2008.2005280

Maghraby, H. A. M., Shwehdi, M. H., & Al-Bassam, G. K. (2002). Probabilistic assessment of photovoltaic (PV) generation systems. *IEEE Transactions on Power Systems, 17*, 205–208.
http://dx.doi.org/10.1109/59.982215

Mahmoud, M. (1990). Experience results and techno-economic feasibility of using photovoltaic generators instead of diesel motors for water pumping from rural desert wells in Jordan. *IEE Proceedings C, 37*, 391–394.

Maleki, A., & Askarzadeh, A. (2014). Comparative study of artificial intelligence techniques for sizing of a hydrogen-based stand-alone photovoltaic/wind hybrid system. *International Journal of Hydrogen Energy, 39*, 9973–9984.
http://dx.doi.org/10.1016/j.ijhydene.2014.04.147

Manninen, M., Lund, P. D. & Vikkula, A. (1990). *PHOTO-A computer simulation program for photovoltaic and hybrid energy systems* (Report TKK-F-A-670). Espoo: Helsinki University of Technology. ISBN 951-22-04657.

Markvast, T. (1997). Sizing of hybrid photovoltaic-wind energy systems. *Solar Energy, 57*, 277–281.

Marwali, M. K. C., Shahidehpour, S. M., & Daneshdoost, M. (1997). Probabilistic production costing for photovoltaics-utility systems with battery storage. *IEEE Transactions on Energy Conversion, 12*, 175–180.
http://dx.doi.org/10.1109/60.629700

Mathew, S., Pandey, K. P., & Anil Kumar, V. (2002). Analysis of wind regimes for energy estimation. *Renewable Energy, 25*, 381–399.
http://dx.doi.org/10.1016/S0960-1481(01)00063-5

McGowan, J. G., Manwell, J. F., Avelar, C., & Warner, C. L. (1996). Hybrid wind/PV/diesel hybrid power systems modeling and South American applications. *Renewable Energy, 9*, 836–847. http://dx.doi.org/10.1016/0960-1481(96)88412-6

Miettinen, K. (1998). *Nonlinear multiobjective optimization*. Boston, MA: Kluwer.
http://dx.doi.org/10.1007/978-1-4615-5563-6

Ming, J., Buping, L., & Zhegen, C. (1995). Small scale solar PV generating system-the household electricity supply used in remote area. *Renewable Energy, 6*, 501–505.

Morgan, T. R., Marshall, R. H., & Brink worth, B. J. (1995). SEU-ARES--A refined simulation program for sizing and optimization of autonomous hybrid energy system. In *Proceedings of the annual meeting of the International Solar Energy Society*. Harare.

Morgan, T. R., Marshall, R. H., & Brinkworth, B. J. (1997). 'ARES'—A refined simulation program for the sizing and optimisation of autonomous hybrid energy systems. *Solar Energy, 59*, 205–215.
http://dx.doi.org/10.1016/S0038-092X(96)00151-X

Muselli, M., Notton, G., Poggi, P., & Louche, A. (2000). PV-hybrid power systems sizing incorporating battery storage: An analysis via simulation calculations. *Renewable Energy, 20*, 1–7. http://dx.doi.org/10.1016/S0960-1481(99)00094-4

Nag, P. K. (2001). *Power plant engineering* (2nd ed.). New Delhi: Tata McGraw-Hill.

Nagata, T., & Sasaki, H. (2002). A multi-agent approach to power system restoration. *IEEE Transactions on Power Systems, 17*, 457–462.
http://dx.doi.org/10.1109/TPWRS.2002.1007918

Nayar, C. V., Phillips, S. J., James, W. L., Pryor, T. L., & Remmer, D. (1993). Novel wind/diesel/battery hybrid energy system. *Solar Energy, 51*, 65–78.
http://dx.doi.org/10.1016/0038-092X(93)90043-N

Nayar, C. V., Thomas, F. P., Phillips, S. J., & James, W. L. (1991). Design considerations for appropriate wind energy systems in developing countries. *Renewable Energy, 1*, 713–722.
http://dx.doi.org/10.1016/0960-1481(91)90018-K

Nehrir, M. H., Wang, C., Strunz, K., Aki, H., Ramakumar, R., Bing, J., … Salameh, Z. (2011). A review of hybrid renewable/alternative energy systems for electric power generation: Configurations, control, and applications. *IEEE Transactions on Sustainable Energy, 2*, 392–403.
http://dx.doi.org/10.1109/TSTE.2011.2157540

Nfaoui, H., Buret, J., & Sayigh, A. A. M. (1996). Cost of electricity generated and fuel saving of an optimized wind-diesel electricity supply for village in Tangier-area (Morocco). *Renewable Energy, 9*, 831–835.
http://dx.doi.org/10.1016/0960-1481(96)88411-4

Nfaoui, H., Buret, J., Sayigh, A. M. M., & Dunn, P. D. (1994). Modelling of a wind/diesel system with battery storage for Tangiers, Morocco. *Renewable Energy, 4*, 155–167.
http://dx.doi.org/10.1016/0960-1481(94)90001-9

Notton, G., Muselli, M., & Louche, A. (1996). Autonomous hybrid photovoltaic power plant using a back-up generator: A case study in a Mediterranean Island. *Renewable Energy, 7*, 371-391.

Ofry, E., & Braunstein, A. (1983). The loss of power supply probability as a technique for designing stand-alone solar electric photovoltaic system. *IEEE Transaction on Power Apparatus and System, PAS-102*, 1171-1175.

Onar, O. C., Uzunoglu, M., & Alam, M. S. (2006). Dynamic modeling, design and simulation of a wind/fuel cell/ultra-capacitor-based hybrid power generation system. *Journal of Power Sources, 161*, 707-722. http://dx.doi.org/10.1016/j.jpowsour.2006.03.055

Papadopoulos, D. P., & Dermentzoglou, J. C. (2002). Economic viability analysis of planned WEC system installations for electrical power production. *Renewable Energy, 25*, 199-217. http://dx.doi.org/10.1016/S0960-1481(01)00012-X

Papaefthymiou, S. V., & Stavros, A. (2014, April). Optimum sizing of wind-pumped-storage hybrid power stations in island systems. *Renewable Energy, 64*, 187-196.

Planning and installing PV system: A guide for installers, architects and engineers. (2005). (pp. 18-24). James & James. ISBN: 1-84407-131-6.

Post, H. N., & Thomas, M. G. (1988). Photovoltaic systems for current and future applications. *Solar Energy, 41*, 465-473. http://dx.doi.org/10.1016/0038-092X(88)90020-5

Protogeropulos, C., Brinkworth, B. J., & Marshall, R. H. (1997). Sizing and techno-economical optimization for hybrid solar photovoltaic/wind power systems with battery storage. *International Journal of Energy Research, 21*, 465-479. http://dx.doi.org/10.1002/(ISSN)1099-114X

Pryor, T., Gray, E., & Cheok, K. (1999, February). *How good are RAPS simulation program?-A comparison of RAPSIM & HYBRID 2.* World Renewable Energy Congress.

PVSYST 4.35. (2009, March). Retrieved 2008, from www.pvsyst.com

Ramkumar, R., Abouzahr, I., & Ashenayi, K. 1992. A knowledge-based approach to the design of integrated renewable energy systems. *IEEE Transaction on Energy Conversion, 7*, 648-659.

Ramkumar, R., Abouzahr, I., Krishnan, K., & Ashenayi, K. (1995). Design scenarios for integrated renewable energy system. *IEEE Transaction on Energy Conversion, 10*, 736-746.

Ramakumar, R., Shetty, P., & Ashenayi, K. (1986). A linear programming approach to the design of integrated renewable energy systems for developing countries. *IEEE Transactions on Energy Conversion, EC-1*, 18-24. http://dx.doi.org/10.1109/TEC.1986.4765768

Rehman, S., Halawani, T. O., & Mohandes, M. (2003). Wind power cost assessment at twenty locations in the kingdom of Saudi Arabia. *Renewable Energy, 28*, 573-583. http://dx.doi.org/10.1016/S0960-1481(02)00063-0

RETScreen. (2009, April 26). *International National Resources Canada.* Retrieved from http://www.retscreen.net/

Richard, N. C. (1989). Development of sizing nomograms for stand-alone photovoltaic/storage systems. *Solar Energy, 43*, 71-76.

Salameh, Z. M., & Safari, I. (1995). The effect of the windmill's parameters on the capacity factor. *IEEE Transactions on Energy Conversion, 10*, 747-751. http://dx.doi.org/10.1109/60.475848

SANDIA. (1995). *Stand-alone photovoltaic systems: A handbook of recommended design practice'SAND87-7023.* Las Cruces, NM: Author.

Sawle, Y., & Gupta, S. C. (2014). Optimal sizing of photo voltaic/wind hybrid energy system for

rural electrification. In *2014 6th IEEE Power India International Conference (PIICON)* (pp. 1-4). Delhi. http://dx.doi.org/10.1109/POWERI.2014.7117758

Sawle, Y., & Gupta, S. C. (2015). A novel system optimization of a grid independent hybrid renewable energy system for telecom base station. *International Journal of Soft Computing, Mathematics and Control, 4*, 49-57. http://dx.doi.org/10.14810/ijscmc

Schaffrin, C. (1998, September). SolSim and hybrid designer: Self optimizing software tool for simulation of solar hybrid applications. In *Proceedings of EuroSun'98.* Portooz.

Seeling-Hochmuth, G. C. (1997). A combined optimisation concet for the design and operation strategy of hybrid-PV energy systems. *Solar Energy, 61*, 77-87. http://dx.doi.org/10.1016/S0038-092X(97)00028-5

Shaahid, S. M., & Elhadidy, M. A. (2003). Opportunities for utilization of stand-alone hybrid (photovoltaic+diesel+battery) power systems in hot climates. *Renewable Energy, 28*, 1741-1753. http://dx.doi.org/10.1016/S0960-1481(03)00013-2

Shaahid, S. M., & Elhadidy, M. A. (2004a). Prospects of autonomous/stand-alone hybrid (photo-voltaic+diesel+battery) power systems in commercial applications in hot regions. *Renewable Energy, 29*, 165-177. http://dx.doi.org/10.1016/S0960-1481(03)00194-0

Shadmand, M. B., & Balog, R. S. (2014, September). Multi-objective optimization and design of photovoltaic-wind hybrid system for community smart DC microgrid. *IEEE Transactions on Smart Grid, 5*, 2635-2643. http://dx.doi.org/10.1109/TSG.2014.2315043

Shang, C., Srinivasan, D., & Reindl, T. (2016). An improved particle swarm optimisation algorithm applied to battery sizing for stand-alone hybrid power systems. *International Journal of Electrical Power & Energy Systems, 74*, 104-117.

Shin, Y., Koo, W. Y., Kim, T. H., Jung, S., & Kim, H. (2015). Capacity design and operation planning of a hybrid PV–wind–battery–diesel power generation system in the case of Deokjeok Island. *Applied Thermal Engineering, 89*, 514-525. http://dx.doi.org/10.1016/j.applthermaleng.2015.06.043

Shrestha, G. B., & Goel, L. (1998). A study on optimal sizing of stand-alone photovoltaic stations. *IEEE Transactions on Energy Conversion, 13*, 373-378. http://dx.doi.org/10.1109/60.736323

Sinha, S. S. (2015) Review of recent trends in optimization techniques for solar photovoltaic–wind based hybrid energy systems. *Renewable and Sustainable Energy Reviews, 50*, 755-769. ISSN 1364-0321. http://dx.doi.org/10.1016/j.rser.2015.05.040

Sinha, S., & Chandel, S. S. (2014). Review of software tools for hybrid renewable energy systems. *Renewable and Sustainable Energy Reviews, 32*, 192-205. http://dx.doi.org/10.1016/j.rser.2014.01.035

Smaoui, M., Abdelkafi, A., & Krichen, L. (2015). Optimal sizing of stand-alone photovoltaic/wind/hydrogen hybrid system supplying a desalination unit. *Solar Energy, 120*, 263-276. http://dx.doi.org/10.1016/j.solener.2015.07.032

Swift, A. H. P., & Holder, M. R. (1988). Design of hybrid energy systems. In *Proceedings of the Seventh ASME Wind Energy Symposium* (pp. 175-181). New Orleans, LA.

Syafaruddin, Narimatsu, H., & Miyauchi, H. (2015). Optimal energy utilization of photovoltaic systems using the non-binary genetic algorithm. *Energy Technology & Policy, 2*, 10-18.

Tiba, C., & Barbosa, E. M. de. S. (2002). Software's for designating, simulating or providing diagnosis of photovoltaic water-pumping systems. *Renewable Energy, 25*, 101-113.

Tina, G., & Gagliano, S. (2010). Probabilistic analysis of weather data for a hybrid solar/wind energy system. *International Journal of Energy Research, 35,* 221–232.

Tomonobu, S., Hayashi, D., & Urasaki, N. (2006). Optimum configuration for renewable generating system in residence using genetic algorithm. *IEEE Transaction on Energy Conversion, 21,* 459–466.

Toroczckai, Z., & Eubank, S. (2005). *Agent-based modeling as a decision making tool* (Vol. 35, pp. 22–27). The Bridge, Publication of the National Engineering Academy.

Torreglosa, J. P., García, P., Fernández, L. M., & Jurado, F. (2014). Hierarchical energy management system for stand-alone hybrid system based on generation costs and cascade control. *Energy Conversion and Management, 77,* 514–526.
http://dx.doi.org/10.1016/j.enconman.2013.10.031

Torres-Hernandez, M. E. (2007). *Hierarchical control of hybrid power systems* (PhD dissertation). University of Puerto Rico, Mayaguez.

Traca De Almeida, A., Martins, A., & Jesus, H. (1983). Source reliability in a combined wind-solar-hydro system. *IEEE Transactions on Power Apparatus and Systems, PAS-102,* 1515–1520.
http://dx.doi.org/10.1109/TPAS.1983.317879

Upadhyay, S., & Sharma, M. P. (2014). A review on configurations, control and sizing methodologies of hybrid energy systems. *Renewable and Sustainable Energy Reviews, 38,* 47–63.
http://dx.doi.org/10.1016/j.rser.2014.05.057

Valenciaga, F., & Puleston, P. F. (2005). Supervisor control for a stand-alone hybrid generation system using wind and photovoltaic energy. *IEEE Transactions on Energy Conversion, 20,* 398–405.
http://dx.doi.org/10.1109/TEC.2005.845524

Valente, L. C. G., & de Almeida, S. C. A. (1998). Economic analysis of a diesel/photovoltaic hybrid system for decentralized power generation in northern Brazil. *Energy, 23,* 317–323.
http://dx.doi.org/10.1016/S0360-5442(97)00094-7

Weiss, G. (Ed.). (1999). *Learning in multi-agent systems.* Cambridge, MA: MIT Press.

The Economic Times. (2009, October 26).

Yang, H. X., Lu, L., & Burnett, J. (2003). Weather data and probability analysis of hybrid photovoltaic-wind power generation systems in Hong Kong. *Renewable Energy, 28,* 1813–1824.
http://dx.doi.org/10.1016/S0960-1481(03)00015-6

Yang, H., Lu, L., & Zhou, W. (2007). A novel optimization sizing model for hybrid solar-wind power generation system. *Solar Energy, 81,* 76–84.
http://dx.doi.org/10.1016/j.solener.2006.06.010

Yang, Z., Ma, C., Feng, J. Q., Wu, Q. H., Mann, S., & Fitch, J. (2006). A multi-agent framework for power system automation. *International Journal of Innovations in Energy Systems and Power, 1,* 39–45.

Yang, H., Wei, Z., & Chengzhi, L. (2009). Optimal design and techno-economic analysis of a hybrid solar-wind power generation system. *Applied Energy, 86,* 163–169.
http://dx.doi.org/10.1016/j.apenergy.2008.03.008

Zaheeruddin, & Manas, M. (2015). Analysis of design of technologies, tariff structures, and regulatory policies for sustainable growth of the smart grid. *Energy Technology & Policy, 2,* 28–38.

Zhou, W., Lou, C., Li, Z., Lu, L., & Yang, H. (2010). Current status of research on optimum sizing of stand-alone hybrid solar–wind power generation systems. *Applied Energy, 87,* 380–389.
http://dx.doi.org/10.1016/j.apenergy.2009.08.012

Permissions

The contributors of this book come from diverse backgrounds, making this book a truly international effort. This book will bring forth new frontiers with its revolutionizing research information and detailed analysis of the nascent developments around the world.

We would like to thank all the contributing authors for lending their expertise to make the book truly unique. They have played a crucial role in the development of this book. Without their invaluable contributions this book wouldn't have been possible. They have made vital efforts to compile up to date information on the varied aspects of this subject to make this book a valuable addition to the collection of many professionals and students.

This book was conceptualized with the vision of imparting up-to-date information and advanced data in this field. To ensure the same, a matchless editorial board was set up. Every individual on the board went through rigorous rounds of assessment to prove their worth. After which they invested a large part of their time researching and compiling the most relevant data for our readers.

The editorial board has been involved in producing this book since its inception. They have spent rigorous hours researching and exploring the diverse topics which have resulted in the successful publishing of this book. They have passed on their knowledge of decades through this book. To expedite this challenging task, the publisher supported the team at every step. A small team of assistant editors was also appointed to further simplify the editing procedure and attain best results for the readers.

Apart from the editorial board, the designing team has also invested a significant amount of their time in understanding the subject and creating the most relevant covers. They scrutinized every image to scout for the most suitable representation of the subject and create an appropriate cover for the book.

The publishing team has been an ardent support to the editorial, designing and production team. Their endless efforts to recruit the best for this project, has resulted in the accomplishment of this book. They are a veteran in the field of academics and their pool of knowledge is as vast as their experience in printing. Their expertise and guidance has proved useful at every step. Their uncompromising quality standards have made this book an exceptional effort. Their encouragement from time to time has been an inspiration for everyone.

The publisher and the editorial board hope that this book will prove to be a valuable piece of knowledge for researchers, students, practitioners and scholars across the globe.

List of Contributors

Bhanu Pratap Soni and Vikas Gupta
Department of Electrical Engineering, Malaviya National Institute of Technology, Jaipur, Rajasthan 302017, India

Akash Saxena
Department of Electrical Engineering, Swami Keshvanand Institute of Technology, Jaipur, Rajasthan 302017, India

Ahmet Kayabasi
Department of Electronic & Automation, Silifke-Tasucu Vocational School, Selcuk University, 33900, Silifke, Mersin, Turkey

Ali Akdagli
Faculty of Engineering, Department of Electrical & Electronics Engineering, Mersin University, 33343, Ciftlikkoy, Mersin, Turkey

Ashok Kumar and Garima Saini
Department of Electronics & Communication, NITTTR, Chandigarh, India

Shailendra Singh
SGIT, Ghaziabad, India

Grzegorz Gralewicz and Grzegorz Owczarek
Department of Personal Protective Equipment, Central Institute for Labour Protection – National Research Institute, Czerniakowska 16, 00-701 Warsaw, Poland

Janusz Kubrak
Vigo SL Sp. z o.o., Poznańska 129/133, 05-850 Ożarów Mazowiecki, Poland

Vikal R. Ingle
B.D. College of Engineering, Sevagram, Wardha, Maharashtra, India

V.T. Ingole
Prof. Ram Meghe Institute Technology & Research, Badnera, Amravati, Maharashtra, India

Esha Gupta and Akash Saxena
Department of Electrical Engineering, Swami Keshvanand Institute of Technology, Office no AC-201, Ramnagaria, Jagatpura, Jaipur 302017, Rajasthan, India

Omveer Singh
Electrical Engineering Department, Maharishi Markandeshwar University, Ambala, Haryana, India

and Ibraheem Nasiruddin
Electrical Engineering Department, Quassim University, Buraydah, Saudi Arabia

M. Deben Singh, Ram Krishna Mehta and Arvind Kumar Singh
Department of Electrical Engineering, North Eastern Regional Institute of Science & Technology, Nirjuli, Arunachal Pradesh 791109, India

Ngasepam Monica Devi and Rajesh Saha
Computer Science and Engineering, National Institute of Technology, Arunachal Pradesh, Yupia 791112, India

Sanjeev Kumar Metya and Santanu Maity
Electronics and Communication Engineering, National Institute of Technology, Arunachal Pradesh, Yupia 791112, India

T.H. Khoa, P.M. Vasant and M.S. Balbir Singh
Department of Fundamental and Applied Sciences, Universiti Teknologi PETRONAS, 31750 Tronoh, Perak, Malaysia

V.N. Dieu
Department of Power Systems, HCMC University of Technology, Ho Chi Minh City, Vietnam

Joanne Mun-Yee Lim
Faculty of Engineering, SEGi University, Jalan Teknologi, Kota Damansara, Petaling Jaya 47810, Selangor, Malaysia

Yoong Choon Chang
Lee Kong Chian Faculty of Engineering & Science, Universiti Tunku Abdul Rahman Sungai Long Campus, Jalan Sungai Long, Bandar Sungai Long, Kajang 43000, Selangor, Malaysia

Mohamad Yusoff Alias
Faculty of Engineering, Multimedia University, Persiaran Multimedia, Cyberjaya 63100, Selangor, Malaysia

Jonathan Loo
School of Science and Technology, Middlesex
University, London NW44BT, UK

Barnam Jyoti Saharia and Munish Manas
Department of Electronics and Communication
Engineering, Tezpur University, Assam 784028,
India

Bani Kanta Talukdar
Department of Electrical Engineering, Assam
Engineering College, Assam 781013, India

**Pandian M. Vasant, Imran Rahman and Balbir
Singh Mahinder Singh**
Department of Fundamental and Applied Sciences,
Universiti Teknologi PETRONAS 32610 Seri
Iskandar, Perak, Malaysia

M. Abdullah-Al-Wadud
Department of Software Engineering, College of
Computer and Information Sciences, King Saud
University, Riyadh, Saudi Arabia

**Yashwant Sawle, S.C. Gupta and Aashish Kumar
Bohre**
Department Electrical Engineering, Maulana Azad
Institute of Technology, Bhopal, India

Index